Lecture Notes in Electrical Engineering

Volume 1265

The book series *Lecture Notes in Electrical Engineering* (LNEE) publishes the latest developments in Electrical Engineering—quickly, informally and in high quality. While original research reported in proceedings and monographs has traditionally formed the core of LNEE, we also encourage authors to submit books devoted to supporting student education and professional training in the various fields and applications areas of electrical engineering. The series cover classical and emerging topics concerning:

- Communication Engineering, Information Theory and Networks
- Electronics Engineering and Microelectronics
- Signal, Image and Speech Processing
- Wireless and Mobile Communication
- Circuits and Systems
- Energy Systems, Power Electronics and Electrical Machines
- Electro-optical Engineering
- Instrumentation Engineering
- Avionics Engineering
- Control Systems
- Internet-of-Things and Cybersecurity
- Biomedical Devices, MEMS and NEMS

For general information about this book series, comments or suggestions, please contact leontina.dicecco@springer.com.

To submit a proposal or request further information, please contact the Publishing Editor in your country:

China

Jasmine Dou, Editor (jasmine.dou@springer.com)

India, Japan, Rest of Asia

Swati Meherishi, Editorial Director (Swati.Meherishi@springer.com)

Southeast Asia, Australia, New Zealand

Ramesh Nath Premnath, Editor (ramesh.premnath@springernature.com)

USA, Canada

Michael Luby, Senior Editor (michael.luby@springer.com)

All other Countries

Leontina Di Cecco, Senior Editor (leontina.dicecco@springer.com)

**** This series is indexed by EI Compendex and Scopus databases. ****

Vančo B. Litovski

Lecture Notes in Analog Electronics

Testing and Diagnosis of Analog and Mixed-Signal Electronic Circuits

Springer

Vančo B. Litovski
Faculty of Electronic Engineering
University of Niš
Niš, Serbia

ISSN 1876-1100 ISSN 1876-1119 (electronic)
Lecture Notes in Electrical Engineering
ISBN 978-981-97-8256-7 ISBN 978-981-97-8257-4 (eBook)
https://doi.org/10.1007/978-981-97-8257-4

This Springer imprint is published by the registered company Springer Nature Singapore Pte Ltd.
The registered company address is: 152 Beach Road, #21-01/04 Gateway East, Singapore 189721, Singapore

Preface

The process of designing, manufacturing, and maintenance of electronic circuits and systems inevitably includes their testing.

It is not possible to deliver a quality and reliable industrial product to the market if it is not thoroughly tested. Hence, the ability to test one's own product is in fact a requirement to market that product. Of course, as is well known, testing does not give the manufacturer new values but only the knowledge that the products he delivers to the market are good as expected.

In some situations both the manufacturer and the maintenance support are facing the need to diagnose the failure which means to find the nature and the location of the fault causing malfunction of a circuit or system. That will be also solved based on testing.

To bring the subject to the reader, in this book, we will visit several topics.

When considering the concepts of testing, functional testing is the first idea that comes in mind. That means one is verifying the functionality of the designed system against the requirements given in advance by measurement. This and many other tests (such as temperature dependence) on the finished circuit or system are done most frequently. The problem is the number of test signals needed to activate the complete functionality of the system and related to this the time needed for testing a single product. Alternatively, one is to select a set of most probable faults and perform structural testing which means one will create test signals which will attack them (in general) only. To the structural testing is devoted the rest of the discussion since functional testing is self-defined.

Most of our attention will be devoted to test signal synthesis for analog and for combinational digital circuits. This issue is almost entirely dependent on the circuit and system designer and is based on knowledge of both the functionality of the system and the failure modes in the life cycle of the product starting with design and ending with field exploitation. Procedures will be advised (and illustrated) leading to selection of test signals accommodated to the structure and the nature of the functionality of the system.

In the heart of test signal synthesis is always the fault model which enables the verification of the synthesized signal and algorithm by simulation (which is an inevitable supportive tool to the test synthesizing engineer). Having in mind the almost unlimited number of possible faults in a circuit or system, effective fault simulation is a must for test synthesis both for analog and for mixed-signal circuits.

Even small systems exhibit problems related to the accessibility to some points inside being needed for monitoring the response. Creating ports enabling accessibility has limited potential since it is bounded by the number of pins of the IC component or the pads of the printed circuit board. That is why the concept of design for testability (DFT) is most frequently adopted. It comprises a change in the structure of the system (mainly complex integrated circuits) which will enable access to many testing points (selected in advance) via a signal bus which is visiting parts of the system (under control of built-in hardware). System synthesis for testability is nowadays standardized which will be discussed below.

A special problem arises in complex analog circuits to create an exciting signal which would activate a part of the whole while isolating the rest. That is solved by the concept of built-in test (BIT) or even the concept of built-in self-test (BIST). That comprises internal reconfiguration of the circuit so that its part being under test generates signals which are analyzed outside of the circuit (BIT) or even inside it (BIST).

The issue of diagnosis is visited next. Here again, the structural view to the problem is considered which means that limited number of diagnostic hypotheses is created and memorized within the so-called fault dictionary. These hypotheses are, of course, related to the most probable faults and are generated by simulation. In that view the process of diagnosis represents a search through the fault dictionary (among the hypotheses) for the fault which produces a faulty response that is the most similar to the measured one. The problem is, of course, that in the context of analog circuits and when many faults are encompassed within the fault dictionary, there is a probability for some faulty responses to be similar. That is why here, among other methods, we strongly recommend the use of artificial neural networks (ANNs), being classifiers by nature, to help distinguish among them.

To diagnose mixed-signal circuits a hierarchical procedure is advised and exemplified.

That completes our proceedings related to testing and diagnosis. What follows is viewing the testing as an industrial activity. In that we will first analyze the nature of the faults. Then we will proceed with the analysis of the lifecycle of an electronic product marking the points where testing and diagnosis should be applied including the nature of them. Test equipment and economy of testing will be part of that analysis too.

We are finishing with a chapter containing examples of test synthesis and diagnosis of some (mostly simple) circuits. Unlike other book of the LNAE series here there will be no chapter devoted to Spice simulation examples since Spice is extensively used throughout the previous chapters.

Niš, Serbia Vančo B. Litovski

Introduction to the Lecture Notes of Analog Electronics (LNAE) Series

The electron is a phenomenon exhibiting divine properties. We all know it exists but none can see it. Obviously, we consider Electronics inherits his divinity.

Electronics is, nowadays, ubiquitous. There is no aspect of human life being not supported by electronics. None would deny this claim. Even so, when comes to studying electronics, and especially analog electronics, it is mostly avoided by many talented students since it is considered difficult to study. Thus, it is our intention to produce a series of books that will be student oriented and, probably not understandable at first glance, electronics oriented. In that way we expect to help attract more young and talented people to study electronics and to further contribute to the human society. In this pamphlet we will consider two issues. We will first try to answer the question as to why electronics is difficult to study, and then we will give the rationale for preparing a series of books related to analog electronics only.

Let's start with the difficulty. Ever since Nikola Tesla introduced the alternating current (AC) circuits, the primary circuit elements became the coil, the transformer, the AC motor, and the transmission line. All these were described by voltage equations which originate from the discovery of the Russian physicist Heinrich Lenz of the directional relationships between induced magnetic fields, voltage, and current when a conductor is passed within the lines of force of a magnetic field. Lenz's law states: *"An induced electromotive force generates a current that induces a counter magnetic field that opposes the magnetic field generating the current."* As a consequence, voltage equations were introduced everywhere including the exclusive use of Kirchhoff's voltage law. That we call electrical engineering approach to the circuit.

The things went so far that even Ohm's law which should read $i = G{\cdot}v$ (currents are on both sides) was transformed into his voltage version $v = R{\cdot}i$. In that, two serious problems were introduced which concern the modern student of electronics. First, in $v = R{\cdot}i$, the independent variable is on the left-hand side of the equation while the function (the consequence) is on the right. So, one is supposed first to reorder the equation to come to the solution. That means that, for evaluation of the current, division is supposed to be performed as opposed to the much more natural arithmetic action of multiplication. Of course, to make things still more difficult, instead of the natural quantity which is conductance, resistance was introduced. For example, if a voltage source v is connected to a resistor whose conductance is $G = 25$ S one would immediately pronounce the value of the current to be $I = 25$ A. The same resistor has a resistance of $R = 40$ mΩ, and to come to the current becomes a serious computational task. The transition was so merciless that no schematic symbol for the conductance exists.

The conflict becomes still more severe when comes to electronic circuits. Namely, all (no exception) electronic components are modeled by current equation (voltage controlled current sources), and, for electronic circuit analysis, Kirchhoff's current low is the most natural means for equation formulation which are known as nodal equations. A student who is trained to think of voltage equations however is immediately frustrated since he/she does not know what to do with the resistor now. Even a well-trained former student, as is the author of this text, is forced to use reciprocals of the resistances all the time and, at the end of the analysis, is forced again to rearrange the obtained expressions to make them tractable.

Of course, modern electronic circuit simulation programs are based on nodal analysis (or the so-called Modified Nodal Analysis) which we would call electronic approach to the circuits. Nevertheless, due to the voltage background, understanding of their functionality is not easy, obstructing their efficient use and, again, is frustrating to many.

Next, the students trying to enter into the subject of electronics are facing the duality in electronics which, as we see it, is twofold. To start with, we will mention the synergy (or the conflict) between the DC and AC signals in electronic circuits. In some analyses, these are independent and in other mutually coupled. The fact that the DC supply is a source of energy which is transferred to the useful signal is part of the explanation on what is going on in the circuit. One should understand, however, why even small AC signals (bearing almost no power) are depending on the value of the voltage of the power supply which in most cases of analysis of incremental circuits is absent from the schematic. On the other side, one is supposed to understand that, in some electronic circuits, voltages larger than the power supply voltage are produced and limit the applicability of some devices.

The other aspect of duality is related to the mutual synergy and antagonism of the time and frequency domain. There is no linear electronic component. That means that the component's current is a nonlinear function of a single or several voltages or currents. So, it is normal, when starting from basic electrotechnics, for nonlinear differential equations to be created to enable the analysis of an electronic circuit. Solving nonlinear differential equations, however, is a task which extremely rarely may be performed in closed form. Practically, there is no time domain electronic circuit analysis without a computer. That is especially true since one has no complete information about the boundary conditions necessary for solving the nonlinear differential equations. That imposes a need for development of incremental models of the electronic devices to allow analysis for small signals. The problem is that, usually, the two versions of the same circuit (the large signal nonlinear and the incremental linear) are not taught to students as a single complete. To learn the analysis of incremental circuits the students are kept in the so-called frequency domain where all of a sudden complex arithmetic is necessary. In that, a complex variable is introduced so that the real frequency of the signal is represented as the imaginary part of the complex frequency. Even the interpretation of the last sentence is confusing if not read several times.

This is not the end, however. The students are often facing textbooks written by teachers that want to promote themselves as the one having strong practical experience. In that slang is used as if they are talking to some repair person. Avoiding universal professional terminology and especially verbs representing phenomena within the circuit invalidates the knowledge delivered and makes the further progress of the student difficult.

Having all these in mind it becomes understandable why not only the student but even the teachers face wild wind-mills as if they are the Spanish fiction hero Don Quixote.

Now, what do we want to do by this series of books? We are not in a position to change the unchangeable. The resistors will stay in the schematics as are everywhere and the imaginary part of the complex frequency will still represent the real frequency of the signal. What we can do is to cover the analog electronics as a whole using the same vocabulary which we consider universal. Also, we would like to enable a start from the very beginning in the sense that only very basic knowledge of the student to be necessary to follow the texts. Finally, we want to give a complete knowledge starting with fundamental physics and ending with simulation and optimization, and, especially, with the sustainability aspects of electronic engineering.

The series will contain the following issues with the probability for some of them to become merged or splitted in two, according to the running inspiration of the author:

Book 1. Low Voltage Electronic Components
Book 2. Electronic Signal Amplification and Linear Oscillators
Book 3. VLSI Circuits Simulation and Optimization (coauthored with Prof. M. Zwoliński)
Book 4. Discrete and Integrated Large Signal Amplifiers
Book 5. Noise in Electronic Circuits and Low Noise Amplifier Design
Book 6. DC Electronic Power Supplies
Book 7. Testing and Diagnosis of Analog and Mixed-Signal Electronic Circuits
Book 8. Artificial Intelligence in Electronic Modelling and Design
Book 9. Sustainable Electronic Design

The titles listed here are subject to refinements, too.

We really hope that these contents will fit in the contemporary views on the subject and will inspire authors to build on or to make improvements in a way they find better than what we did.

Contents

About the Author

Prof. Vančo B. Litovski was born in 1947 in Rakita (alleged birthplace of Olympias of the Molossians, the mother of Alexander, the Great, of Macedonia), South Macedonia, Greece. He graduated from the Faculty of Electronic Engineering in Niš in 1970 and obtained his M.Sc. in 1974 and his Ph.D. in 1977. He was appointed as a teaching assistant at the Faculty of Electronic Engineering in 1970 and became a Full Professor at the same faculty in 1987. He was elected as Visiting Professor (honoris causa) at the University of Southampton in 1999. From 1987 until 1990, he was a consultant to the CEO of Ei, and was Head of the Chair of Electronics at the Faculty of Electronic Engineering in Niš for 12 years. From 2015 to 2017, he was a researcher at the University of Bath. He has taught courses related to analogue electronics, electronic circuit design, and artificial intelligence at the electro-technical faculties in Priština, Skopje, Sarajevo, Banja Luka, and Novi Sad. He received several awards including from the Faculty of Electronic Engineering (Charter in 1980, Charter in 1985, and a Special Recognition in 1995) and the University of Niš (Plaque 1985). Prof. Litovski has published 6 monographs, over 400 articles in international and national journals and at conferences, 25 textbooks, and more than 40 professional reports and studies.

His research interests include electronic and electrical design and design for sustainability, and he led the design of the first custom commercial digital and research-oriented analogue CMOS circuit in Serbia. He has also headed 8 strategic projects financed by the Serbian and Yugoslav governments and the JNA and has participated in several European projects funded by the governments of Germany, Austria, UK, and Spain, and the EC as well as the Black See Organization of Economic Cooperation (BSEC).

7.1 Basics

Abstract The basic ideas of testing and diagnosis are introduced. We start with the existence and physical nature of faults within the electronic circuit to establish the feeling on the necessity of testing and diagnosis. We explain here why under "testing" in this book we mainly understand test signal synthesis. We start with the fault and introduce the inductive fault analysis relating the faults and the possible failures. Functional and structural testing are contrasted and rationale for implementation the one and the other are given. Specialist's terms such as testability, fault sensitization, detectability, observability, and similar are introduced. To diagnosis special attention is paid including a set of algorithms used for analog and mixed-signal circuits. We also introduce the notion on the importance of testing in conjunction with the price paid for it including the test sequence(s) design, implementing on testing hardware, and optimizing the speed of testing in mass production environment.

7.1.1 Introduction

Within the design and production of electronic circuits, testing occupies one of the most important places. The task of testing is to establish whether a project, process, technology, component, or a system is functioning correctly. In the following text, in order to avoid long references, we will talk about the testing of electronic circuits, which will mostly cover all the above-mentioned results of the production phases. The term "correct" implies that there should be a definition of correct operation of the circuit so that testing is reduced to verifying that the response of the circuit fits within the definition of correct operation.

The cause of a possible failure of an electronic circuit or system can lie in a design error, technological failure, and exploitation conditions (excess temperature, excess current or voltage, ionizing radiation, mechanical shock, stress or impact, contamination, aging, etc.), which includes the storage and transportation of finished products. Both can lead to defects. This may mean that a part of the circuit will be faulty. This, again, may mean that the behavior of the system in which the circuit is embedded will not fit within the definition of correct operation. In such a case, the

V. B. Litovski, *Lecture Notes in Analog Electronics*, Lecture Notes in Electrical Engineering 1265, https://doi.org/10.1007/978-981-97-8257-4_1

defect is mapped into a failure. For example, one of the connections in the circuit may be broken. As a result, it is likely that at least one function of the circuit will be disrupted at least at one time instant. In this regard, the testing task would be as follows:

- establish a signal (waveform, time scale) that will enable the response of the faulty and the correct circuit to differ when measuring the faulty circuit and
- among all such signals, choose the one that will enable testing in the shortest time, that is, which will make testing the cheapest.

It is easy to draw conclusions about the complexity of electronic devices (testers) that perform the very function of testing. They need to meet technological performance in terms of speed, reliability, and productivity that fundamentally exceeds the performance of the component under test (CUT). Of course, the price of the tester used also comes from this. More will be said about the tester and its properties later.

Testing of finished products is performed during production, in the plant, and during exploitation, in laboratories or service centers. When it comes to production, it should be kept in mind that with large batches, the duration of testing is multiplied by the number of components or systems that are produced. This means that the total testing of one series can take time, even when using the most sophisticated equipment and when the most favorable test signal is implemented.

All in all, testing is an expensive activity. It does not add value to the product because it stays just as valuable as it was before testing, and it significantly raises its price. In some situations, the cost of testing exceeds the cost of production. In exceptional cases, even more than once. In turn, the result of the test is information that the circuit or system is correct and can be incorporated into the device as such. Avoiding testing directly increases the likelihood of device malfunction. In this regard, the price of production should also include the repair of untested devices outside the plant where they were produced, so that avoiding testing is usually not worth it. The economics of testing will be discussed separately later in this book.

From a design point of view, the cost of testing, and thus the price of the product, can be reduced in several ways.

First of all, a correct project is the first condition for a correct product. This means that, during design, verification of each design phase should be systematically applied in order to avoid errors in the design. Among such activities, we can easily recognize simulation for verifying the synthesized schematic, checking the design rules on the completed layout of the integrated circuit or printed circuit board, and simulation after designing the layout (with data extracted from the layout) for verification the completeness (integrity) of the layout and evaluation the delay times in critical circuit connections (which are difficult to be estimated during the pre-layout design phase).

Another very significant aspect of the influence of the design phase on testing is related to the application of the concepts of "design for testability" (DFT). In this case, architecture of the electronic circuit, printed circuit board or system is created to enable easy testing. In doing so, sometimes the concept of "boundary scan" is used, which is sanctioned by IEEE standard 1149.1 [1, 2] or 1149.4 [3] and their supplements, and more recently, the concept of "built-in self-test" (BIST).

Finally, one of the most significant aspects of the influence of the design phase on the testing phase is the synthesis of the test signal. In this phase, the designer is faced with the task of generating signals that the tester will bring to the input of the circuit (or system) being tested, as well as signals of the desired response that the tester will compare with the measured ones. As already mentioned, this set of signals should be such that the testing is performed in the simplest and fastest possible way, without leaving any doubts about the correctness of the device being tested.

The concepts of testing digital and analog circuits are different. Due to mass production, due to the very high scale of integration as well as due to the specific waveform of the signal used in digital circuits, the problem of testing digital circuits has been given incomparably greater attention until now, and it can be said that this problem is practically solved from the point of view of industrial application. Namely, there are developed comprehensible and programmable algorithms, and thus also software aids, which enable practically automatic generation (synthesis) of test signals.

The situation with analog integrated circuits is not the same. First, the integration of complex analog or mixed-signal (analog and digital) electronic circuits is more recent. Then, until recently, a series of such products were not large (when dealing with complex analog functions). Finally, the waveforms and the nature of the quantities that can be measured are numerous, making automatic synthesis much more difficult. Dilemmas arise about whether to measure voltages, currents, or powers. Is it more convenient to perform measurements in the time or frequency domain? Or, is it enough to establish knowledge on the DC values. If it is time domain, which waveforms are the most suitable? What are the amplitudes, edge durations, and the like? If it is about the frequency domain, should we measure amplitudes or phases? It is of particular importance to choose the measurement (testing) points. What are the quantities (voltages, currents, or powers) in the circuit in which, from the point of view of testing, the fault effects that we want to test are mapped in the most favorable way? Bearing in mind that with encapsulated integrated circuits only those points that are connected to the circuit pins are available, the selection of measurement points and their connection to the corresponding pins becomes a key problem from the testing point of view. Finally, the components or elements that are built into an analog electronic circuit always have tolerances. The question arises, what is the limit up to which it can be said that the response deviates from the desired one due to the final tolerances of the elements, and not due to a fault in the circuit? What tolerance is tolerable and is not a defect?

Regardless of whether analog or digital circuits are treated, the rationalization of testing is most often based on the reduction of the number of measurements. Namely, according to the experiences related to the technology implemented, it is possible to say in advance which of the possible faults are the most likely to happen. Based on a study [4], one can claim that in classical CMOS technology of digital circuits, 75% of all circuit faults are open and short circuits. In that, according to [5] more than 65% of the open-circuit faults exhibit a resistance larger than 1 GΩ while the rest are evenly distributed over a wide range of (large) resistances.

That is why when compiling the test, signals are created that are tasked to enable the detection of certain probable faults. If we do so, we will face the problem of test verification. Namely, if the test or signal is composed so that the effect of the given fault is reflected on the output, then before that test is actually applied on the tester, it should be checked whether it is really well composed. A fault simulator is used for this purpose.

A fault simulator is a software engine whose task is to simulate a circuit with a fault and to generate the output signal that such a circuit would have. In order for this machine to work efficiently, it is necessary to solve two problems at the same time. The first one refers to the insertion of the fault into the fault-free circuit. Namely, in the description of the fault-free circuit, which, of course, has already been simulated, a fault should be incorporated and the simulation repeated. This procedure should be repeated as many times as there are faults whose test should be verified. Just as all components or logical operators are to be modeled in order to be incorporated into the description of the electronic circuit or system being simulated, the fault should be modeled as well. Fault modeling is one of the most important activities in the process of developing a simulator that will serve for test (signal) verification. On the other hand, when evaluating the effectiveness of the fault simulation, it should be kept in mind that any simulation, be it logic or electrical (circuit), takes considerable time. If our aspirations are such that the testing includes a large number of possible faults, the number of simulations will be very large. Therefore, the test verification procedure will become ineffective. That is why it is important to develop procedures that will enable simultaneous or, what is the same, concurrent, simulation of several faults. The basic guiding idea used here is the fact that the original (or, fault-free) circuit and the faulty one differ little, at least in a topological sense. It makes sense, especially with digital circuits, that their responses do not have to be very different.

Before giving an overview of the scope of this book, we will mention another aspect that makes fault simulation a significant activity in the field of designing and manufacturing electronic circuits and systems in general. One of the results of fault simulation is the so-called fault dictionary. Actually, it is a table containing (fault code—circuit response) pairs. Each fault is associated with the response of the circuit which is obtained by simulation for the case when that fault is embedded into the circuit. This information is of inestimable value from the point of view of diagnostics of electronic circuit malfunctions.

Diagnostics of circuits and systems is an activity which, in its content and objectives, is much broader than testing. While the task of testing is only to establish the correct or incorrect operation of the circuit, there are two other tasks to be fulfilled by diagnosis. It should make it possible to establish the nature and the location of the fault in the circuit. Based on correct diagnostics, it is possible to intervene in the correction of the production process and, of course, it is possible to intervene in the replacement of a faulty component during the maintenance of a system that is in operation. The fault dictionary, during diagnostics, is used by comparing the (single) measured response of the faulty system successively with the individual faulty responses obtained by simulation. When the two responses match, the fault corresponding to the pair in the dictionary is extracted.

In the following text, the problem of automation of test generation for analog (linear and nonlinear) circuits will first be discussed in more detail. In this case the presentations will be mostly based on literature [6–7]. Then, methods for efficient fault simulation in analog linear and nonlinear circuits will be introduced. The procedures for generating tests in digital circuits will be shortly visited to enable mixed-signal test generation and, what is very important, to introduce the concept of design for testability. Mixed-signal circuits are frequent in modern CMOS integrated electronics. Testing such circuits represents a specific challenge. Only a limited set of mixed-signal functions will be visited in the next so that a part only of the vast family of these circuits will be considered. Having almost complete information on test synthesis, the testing will be considered in the next as an industrial action. Specific circuit design concepts to improve testability will be proposed which are known as built-in test (BIT) or, more specifically, built-in self-test (BIST). To complete the subject the concept of diagnosis as such will be discussed first. Then diagnosis of analog and mixed-signal circuits based on inference mechanism supported by artificial neural networks (ANNs), nowadays known as deep learning algorithms or even as artificial intelligence (AI), will be introduced. The reader is advised to consult [8, 9] as supportive reading in diagnosis.

As usual in the LNAE series the book contains a set of solved problems.

In this chapter, the concept of faults, causes of faults, and fault effects will first be introduced. Then, the basic concepts of testing will be presented in a simple way in order to see the fundamental procedures of test generation.

7.1.2 Faults and Fault Effects

As already said, any deviation of the actual electrical schematic of the circuit from the one required to perform its function correctly can be considered as a fault. Such deviations can be lack of some element of the layout, interruption of some connection, short circuit between two connections (these can be transistor electrodes), drastic narrowing of the connection, narrowing or expansion of a particular active area (diffusion, polysilicon or metal gate, implantation), damage (puncture) of certain circuit parts, etc. The causes of such faults can be by design or due to technological imperfections. Design errors can be twofold. The first category includes errors related to the lack of an element (function or connection) and building in redundant elements. The second category, which is encountered more often, includes dimensioning errors. For example, a metal strip that is too narrow, during the operation of the component, if it happens that large current flows through it, can break due to electromigration. Too closely spaced transistors or protective rings can form a four-layer structure (thyristor) that conducts at lower voltages than expected. At the same time, we are not talking about violations of design rules, but about inappropriate dimensioning, which is particularly characteristic to analog circuits. Dimensioning errors are especially common with components. It is easy to make mistakes when sizing resistors, capacitors, transistor widths, etc.

Imperfections of the technological process are one of the most common causes of faults. The most easily understood fault is excessive metal etching which is characteristic for both, integrated circuits when it comes to aluminum etching, and printed circuits when it comes to copper etching. The number of processes involved in making integrated circuits is large, and the number of different materials and chemical compounds used is also large. Each of the processes can, similarly, to etching, lead to faults. Unwanted impurities in any part of the component or circuit, for example, can be mapped into a fault in one way or another.

Based on the nature of faults, they can be divided into two basic categories: catastrophic (hard) or parametric (soft). By catastrophic we mean those that lead to a change in the topology of the circuit (although, in principle, they do not have to change the function), and by parametric we mean those in which the change is only in the value of one of the parameters (which, sometimes, can change the function).

In digital circuits, the modeling of faults is greatly simplified. First, the so-called stuck-at faults will be mentioned. There are stuck-at-1 and stuck-at-0. These conditions are attached to the signals (node voltages which are represented by logic states) and represent the consequences of actual faults in the circuit. The simplest fault that is modeled by the stuck-at-1 fault is a short circuit between the signal and the supply voltage. On the other hand, the simplest fault that is modeled by the stuck-at-0 fault is a short circuit between signal and ground. In addition to these two states, a permanently open (stuck-at-open) and permanently short-circuited (stuck-at-short) circuit is introduced. Finally, when modeling faults in digital circuits, the delay deviation model is also used, that is, the delay fault model.

When modeling faults in analog circuits, in addition to all these models, it is necessary to use the parameter increment model. It should be borne in mind that when the fault is parametric, the fault simulation is in fact a tolerances analysis or, in other words, sensitivity analysis for large increments of the parameters. At the same time, both with digital and analog circuits, when the fault simulation is performed for the purpose of testing, it most often includes only one fault. Of course, it is possible to introduce and simulate several faults at the same time, but this is considered irrational because it has been shown that when the test detects each fault individually, it will, with a very high probability, also detect faulty operation of circuit with multiple faults that exist simultaneously.

When multiple faults in linear circuits are to be considered the k-fault testability is invoked [10]. The technique transforms the fault element parameters into fault currents, with the constraint that the number of faults in the network does not exceed k. Consequently, the key parameter for validation is the fault current. A fault current of 0 indicates an absence of faults, while any other value suggests the presence of a fault. This approach leverages the alterations in fault element parameters and the correlation between voltage and current changes at detectable nodes. Numerous k-fault diagnosis approaches have been proposed, including the branch method, node method, loop method, mesh method, and cut set method.

Based on these considerations the importance of the concept of the fault effect may be recognized. For example, in a simple two NMOS transistor amplifier, short circuiting the gate and drain of the NMOS transistor used as a load (to the amplifying NMOS transistor) changes the electrical schematic, but the amplifier retains its function. That same short circuit, however, when it occurs at the PMOS transistor in a CMOS inverter, is actually a short circuit between the input and the output and fundamentally changes the circuit's function. In the first case, the fault effect was significant, but not decisive, and in the second case it was decisive.

In digital circuits, the fault effects can be expressed as an error in the state of the output at a given moment (instead of unity there is a logical zero or vice versa) or as an error in the time diagram, i.e., signal delay.

In analog circuits, the fault effects can be expressed in countless ways. Among them we will name only a few. When we talk about DC modes, the fault effects are usually expressed in DC voltage and current levels in the circuit. Among them, the most important are the voltages and currents at the external terminals, including the power supply voltage terminal. The current of the power source always contains aggregate information about the operation of the circuit considering that it represents a superposition of the currents of all branches connected to the power supply.

When talking about the fault effect in the frequency domain, different results are possible. As a result of a fault, hard or parametric, the value of the nominal gain of the circuit can be changed, the value of the nominal phase of the circuit can be changed, the slope of the frequency characteristic at the bandwidth edge can be changed, the gain value within a frequency range can be disturbed, the bandwidth may be shifted along the frequency axis, etc.

Many of these effects can be present simultaneously as a result of the same fault. In the frequency domain, the catastrophic consequences of a non-catastrophic (soft) fault can easily become noticeable. Namely, if due to the increase of one (critical) parameter in the circuit, there is an unacceptable change in the cut-off frequency of the circuit or system, it becomes unusable even though the topology of the circuit is not disturbed.

The fault effect in the time domain can also be mapped into different consequences for the output signal. It is possible to fundamentally change the waveform (of output voltage, output current, or supply current) or to change the response parameters such as rise time, fall time, delay time, overshoot, undershoot, frequency, etc.

Finally, in recent times there are more and more integrated circuits that combine both analog and digital functions. The fault effects in such circuits can manifest itself in a way that is related to the functionality of the circuit. When dealing with A/D or D/A converters, for example, the effect of a soft fault would be expressed in a conversion error that is larger than planned. A comparator may be switching with the wrong threshold or different hysteresis. A phase-locked-loop circuit may not converge, etc.

All this speaks of the complexity of mapping a fault into the fault effect, and therefore of the complexity of the work that would be called the synthesis and verification of the test for analog circuits.

Figure 7.1.1a shows the layout of a simple integrated circuit (standard cell) and indicates some of the possible faults (physical faults) according to the following list

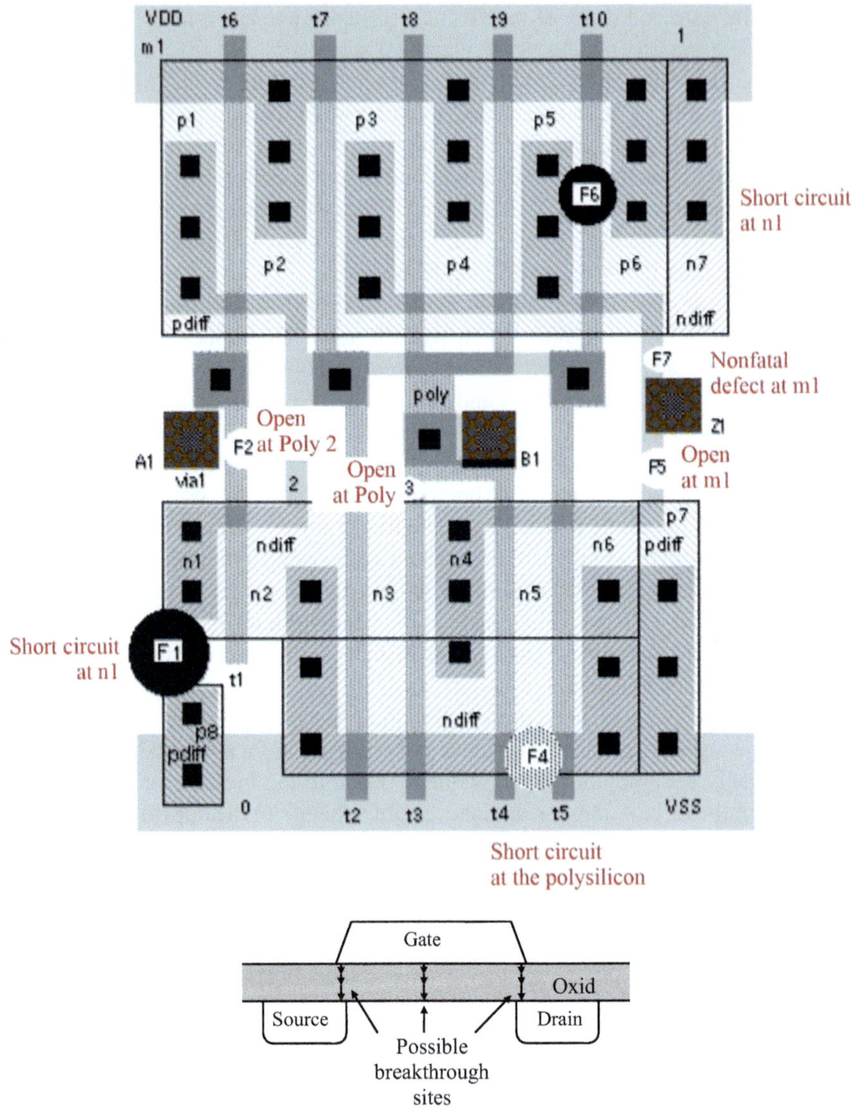

Fig. 7.1.1 **a** Layout of a simple integrated circuit with indications of some faults. **b** Cross-section of a MOS transistor and possible short circuits due to breakdown of thin the oxide

- F1 is a short circuit between connections at the m1 level (metal 1) and connects node n1 to VSS.
- F2 is an open circuit in the polysilicon layer and separates the gate of transistor t1 from the rest of the circuit.

- F3 is an open circuit in the polysilicon layer and separates the gate of transistor t3 from the rest of the circuit.
- F4 is a short circuit on the polysilicon layer and connects the gates of transistors t4 and t5.
- F5 is an open circuit in m1 and separates node n4 from output Z1.
- F6 is a short circuit on m1 and connects nodes p5 and p6.
- F7 is a non-fatal fault that only increases the resistance of the connection to m1.

The very procedure of associating a functional fault with a physical defect is often called *inductive fault analysis* [5, 11–13]. First, a set of "dangerous" physical faults is established that can be assumed to be extractable from the mask information used to build the circuit. An extended list of possible physical failures is given in Table 7.1.1. Then, based on the structure of a given set of masks (for a given integrated circuit), based on knowledge of the technological process, the probability of the existence of those failures is generated. This creates a hierarchical list of faults. In that list, the most significant are the failures that are the most likely. Finally, physical defects are joined by faults in the way we discussed above. This is how an inductive list of faults is created.

Table 7.1.1 Physical defects

F1	Metal break
F2	Polysilicon break
F3	Break of diffusion
F4	Break at the metal-polysilicon junction
F5	Break at the metal-diffusion junction
F6	Break at the polysilicon-diffusion junction
F7	Break at the metal-polysilicon-diffusion junction
F8	Short circuit between metal connections
F9	Short circuit between polysilicon bonds
F10	Short circuit of diffusions
F11	Short circuit between metal and polysilicon
F12	Short circuit between metal and diffusion
F13	Short circuit between polysilicon and diffusion
F14	Short circuit between metal, polysilicon, and diffusion
F15	New transistor
F16	The transistor constantly in conducting state
F17	Transistor constantly cut-off
F18	A transistor is missing
F19	Change of the transistor's *W*/*L*
F20	A new junction has been created
F21	Simultaneous creation of a new transistor and a metal-diffusion break

Of course, in addition to these faults, there are others that this procedure does not cover. One of them is a gate breakthrough which we will discuss in more detail later. In addition, "faults" whose importance increases with the reduction of the dimensions of the components are the so-called global defects affecting a large part of the chip surface. Among these are fluctuations in oxide thickness, too strong or too weak etching on one part of the surface, etc.

A careful analysis establishes the consequences of a physical defect on the electrical properties [14]. At the same time, the number of possible physical defects is significantly higher than the number of faults modeled in the electrical circuit's schematic. For example, electrical disconnection is a fault that models each of the following physical defects: lack of metal in the first layer of bonding metal; lack of metal in the second metal layer for bonding, and lack of polysilicon for bonding. Similar mappings are performed with other defects.

In the following text, for the sake of illustration, an example of mapping of a technological defect (breakdown) into a fault effect will be briefly described [15]. It is a breakthrough of the gate oxide of a MOS transistor. Breakthrough can be caused by electrostatic discharge or electric shock. It has the effect of establishing an electrical connection from the gate to the silicon surface of the MOS transistor. As shown in Fig. 7.1.1b that connection can lead from the gate to the substrate (p- or n-wells), from the gate to the source, and from the gate to the drain.

Figure 7.1.2 shows the output characteristics of the MOS transistor before (**a**) and after (**b**) gate oxide breakdown. It can be seen from figure that the transistor has not lost its basic property. From the point of view of these characteristics, the most important difference between a correct and a defective transistor is in the reduction of its transconductance. Additional measurements showed that the damaged components have an average of 29% lower transconductance than the correct ones. What is not shown, as usual, is the gate current of the MOS transistor. Namely, due to the breakthrough from the gate to the substrate, there is a finite input resistance of the gate transistor and a finite gate current. These considerations (transistor transconductance change and decrease of its input resistance) can be used to create a fault model when simulating the faults of an analog circuit that would contain a MOS transistor with a broken gate.

For a complete characterization of the fault effect, however, a more complex experiment was performed in [15]. One transistor, whose gate breakthrough was deliberately caused by a laser beam, was built into a circuit as shown in Fig. 7.1.3. It is a PMOS transistor in the second inverter. Figures 7.1.4 and 7.1.5 represent the measured responses of this circuit.

Figure 7.1.4 shows the measured dependence of the input and output voltage of the second inverter in the circuit of Fig. 7.1.3. It can be seen that the value of the input voltage (output voltage of the first inverter), due to the finite input resistance of T_4, reaches only six volts, which is enough for the three-stage inverter to preserve its function (despite the fact that some of the characteristics have being deteriorated).

On the other hand, Fig. 7.1.5 represents the overall transfer characteristic and power supply current of the three-stage inverter. Here it can be seen that the voltage transfer characteristic has an almost perfect shape. Eventually (imperceptibly), its

Fig. 7.1.2 Characteristics of **a** correct and **b** damaged transistor

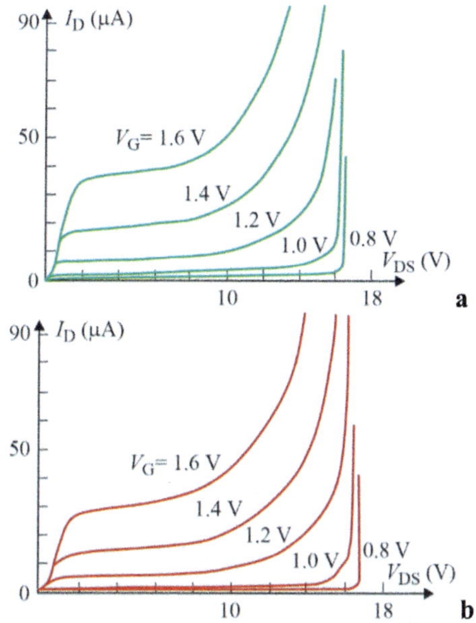

Fig. 7.1.3 Three-stage inverter circuit in which one of the transistors is faulty

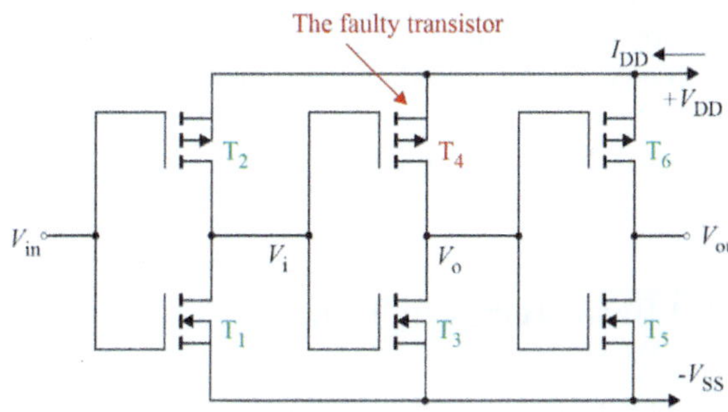

Fig. 7.1.4 Input (V_i) and output (V_o) voltage of the second inverter circuit from Fig. 7.1.3

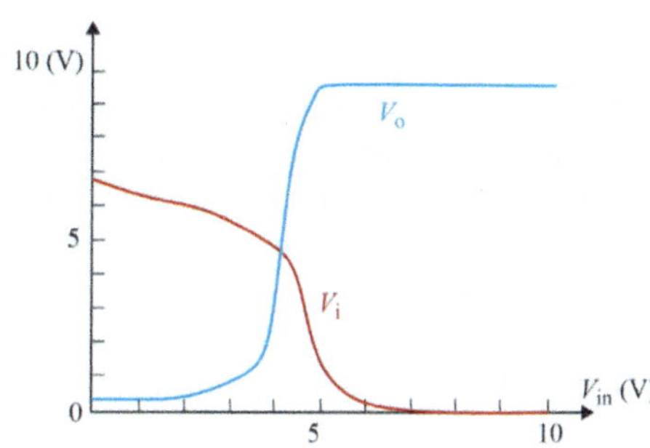

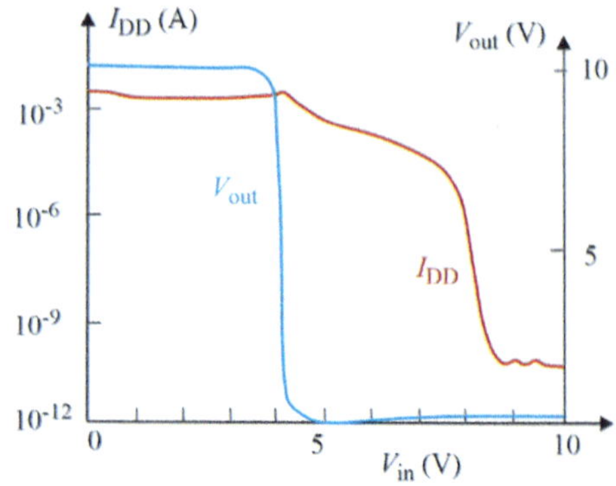

Fig. 7.1.5 Transfer characteristic (V_{out}) and supply current (I_{DD}) of the circuit from Fig. 7.1.3

slope deviates from the slope of the transfer characteristic of the correct circuit. However, this does not apply to the supply current. Namely, it is known that with CMOS inverters, the supply current is practically equal to zero all the time, except for a short interval when the output voltage changes state. In this case, however, when the gate voltage of the faulty transistor increased, it drew a current of as much as 2.7 mA.

The result of this study exposes the mapping of the fault into the fault effect and the importance of measuring the supply current from the point of view of testing and diagnosing faults and, at the same time, about a possible test signal that would enable the detection of a gate-to-substrate breakdown type defect. That test signal is therefore a logical unit on the gate of the faulty PMOS transistor.

7.1.3 Basic Terms in Testing

Based on the previous presentations, an impression was gained about the origins of the faults in modern electronic circuits and the importance of testing as well as the goals that should be achieved with it. In this section, the basic concepts applied in fault modeling as well as in the generation (synthesis) of test signals will be listed.

In order to get an idea of the basic problems with which the generation of test signals collides, let us imagine that in a combinational circuit there is an NAND-gate with a fault so that its output is stuck-at-1 as in the case of the circuit marked with "1" in Fig. 7.1.6.

In order for the fault effect to become observable at the input of the NAND-gate marked with "2", it is necessary that the signals *B* and *C* are such that they force a

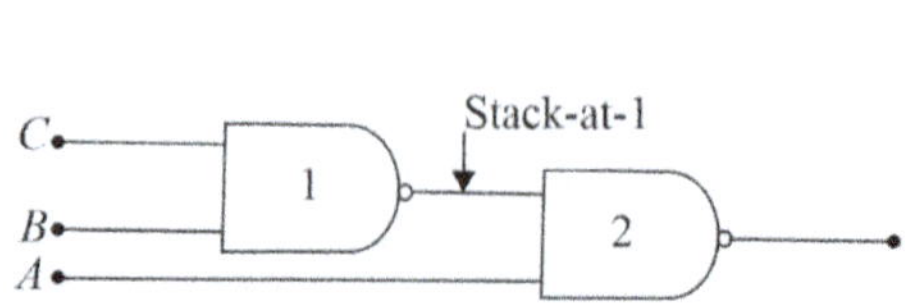

Fig. 7.1.6 Output of the NAND circuit denoted "1" is stuck-at-1

logical zero at the output of circuit "1". A logic 1 that is constantly present due to a fault will be different from the desired value of the signal which allows the *detection* of the existence of the fault.

It is said that in this way the *detectability* of the defect we are considering is ensured. If there is a set of input signals that will force a given node into a state opposite to that caused by the fault, we say that the defect is detectable. *Controllability* is sometimes used as a measure of the detectability of digital circuits. This means either the number of primary inputs that should be controlled in order to detect a given fault or the number of logical operators (gates) from the primary input to the fault location, which should be set in appropriate states to allow detectability.

With analog circuits, the notion of detectability is broader. Namely, from the example of the inverter from Fig. 7.1.3 it was seen that for a given input signal, the output voltage did not show the fault effect, which does not mean that detectability is not achieved with the given input signal. The same input signal can be used to achieve the needed fault effect, but the supply current should be observed, not the output voltage.

This discussion once again confirms that the problem of testing analog circuits is more complex. After all, there may be some waveform of the input voltage for which the fault effect is also mapped into the output voltage. Furthermore, if in a circuit a fault of the stuck-at-open type is tested in the branch where a capacitor is located, a DC signal cannot be used as a test. A time-varying signal, however, such as a sinusoid, for example, can prove to be very convenient. In that, the notion of controllability as expressed for digital circuits loses its importance. This resulted in the fact that today most often, controllability is understood as detectability. This is especially true for discussions on analog testing.

Compiling a signal (test) that ensures detectability does not mean that the work of generating the test is finished. It is still necessary to ensure that the fault effect is mapped onto the output. In the circuit of Fig. 7.1.6 this is achieved by setting the signal *A* to logic 1. This, in effect, makes it neutral and allows the upper input terminal of circuit "2" to control the output. It is said that the path from the fault site to the exit is made sensitive to the fault (*sensitized*) and that the "observability of the effect of the fault," or simply "*observability*," is enabled. If both detectability and observability are achieved, the defect is *testable*. If, by any chance, due to the configuration of the circuit it was not possible to set the signal *A* to logic 1, it would not be possible to compile a complete test, and we would speak of an untestable fault. In such an example, the path from the fault to the output would be insensitive to the fault. If testability is insisted upon, in such a situation, the project or logic schematic would have to be changed. Such an activity is known as "designing for testability" or in short DFT. With analog circuits, the question of observability is posed in a different light. Namely, with analog circuits, observability is always present, but this does not solve the problem. The fault effect, measured at the output, can be so small that it is practically immeasurable. In that case, observability is lost. Increasing the observability can be achieved in several ways: by choosing the output quantity that is observed (voltage, current, supply current), by choosing the waveform that is used (direct current, sine, pulse, and other waveforms), as and by choosing suitable values of the signal parameters used. For example, bearing in mind the frequency dependence of the sensitivity of the output voltage to parameter changes, one should choose that value of the signal frequency for which the sensitivity is the highest and the observability is achievable in the best. In the case of nonlinear circuits, it is necessary to simultaneously consider both the DC components that provide the working conditions and the time-varying components that are processed.

Although the answer has already been partially indicated, the question about the concept of choosing a testing procedure deserves more attention. Namely, the issue of testing can be viewed in the following way. A circuit is given that is designed to fulfill a function. Checking whether it fulfills that function is, in fact, testing the circuit. This approach to testing is known as *functional* or *specification-based testing*. The problem of generating the test signal is solved per se because it is contained in the project specifications. The following problems, however, have been observed with this concept.

When dealing with a simple combinational circuit, the use of a truth table (hence an exhaustive functional check) seems tempting. By bringing all possible combinations of input signals, we apply an exhaustive test that can establish any deviation of the response of the measured circuit from the response of the correct one. In connection with the exhaustive test of combinational circuits, however, two problems can be observed. First, it is not easy to develop a truth table for a circuit of arbitrary complexity. Such a table is usually not submitted to the designer who deals with testing.

Second, a large, printed circuit board or integrated circuit may have dozens or hundreds of input terminals. In addition to the difficulty of generating a truth table for such a system, an exhaustive test is not at all practical. Namely, for a circuit with n inputs, the truth table contains 2^n lines. If the required testing time is only one microsecond for each row in the table, a circuit with $n = 30$ will take 18 min, and a circuit with $n = 50$, all 36 years. For sequential circuits, the concept of an exhaustive test is even more difficult to apply.

It is particularly interesting to consider the influence of the method of expressing the function of complex integrated circuits on test generation. Although the function is basically easy to define, that definition is difficult to translate into a set of test signals whose composition and time order are precisely determined. A microprocessor, for example, can be defined as a component that executes each instruction in an instruction set. An exhaustive test set of instructions, however, would process every instruction with all possible operands and combinations of operands. This would require practically infinite time, as with combinational circuits. In analog circuits, as shown with the faulted gate in the circuit of Fig. 7.1.3, the functional performance of the circuit can be largely satisfied, while some others (such as the value of the supply current, i.e., the total dissipation of the circuit) are unacceptable.

Nevertheless, when we consider functional testing, we can say that despite the limitations, some elements of the concept of functional testing find application in the automatic generation of a test sequence, although an exhaustive test, except for small circuits, is never applied.

Instead of checking that the circuit is working properly, which corresponds to a functional test, it is possible to check whether the circuit is faulty. Test programs based on this concept are known as *structured test* programs. They try to detect individual structural faults.

The goal of testing based on the structural concept is to construct such a test signal that will ensure that the *faulty* circuit or system, when given the appropriate test, will perform incorrectly. We will abbreviate this activity as GTS, which stands for Generating Test Signals. Here, as with functional testing, the danger of the curse of dimensionality is constantly lurking. Namely, the number of possible faults in one complex circuit is very large. Putting together a test signal that will detect every possible fault, per se, is a very extensive job. To test the circuit in such a way as to check the presence of each possible fault individually (in the industrial conditions of high-series production) is an even more extensive and expensive job. Therefore, as already mentioned earlier, from the total set of theoretically possible faults, a "realistic set of faults" is formed by applying appropriate criteria. Some of those criteria are based on the statistical properties of certain faults that were previously established in the production process, and some are related to the circuit layout and possible faults that arise from the mutual special relationship of the circuit parts. For such a formed realistic set, a test sequence is generated, which will determine that, with some probability, the circuit will work correctly.

Figure 7.1.7 shows the flow of the basic GTS procedure based on the structural concept. It is assumed that there is a preprepared list of faults (usually based on the inductive fault analysis) for which the circuit should be tested. Each test signal (TS) is composed to cover one of the faults, the first in line from the list of faults. Very often it will turn out, however, that the given test will also cover some other faults from the list. As a result, the number of tests required in the final test program will be significantly less than the number of faults in the fault list. Therefore, to avoid unnecessary repetition, it is checked which faults are covered by the given test, and the covered faults are deleted from the list regardless of the fact that no separate test has been written for them. Finally, it is checked whether the test for each fault has been composed, i.e., whether all faults from the list of faults have been covered. Now, once again, it can be confirmed that the list of faults for which we compose tests is shorter than the total list of faults for reasons of economy. Otherwise, the number of cycles in the procedure described in Fig. 7.1.7 would be very large even for not very complex circuit.

Prepare a list of faults.
While all faults from the list are **exhausted** {
  Select the next fault.
  Compose a test for the selected fault.
  For **all other elements** in the fault list {
    Delete from the list of faults those that are covered by the composed test.
  }
}

Fig. 7.1.7 Generating test signals based on the structural concept

References

1. Bleeker H, van den Eijnden P (1993) Boundary-scan test, a practical approach. Kluwer Academic Publishers, Dordrecht
2. Bleeker H, van den Eijnden P (1993) IEEE standard test access port and boundary-scan architecture. IEEE, New York
3. Bleeker H, van den Eijnden P (2010) IEEE draft standard for a mixed-signal test bus. IEEE P1149.4/D2, pp 1–113
4. Milovanović DP (1991) Modeling and Simulation of faults in CMOS integrated circuits using a modified concurrent method. PhD thesis, Elektronski Fakultet, Niš, Serbia (in Serbian)
5. Sachdev M, de Gyvez JP (2007) Defect-oriented testing for nano-metric CMOS VLSI circuits. Springer, Berlin
6. Spina R, Upadhyaya S (1997) Linear circuit fault diagnosis using neuromorphic analyzers. IEEE Trans CAS II Analog Digit Signal Process 44(3):188–196
7. Soma M (1996) Automatic test generation algorithms for analogue circuits. IEE Proc Circuits Devices Syst 143(6):366–373
8. Fenton WG, McGinnity TM, Maguire LP (2001) Fault diagnosis of electronic systems using intelligent techniques: a review. IEEE Trans Syst Man Cybern Part C Appl Rev 31(3):269–281
9. Stratigopoulos H-G (2023) Machine learning support for diagnosis of analog circuits. In: Machine learning support for fault diagnosis of system-on-chip. Springer Nature, Berlin, hal-03872972
10. Liao W, Liu J (2008) Research on k-fault diagnosis and testability in analog circuit. WSEAS Trans Electron 5(10):417–426
11. Corsi F, Morandi C (1991) Inductive fault analysis revisited. IEE Proc Part G 138(2):253–263
12. Jee A, Ferguson FJ (2003) Carafe: an inductive fault analysis tool for CMOS VLSI circuits. Digest of papers, Eleventh annual 1993 IEEE VLSI test symposium, Atlantic City, NJ, USA
13. Harvey RJA, Richardson A, Kerkhoff HG (1995) Defect oriented test development based on inductive fault analysis. Collection of papers presented at the International mixed-signal testing workshop, Grenoble, pp 2–9
14. Maly W (1987) Realistic fault modeling for VLSI testing. In: Proceeding of the 24th ACM/IEEE design automation conference, Miami Beach, Florida, pp 173–180
15. Soden JM, Hawkins CF (1986) Test considerations of gate oxide shorts in CMOS ICs. IEEE Des Test Comput 3(4):56–64

7.2 Testing Analog Circuits

Abstract Synthesis of a sequence of test signals for a rather complex analog circuit is a serious professional challenge which is mainly related to the difficulties to ensure both controllability and observability of the fault. Usually, attempts are made to use alternatively the DC, frequency, and time domain so exploiting the specific properties of the circuit. To do that, in this chapter we introduce several concepts including correlation of the time domain responses, multifrequency analysis, and most frequently DC studies. All these will be elaborated and exemplified in the next. In addition, attention will be paid to the testing economics, i.e., to find tests that need simple equipment, need less time to be implemented, and (possibly) cover as many faults as possible. An algorithm for creation of an optimal testing sequence will be described too.

7.2.1 Introduction

Analog circuits are not so complex as digital ones when weighed from the point of view of number of elements. From the point of view of testing, however, specific problems arise that make testing more complex. Testing is often performed over large ranges of circuit parameters, with large ranges of ambiental parameters as well as over long frequency and time intervals. This makes testing uneconomical and impractical. The accuracy of the measured values depends largely on the resolution of the test equipment and the quality of input excitations created for testing purposes. On the other hand, special measures are often taken to make the response insensitive to some parameters, when synthesizing analog circuits. Thus, in the middle of the bandwidth, the gain of the feedback system does not depend on the gain of the amplifier in it. This, however, makes it difficult to test the correctness of that part of the circuit. Furthermore, some parts of the circuit, such as the temperature compensation or short-circuit protection circuit are completely inactive (latent) until extremely specific operating conditions are established. All of this speaks in favor of applying the structural test concept to generation of the analog circuit test.

V. B. Litovski, *Lecture Notes in Analog Electronics*, Lecture Notes in Electrical Engineering 1265, https://doi.org/10.1007/978-981-97-8257-4_2

A special problem when testing a mixed-signal circuit may become the mutual influence of the analog and the digital part due to interference since they are built on the same silicon die (or, in some cases on the same printed circuit board). For example, switching noise in the digital part may be considered as a (strong) noise for the analog which may disturb its function.

It does not take much to prove that in the practical implementation of the test generation procedure we always rely on both concepts. Functional testing does not have to be complete just like structural testing is not. Applied together, these concepts can lead to highly effective and reliable results.

In this chapter, we will first consider the basic aspects of the problem of choosing test signals and their parameters. The basic concepts of testing will be presented through the presentation of three signal-generation procedures for testing analog circuits: testing based on the supply current, multifrequency analysis, and testing in the time domain. In doing so, the review work [1] as well as [2] will be used as a basis. Regardless of the methods that will be listed here, it should always be remembered that a functional test is also applied at the same time. For a filter, for example, it is natural to first perform system measurements out of which it will be easy to conclude about the correctness of the basic properties of the circuit (gain at medium frequencies, cut-off frequencies, etc.).

7.2.2 Testing Based on the Supply Current

The supply current is, as has been emphasized several times, a picture of all the events in the circuit, and as such it is very often used for testing purposes. As a method for testing integrated circuits, it was first introduced in 1981 [3]. At the same time, when talking about testing based on the DC component of the supply current one uses different labels such as I_{DDQ}-, I_{DDX}-, or simply I_{DD}-test. In addition to steady-state supply current analysis, transient supply current analysis [4], called I_{DDT} testing and RMS supply current analysis (I_{RMS} testing) [5], can be used to test electronic circuits. Finally, as a succession, attempts have been made to develop test methods based on transients of the supply voltage [6], referred to as V_{DDT} testing.

The special feature of testing using the supply current refers to the fact that in this case, in principle, it is not necessary to apply special measures in order to achieve observability, i.e., measures that would allow the fault effect to spread from the faulty site to the exit. The output, in this case, the supply voltage terminal, is always easy to reach. This property of the I_{DD} test is sometimes called "automatic observability [7]."

The presentations in this section will be largely based on the results described in [8].

There are three basic ways of using the power supply current for testing purposes.

DC supply current has always been an essential part of component testing. It is a measure of the total dissipation in the static mode and as such must be known in advance. Its value, however, depends on the value (state) of the input signals. With digital circuits, it is possible to compile a dictionary that contains, on the one hand, a set of input words, and on the other, the values of the direct current of the power supply. With analog circuits, it is possible to generate the transfer characteristic: input_voltage–supply_current. A fault in the circuit may or may not affect the value of the DC supply current.

Thus, for the fault described in Fig. 7.1.1, when a faulty transistor is used as in Fig. 7.1.3, at some values of the input signal (which can be seen from Fig. 7.1.5) the supply current is noticeably different from that in the fault-free circuit. For input voltage values higher than 8.5 V, however, it cannot be said that a fault or breakdown in the gate of the PMOS in the second inverter is recognized in the supply current. It should be emphasized here that the entire logic function (or the gain, if the chain is considered as a three-stage CMOS amplifier) of the circuit from Fig. 7.1.5 is not disturbed by the given fault, i.e., which means it is logically redundant. Nevertheless, the I_{DDQ} test reveals such a fault.

The presence of multiple faults can sometimes lead to a wrong conclusion about the correct operation of the circuit. Namely, it can happen that the "next" fault in the direction of signal propagation "masks" the effect of the previous one and thus leads to a wrong picture of the correctness of the circuit. The I_{DDQ} test, however, accumulates the faulty currents (accepts them as parallel information) and masking is practically impossible. This is especially so if it is possible to separately monitor the supply currents of individual parts of the circuit [9], which is convenient for printed circuit boards. Finally, there is no other way to detect faults such as a shifted value of the threshold voltage of the transistor, a short circuit of the gate, an interruption of the bilateral switch, and some others.

We should note here that, as described in [10], as "input voltage" one may use the power supply voltage and create proper dependence of any circuit response (e.g., output voltage of the circuit) on it. The authors claim "When the power supply is discretely ramped up, all transistors in the circuit pass through several regions of operation, e.g., subthreshold (region A), linear (region B), and saturation (region C). The advantage of a transition from region to region is that defects can be detected with distinct accuracy in each of the operating regions…"

In some situations, however, it is not enough to know only the DC component of the supply current to detect the fault. To illustrate this, consider a transient in a circuit containing operational amplifiers. The current that the operational amplifier "draws" from the power source is mainly determined by the constant current source in the differential amplifier. However, during the transient, a large part of the supply current is the current of the output stage, which is also a measure of the output voltage of the operational amplifier. This effect can be amplified if the operational amplifier is additionally loaded. Therefore, by measuring the supply current in transient mode, we can establish a possible fault in the output circuit.

When a time-varying signal is used for fault detection, the obtained result can be used in two ways. One of them is to analyze the supply current spectrum (I_{DDT}), and the other is to measure its effective value (I_{RMS}).

Recognizing the presence of distortion is most effectively achieved by comparing the spectra of the supply current of the fault-free and faulty circuit. At the same time, for more efficient work and considering that the waveforms are given as a table of values, spectral analysis is performed with a spectrum analyzer that applies a fast Fourier transform (FFT).

Comparing the thus obtained spectra of the responses of the fault-free and faulty circuit is a specific task. It is based on the calculation of the autocorrelation of the spectrum of the response of the fault-free circuit:

$$R_{i_{\mathrm{DD}}i_{\mathrm{DD}}}(\tau) = \frac{1}{T}\int_{0}^{T} i_{\mathrm{DD}}(t)i_{\mathrm{DD}}(t+\tau)\cdot \mathrm{d}t, \tag{7.2.1a}$$

and the correlation of the response spectra of the fault-free and faulty circuits:

$$R^{\mathrm{F}}_{i_{\mathrm{DD}}i^{L}_{\mathrm{DD}}}(\tau) = \frac{1}{T}\int_{0}^{T} i_{\mathrm{DD}}(t)i^{\mathrm{F}}_{\mathrm{DD}}(t+\tau)\cdot \mathrm{d}t. \tag{7.2.1b}$$

The superscript F is used to mark the signal in the circuit with a fault (faulty) while the values for the correlation coefficient depend on the sampling interval used to calculate the integral and on the duration of the correlation delay τ. A suitable choice of these quantities can affect the accuracy of the calculation and the efficiency of the procedure [11].

Index of functionality (I_{f}) [12] is introduced which tells how similar the spectra of a fault-free and the faulty circuit are:

$$I_f = \left[1 - \frac{\sum\left(\left|R_{\mathrm{xx}} - R_{\mathrm{xy}}\right|\right)}{\sum\left(\left|R_{\mathrm{xx}} - R_{\mathrm{xy\text{-}wc}}\right|\right)}\right]\cdot 100\%, \tag{7.2.1c}$$

where R_{xx} is given by (7.2.1a), R_{xy} by (7.2.1b), and "wc" denotes the worst case for all circuits and refers to a hard fault which is usually taken when the output is permanently at ground. The summation in (7.2.1c) is performed over all faults.

It can be expected that each fault will distort the supply current in its own way, that is, it will imprint itself in its spectrum. This, however, as in other situations, does not have to be sufficiently expressed. That is why sometimes the measurement of the supply current is combined with measurement of the response. Using both pieces of information makes fault detection even more reliable. For this purpose, the correlation function of two responses is used, which will be briefly discussed here.

Let $v^{\mathrm{G}}_{\mathrm{out}}(t)$ denote the voltage at the output of the fault-free (good) circuit, and $i^{\mathrm{G}}_{\mathrm{DD}}(t)$ its supply current. Let, in the presence of a fault, it be measured $v^{\mathrm{F}}_{\mathrm{out}}(t) = v^{\mathrm{G}}_{\mathrm{out}}(t) + \Delta v_{\mathrm{out}}(t)$ and $i^{\mathrm{F}}_{\mathrm{DD}}(t) = i^{\mathrm{G}}_{\mathrm{DD}}(t) + \Delta i_{\mathrm{DD}}(t)$. For the correlation coefficient of the signal in the faulty circuit, we have

Fig. 7.2.1 Impedance converter. Z_1, Z_2, Z_3, and Z_4 are resistances, and Z_L is an inductance

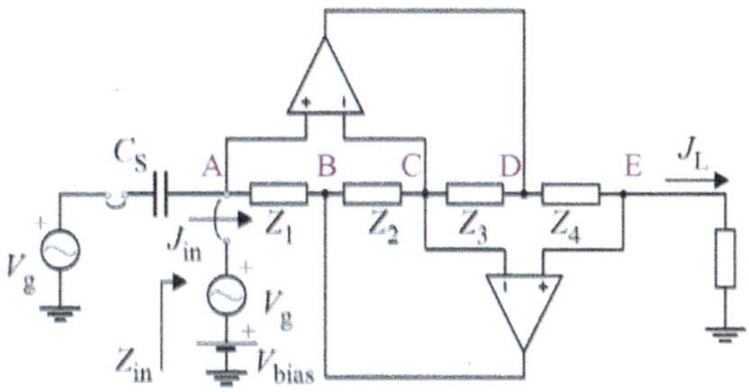

$$R^{F}_{v_{out}i_{DD}}(\tau) = \frac{1}{T}\int_{0}^{T} v^{F}_{out}(t)i^{F}_{DD}(t+\tau)\cdot dt$$
$$= R^{G}_{v_{out}i_{DD}}(\tau) + R_{\Delta v_{out}\Delta i_{DD}}(\tau) + R_{v_{out}\Delta i_{DD}}(\tau) + R_{\Delta v_{out}\Delta i_{DD}}(\tau). \quad (7.2.1d)$$

It can be seen that the correlation coefficient between the output voltage and the supply current of the faulty circuit depends on the four components that arise after the substitution of $v^{F}_{out}(t)$ and $i^{F}_{DD}(t)$ into (7.2.1). The obtained results as reported in [6] show better fault coverage than when the output signal and supply current measurements are analyzed individually.

Finally, the presence of a fault is reflected in the effective (RMS) value of the supply current. That is why the effective value, which is easier to measure, can be used for testing purposes.

Figure 7.2.1 shows an impedance converter circuit where the impedances Z_1 to Z_4 are resistors (100 kΩ), and the impedance Z_L is inductive ($L = 50$ mH). We first use the upper branch for excitation with the coupling capacitor used as $C_S = 0.01$ F. The schematic of the amplifiers used is shown in Fig. 7.2.2a [13]. In this amplifier all P-channel transistors have $W/L = 96.94$ with $L = 55 \times 10^{-6}$ m, and all N-channel transistors have $W/L = 1487.41$ with $L = 20 \times 10^{-6}$ m. Exception is T_7 for which $W/L = 1.38696$ with $L = 3.605 \times 10^{-6}$ m is valid.

The power supply voltages are $V_{DD} = -V_{SS} = 10$ V. Figure 7.2.2b depicts the frequency domain response of the operational amplifier (OA) used, while Fig. 7.2.2c represents two impedances in the circuit.

Here, the blue line represents (scaled) modulus of the (capacitive) input impedance of the circuit of Fig. 7.2.1 and the red line the impedance (inductance) of the load.

If the circuit is excited by (V_g) a sinusoidal signal of amplitude 5 V (Fig. 7.2.3a) at supply voltages of ±10 V, the waveform of the supply current that is shown in Fig. 7.2.3b is obtained. It can be observed that a non-sinusoidal waveform is obtained which, for periodic excitation, has periodic properties.

For a 10 kHz signal (as in Fig. 7.2.3a where $T = 100$ μs), the DC value of the supply current is $I_{DDQ} = 94.83$ mA, and its RMS value is $I_{DRMS} = 95.15$ mA. The presence of a fault will be reflected in the distortion of the waveform from Fig. 7.2.3b. If the channel width of transistor T_5 is changed by 100% (doubled) in the first (left) operational amplifier, the supply current waveform shown in Fig. 7.2.3c is obtained.

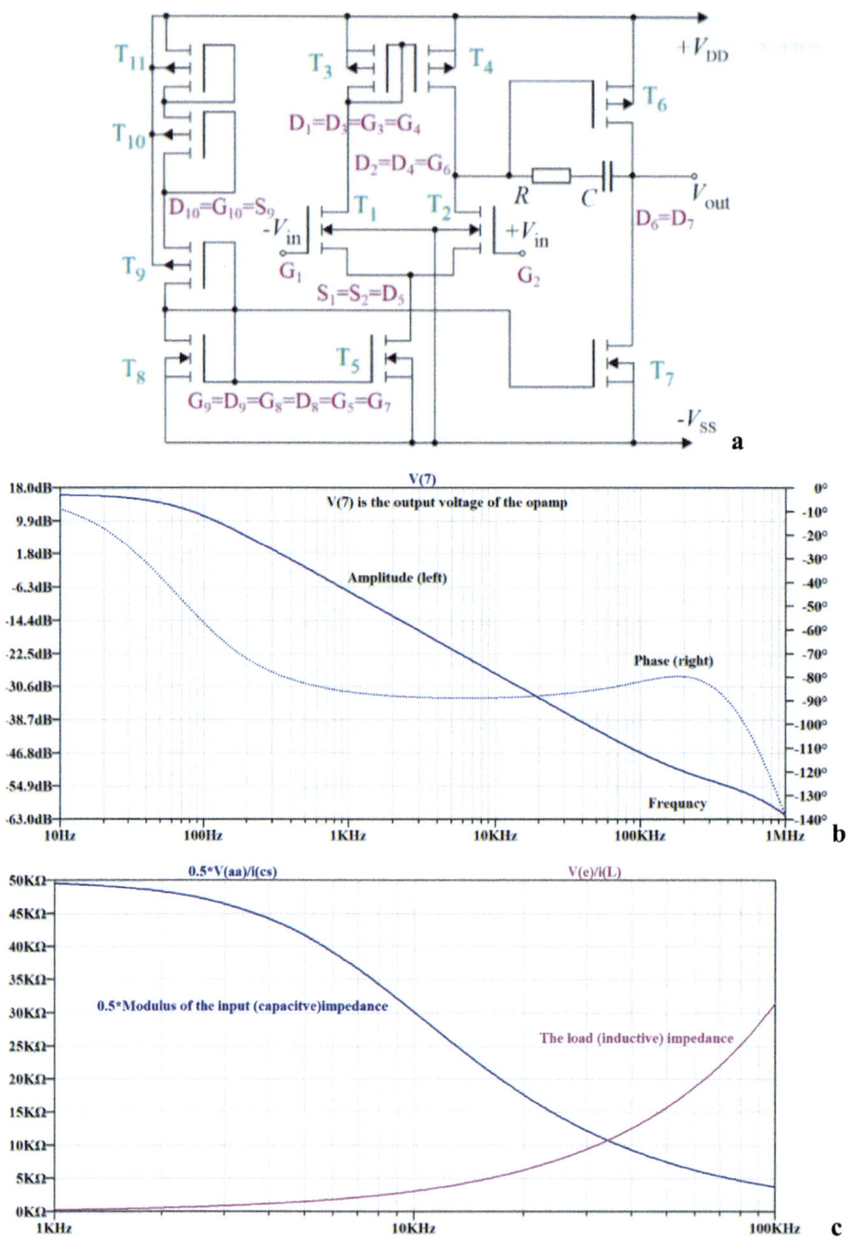

Fig. 7.2.2 Operational amplifier used in the circuit of Fig. 7.2.1. **a** Schematic of the OA, **b** its amplitude and phase characteristics, and **c** impedances in the circuit of Fig. 7.2.1

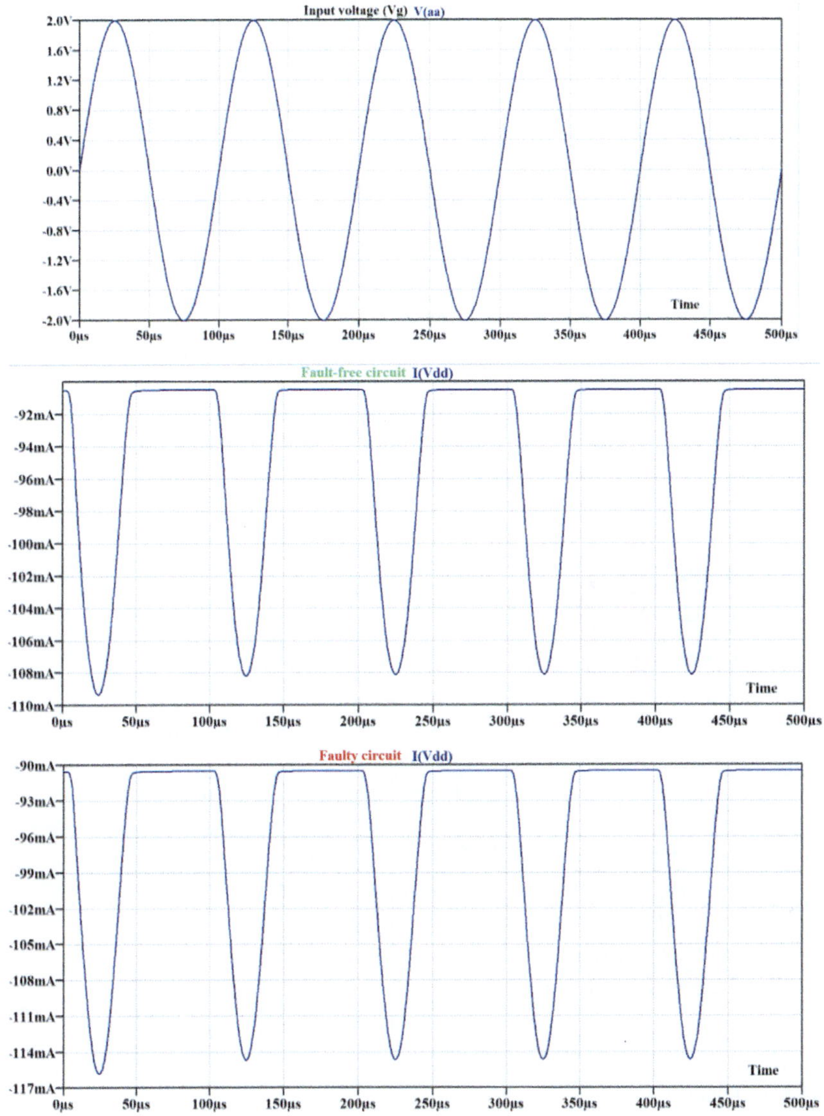

Fig. 7.2.3 Waveforms of input voltage ($V_{in} = V(\text{a})$) and supply current (I_{DD}) of the circuit from Fig. 7.2.1. **a** Input voltage, **b** supply current of the fault-free circuit, and **c** supply current of the faulty circuit as a function of time. Spectra of the supply current of the circuit from Fig. 7.2.1. **d** Fault-free circuit, **e** faulty circuit, and **f** comparison table

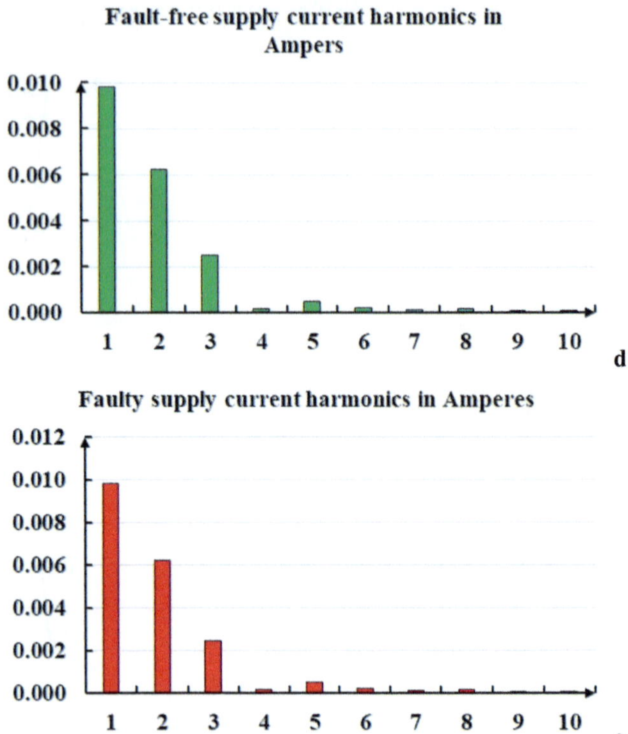

Parameter	Fault-free circuit	Faulty circuit	Common parameter
I_{DDQ}	94.83 mA	96.19 mA	
I_{DRMS}	95.15 mA	96.676 mA	
THD	66.41 %	69.24 %	
Autocorrelation ($R_{i_{DD}i_{DD}}$)	0.0352 A^2	0.0352 A^2	
Mutual correlation ($R_{i_{DD}i_{DD}^{L}}$)			0.02765 A^2
Index of functionality (I_f)			78.47 %

f

Fig. 7.2.3 (continued)

In this case, the DC value of the supply current is $I_{DDQ}' = 96.19$ mA, and its effective value is $I_{DRMS}' = 96.676$ mA.

By spectral analysis of the signals from Fig. 7.2.3b, c, the results shown in Fig. 7.2.3d, e respectively, are obtained.

The most noticeable difference is in the fundamental harmonic (10,000 Hz). The following mean values are obtained for these signals (τ is changed from 0 to T):

$$R_{i_{DD}i_{DD}\_av} = 3.52321 \times 10^{-1}\ A^2$$

and

$$R_{i_{\mathrm{DD}} i^{\mathrm{F}}_{\mathrm{DD\_av}}} = 2.76478 \times 10^{-2}\ \mathrm{A}^2.$$

If we assume $R_{i_{\mathrm{DD}} i^{\mathrm{F}}_{\mathrm{DD\_wc}}} = 0$, we get $I_{\mathrm{f}} = 78.47\%$. In that the sums in (7.2.1c) have only one summand since only one fault is considered.

Before proceeding we will recapitulate what we learned from this example. The results are summarized in a form of table in Fig. 7.2.3f.

By studying this table we may conclude how sensitive the supply current is to a single parametric change (fault) in a relatively complex circuit. Both the I_{DDQ} and I_{DRMS} are noticeably different. That also stands for the THD. The value of the mutual correlation and the index of functionality are even more conceivable expressing the difference in the behavior of the fault-free and the faulty circuit.

This would exhaust the description of the example.

The RMS value of the supply current is a function of the DC component, of the amplitude, and of the frequency of the input signal. In that, for testing purposes, the DC component of the input voltage should be chosen in such a way as to ensure maximum sensitivity (or maximum deviation) of the RMS value to changes in the signal frequency. To make it clearer, we can say that it can be expected that with the same fault, the change in RMS value will be different for different values of the DC component of the input test signal. The amplitude of the alternating component should be large enough to achieve activation or controllability of the fault, and at the same time small enough not to disturb the operating mode of the correct circuit. It usually does not change during the measurement. Which values of the DC component of the input signal (usually two or three) and which frequency values (also two or three) enable the detection of faults will be determined by simulations. The final result contains a table that has the form of Table 7.2.1. The idea was first proposed in [2].

Table 7.2.1 Test vectors to be used for testing the listed set of faults in the left operational amplifier of Fig. 7.2.1

Type	Test vector			FF	Faulty	Difference (mA)
	V_{g} (V)	V_{bias} (V)	f (Hz)	I_{DRMS} (mA)	I_{DRMS} (mA)	
D1-at-S1	0.1	−3	1000	196.26	120.61	75.65
G1-at-S1	0.1	0	1000	192.97	90.489	102.481
G2-at-S2	0.1	−3	1000	196.26	129.77	66.49
D4-at-S4	0.1	−3	1000	196.26	120.6	75.66
D4-at-G4	0.1	0	1000	192.97	142.07	54.2
D5-at-S5	0.1	3	1000	120.61	1147.1	−1026.5
G7-at-S7	0.1	−3	100	196.26	98.465	97.795
D7-at-S7	0.1	−3	1000	196.26	120.61	75.65
G9-at-S9	0.1	3	1000	120.61	309.68	−189.07

It was implemented here on the circuit of Fig. 7.2.1, and faults were introduced in the first to the left operational amplifier. The excitation was changed, however. A DC offset (V_{bias}) was added as shown in the figure. A list of faults was created according to Table 7.2.1 where, as usual x-at-y means short circuit between nodes x and y. The following values were associated with the excitation signal: The amplitude of the AC signal component $V_g \in \{0.1; 1; 2.5; 5; 7.5\}$ V, the DC offset of the input signal $V_{\text{bias}} \in \{-3; 0; 3\}$ V, and the frequency of the input AC signal $f \in \{100; 1000\}$ Hz. This means that ten simulations per fault were performed in order to find the test vector $\boldsymbol{X} = \{V_g, V_{\text{bias}}, f\}$ which leads to maximum deviation from the fault-free response of the circuit. After selection, Table 7.2.1 was obtained.

By analyzing this table one should mention first that it appears that the signal amplitude has no influence on the I_{DRMS}. That in fact is not true. Simply, the amplitude given in the table leads to the largest difference between the faulty and the fault-free value of I_{DRMS}.

Looking to the table as a whole, for the selected list of faults, one may say that noticeable difference in the value of I_{DRMS} may be created allowing proper testing. In addition, one may notice that the same test vector covers several faults which may be exploited for reducing the number of tests. On the other side, all but three differences differ suggesting that these results may be used for diagnostic purposes too.

When analyzing this table one should not forget that I_{DDQ} is also affected by the fault and as can be seen from the simulation results, in most cases, its change dominates the change of the RMS value.

Before proceeding we will shortly recapitulate the I_{DDQ} and I_{DRMS} testing idea in order to estimate the complexity of the task of test signal synthesis. Namely, for a single soft fault, when I_{DDQ} test was considered, we had to perform two simulations and to calculate the correlation coefficients and the index of functionality. In that, a simulator and software for post-processing must be used alternatively.

That means, if thorough testing is planned, a huge amount of computations must be performed including careful fault insertion and data management. Similarly, for every catastrophic fault of Table 7.2.1 (containing only reduced set of faults in one of the operational amplifiers) ten simulations were to be performed in order to get the test vector. To conclude, generation of a proper test is a laborious and time- and resource-consuming task which makes testing expensive and, frequently, suggests reduction of the set of possible faults to allow faster time to market.

7.2.3 Analog Testing in the Frequency Domain

Analog circuits are most frequently realized as linear amplifiers and filters. In this case, usually, for testing purposes, one may observe the amplitude characteristic (nominal gain and cut-off frequencies) and the phase characteristic (phase shift at a given frequency) in the frequency domain. These may all be measured and classified after fault simulation.

When large deviations or catastrophic faults are present, however, circuit supposed to be linear exposes nonlinear behavior due to the rough change of the quiescent operating point(s) of the active element(s). Change of the location of the quiescent operating point especially affects the nonlinear capacitances of the active elements so simultaneously affecting the DC and the AC behavior of the circuit.

To get as a broad picture as possible to the aspects of testing in the frequency domain we will first study the testing of a simple NMOS Class A voltage amplifier. In that, in order to make the influence of the parameters more recognizable, i.e., to make it easier to test, the amplifier, depicted in Fig. 7.2.4a, is loaded by an inductor (L) to produce resonance. Following are the circuit element values: $R_1 = 13$ kΩ, $R_2 = 7$ kΩ, $R_3 = 1$ MΩ, $R_S = 350$ Ω, $R_D = 1$ kΩ, $C_S = 10$ μF, $C_E = 100$ μF, $L = 1$ μH, and $V_{DD} = 12$ V.

The simulation results are listed in Table 7.2.2. Here $+\Delta R_x$ means doubled value of R_x and $-\Delta R_x$ means halved value of R_x.

By inspection of Table 7.2.2 one easily conclude that in this way all faults are testable the difference to the fault-free case being in the value of the resonant frequency, the gain, or both. Again, with exception of the $-\Delta R_1$ and $+\Delta R_2$ faults, diagnosis is enabled.

Similar conclusions may be drawn from the next example which represents a Class C power amplifier discussed in LNAE_Book 4. Power amplifiers are working under stressful conditions and are more susceptible to failures. Hence the special interest. The circuit parameters are $V_{CC} = 20$ V, $V_{BB} = 1$ V, $R_B = 10$ kΩ, $R_L = 1$ MΩ, $C_S = 100$ μF, $L_S = 10$ mH, $C = 100$ pF, $C = 100$ pF, and $L = 0.1$ mH.

Here, in addition to the variation in the biasing circuit related to the base, there is interest on the influence of the power supply voltage to the performance of the circuit not only because of the power delivery and efficiency but also because of the change of the values of the transistor's capacitances. One is not to forget that power transistors exhibit larger (voltage-dependent) capacitances. Of course, variations in the values of the reactive part (parallel resonant circuit) of the amplifier are also of great importance.

The simulation results represented in a short fault dictionary are listed in Table 7.2.5. The frequencies listed in the second column of this table may be considered as the main attribute of the test signal for a given fault. All passive elements marked with "+" are equal to doubled nominal values while the ones marked with "−" are halved. The voltage sources are incremented by ±20%.

By analysis of this table we first conclude that the base biasing circuit has not much of an influence and faults related to it cannot be tested in an easy way. The power supply voltage, however, influences the gain at the resonance while not changing its frequency. One may observe the opposite signs of the increments of both the frequency and the maximum amplitude due to variations of the inductance and the capacitance in the parallel resonant circuit which means that these faults are detectable but cannot be used as diagnostic data. Finally, the shorts between the transistor terminals are easily identified (Table 7.2.3).

The dependence of the sensitivity of the output quantity (voltage or current) of the *linear* electronic circuit on frequency can be easily used to generate the test. The

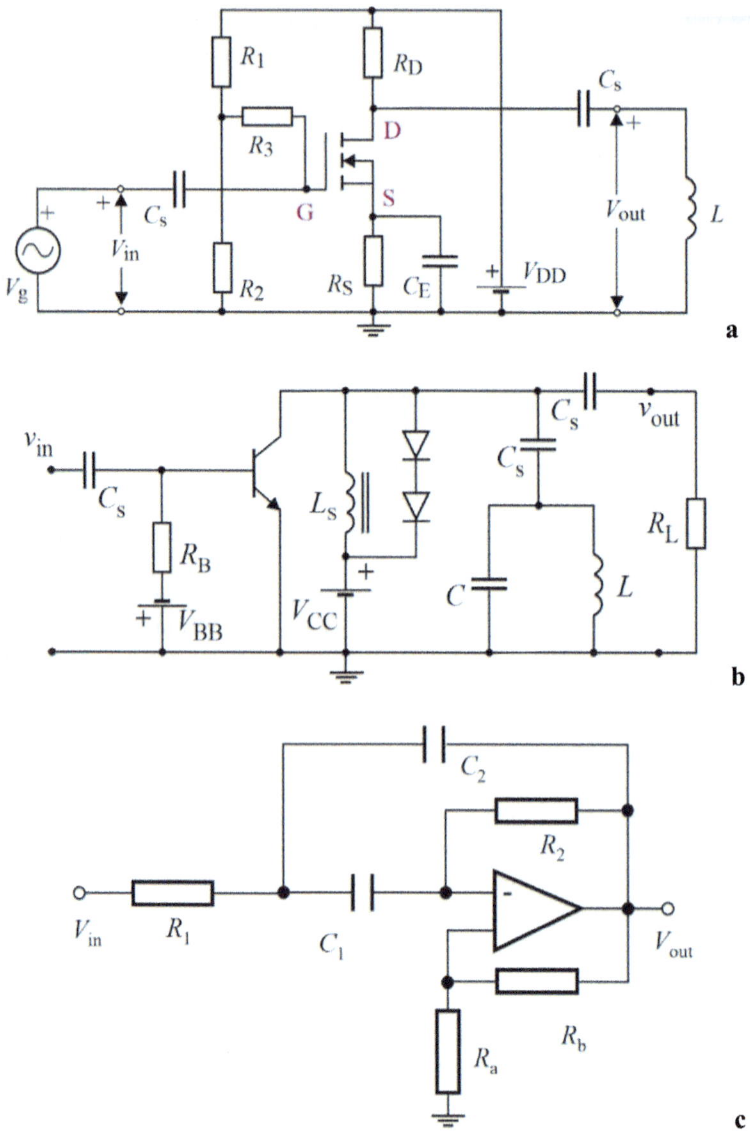

Fig. 7.2.4 Example circuits. **a** Class A amplifier, **b** class C amplifier, and **c** a bandpass filter

basic idea exploited here refers to finding those points on the frequency characteristic where the sensitivity of the output quantity to changes in some circuit parameter is the highest. This usually refers to soft defects that can be divided into small and large increments.

Table 7.2.2 Fault dictionary for the circuit of Fig. 7.2.4a (Part 1)

Type	Location of the maximum f (MHz)	Maximum amplitude V_{out}/V_{in} (dB)
FF	3.8998	36.2
$+\Delta R_1$	4.4019	16.9
$-\Delta R_1$	300.47	−56.4
$+\Delta R_2$	300.47	−56.4
$-\Delta R_2$	4.4042	16.9
$+\Delta R_S$	4.1438	31.9
$-\Delta R_S$	3.2800	24.9
$+\Delta R_D$	2.9984	36.6
$-\Delta R_D$	4.1351	31.8
D-at-G	0.0711	35.1
D-at-S	0.0528	−38.8
G-at-S	4.4415	10.1

Table 7.2.3 Fault dictionary for the circuit of Fig. 7.2.4a (Part 2)

Type	Location of the maximum f (MHz)	Maximum amplitude V_{out}/V_{in} (dB)
FF	1.5191	21.69
$+\Delta R_B$	1.519	21.69
$-\Delta R_B$	1.519	21.69
$+\Delta L$	1.0795	18.77
$-\Delta L$	2.143	24.57
$+\Delta C$	1.1015	18.92
$-\Delta C$	2.050	24.22
$+\Delta V_{CC}$	1.519	16.55
$-\Delta V_{CC}$	1.519	22.22
$+\Delta V_{BB}$	1.519	21.66
$-\Delta V_{BB}$	1.519	21.71
C-at-B	2.268×10^{-3}	11.45
C-at-E	0	0
B-at-E	0	0

The essence of the difference in the value of the sensitivity stems from the property of electronic circuits, even linear ones, that the dependence of the response of the circuit is in almost all cases a nonlinear function of the circuit parameters.

Namely, we say that some electronic circuits are linear because, in them, linear functions between currents and voltages in the circuit arise, but this does not mean that there are linear functions between the currents and the resistances in the circuit. The latter dependences are always nonlinear. Thus, when the increments of the parameters are small, the method of adjoint networks [14] can be used to calculate the sensitivity of the output quantities and in this way, for a given frequency, all sensitivities in the circuit can be determined with only two analyses.

When the parameter increments are large, even for linear circuits, it is necessary to separately analyze the faulty circuit for each fault (for each parameter value) individually, at a given frequency. In order to overcome this difficulty, one often proceeds by specifying the function of the circuit in a symbolic form. For complex circuits with a large number of parameters, this procedure, even with the use of a computer, is not so clear, but, all the same, it can be automated. The essence of the application of the symbolic function is that it eliminates the need to re-factorization the circuit matrix for each fault and for each frequency. In this way, considerable time savings may be achieved.

If one restricts its scope to linear circuits meaning that the linearity is not disturbed even for catastrophic faults, one may implement the so-called multifrequency analysis. The idea was introduced in [15]. In the example we present here we will use the circuit of Fig. 7.2.4c which is a simple second-order RC active bandpass filter. The nominal values of the circuit elements are $R_1 = R_2 = 10\ \text{k}\Omega$, $R_a = 10\ \text{k}\Omega$, $R_b = 100\ \text{k}\Omega$, and $C_1 = C_2 = 10$ nF. The operational amplifier is considered ideal with infinite input impedance, zeroth output impedance, and infinite gain. Its amplitude and phase characteristic in absence of faults is depicted in Fig. 7.2.5a. The excitation signal amplitude was 1 V while the response has a maximum of 579 mV at 1.593 kHz.

A very detailed study of the sensitivity of the output voltage was carried out. In doing so, only the elements that are outside the operational amplifiers were observed. For them, sensitivity to small increments; sensitivity to large increments of +40% and −40%; sensitivity to element short-circuited; and sensitivity to element open-circuited were established. All this for a frequency range from zero to 20 kHz. Some of the obtained results are reproduced in Fig. 7.2.5b, c. They refer to soft faults.

The results of the study are presented in the form of Table 7.2.4. It contains the values of the sinusoidal excitation frequencies for which the maximum sensitivity to the given parameter is obtained. Thus, it is most convenient to use a signal whose frequency is 1.43 kHz for testing the small increment of the resistor R_1. At that frequency, the sensitivity is the highest. This means that if there is a fault, it will be reflected in a larger increase in the output voltage than at other frequencies. For testing a large positive increment of the same resistor, a signal whose frequency is 1.98 kHz should be used while for the large negative increment the most convenient frequency would be 1.59 kHz. Open-circuited R_1 suppresses the transfer of the signal meaning that the output voltage will be equal to zero at all frequencies. Finally, short-circuited R_1 leads to an output voltage of 11 V at "infinite" frequency.

Table 7.2.4 can be used to test the circuit of Fig. 7.2.4c as follows. A set of signals (sinusoids) is formed, which will be used as test signals. The set consisting of numerical values of the frequencies of those sinusoids is called the test vector. As can be seen from Table 7.2.4, depending on the types of the faults we are testing, we also determine the size of the test vector. If we cover only soft faults with small parameter increments, the number of elements of the test vector will be the smallest.

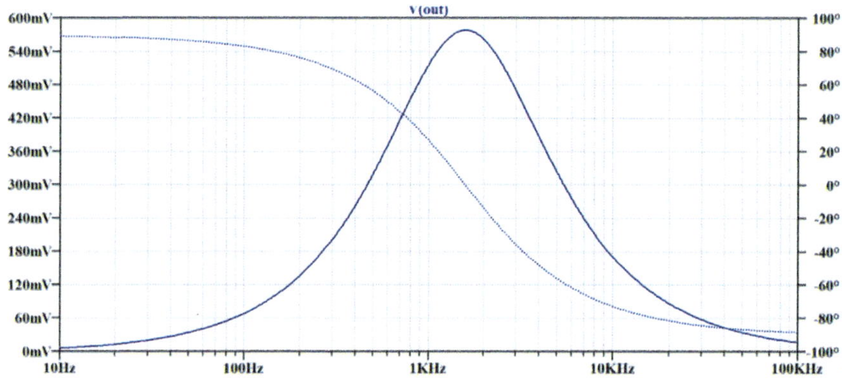

a

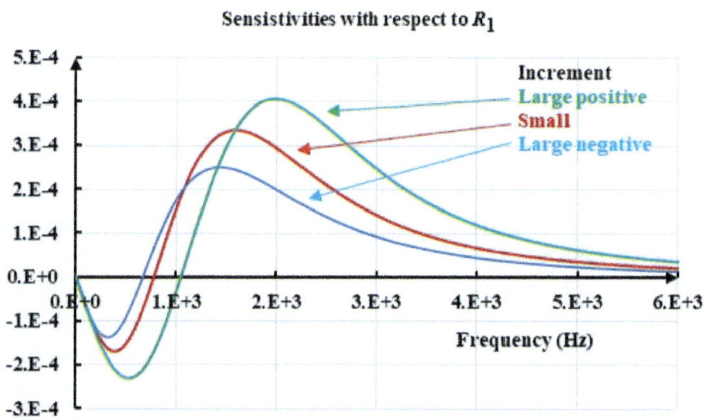

b

Fig. 7.2.5 Sensitivity to changes in circuit parameters. **a** The amplitude and phase characteristic of the fault-free circuit, **b** sensitivities with respect to R_1. Sensitivity to changes in circuit parameters. **c** Sensitivities with respect to C_2

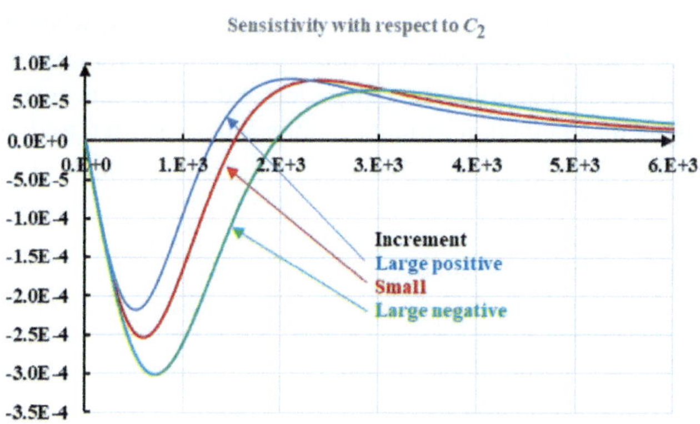

c

Fig. 7.2.5 (continued)

Table 7.2.4 Multifrequency analysis of the circuit of Fig. 7.2.4a

Element	Small deviation f (kHz)	Large deviation + 40% f (kHz)	Large deviation − 40% f (kHz)	s-at-open f (kHz)	s-at-open V_{out} (V)	s-at-short f (kHz)	s-at-short V_{out} (V)
R_1	1.59	1.43	1.98	Any	0	∞	11
R_2	0.97	0.95	0.98	0	11	Any	0
R_a	1.17	1.18	1.16	Any	1	1.59	0.5
R_b	1.17	1.16	1.18	1.601	0.5	Any	1
C_1	0.77	0.68	0.87	Any	0	0	1.22
C_2	0.6	0.52	0.71	∞	1.22	Any	0

Table 7.2.5 Parameters of the response of the fault-free circuit of Fig. 7.2.7a

Test	T_r (μs)	τ_d (μs)	τ_r (μs)	γ (%)	V_{sat} (V)
1	0.001	129.9	163.2	16.3	1
2	1000	596.61	771.19	13.4	1

An example of such a vector is: $f_{kHz} = \left[1.59\ 0.97\ 1.17\ 0.77\ 0.6\right]$. The numerical frequency values in this vector are taken from the third column of Table 7.2.4.

If soft faults with large parameter increments are also to be covered, the test vector should be extended by the following vector: $f_{kHz} = \left[1.43\ 0.95\ 1.18\ 1.16\ 0.68\ 0.52\right]$ for the case of large positive increments, and the vector $f_{kHz} = \left[1.98\ 0.98\ 1.16\ 1.18\ 0.87\ 0.71\right]$ for large negative increments.

Finally, if catastrophic faults are to be covered, signals whose frequencies are in the next vector should also be taken $f_{kHz} = \left[0\ 1.59\ 1.601\ \infty\right]$.

The analysis shows that some components share the same test frequency. However, this does not bother us as long as our goal is to determine whether the circuit is working properly or not. Simply put, the test will activate one or the other fault if either is present in the circuit. For diagnostic purposes, that is, when we want to isolate one fault, it will be necessary to perform several measurements with signals of different frequencies and, probably, application of different concepts.

When applying this method, maximum caution should be exercised in order not to make a mistake. Namely, as explained in [16] starting from the results described in [17], during the analysis in the frequency domain we use linearized models of active components for a given operating point. However, some of the faults may affect the positions of the quiescent operating points of the transistors in the circuit, thus rendering the analysis for AC signals of the defective circuit unusable. In the example from Table 7.2.4, there were no such elements.

7.2.4 Analog Testing in the Time Domain

We addressed testing in the time domain partially when we considered the I_{DDT} test. As a reminder, by the method introduced for the I_{DDQ} test, to qualify a signal as a testing one, one needs to perform two time domain analyses for every fault, to calculate the autocorrelation of the fault-free signal and the mutual correlation of the fault-free and the faulty signal, and finally to come out with the index of functionality. Proper value of the latter qualifies the proposed signal to be accepted as test signal. As already mentioned, this is a very time-consuming and tedious procedure if serious fault-coverage is to be achieved.

One has to have in mind that the same procedure as implemented to I_{DDQ} may be used for any time varying response of the circuit but we will here consider that the method was already outlined and there is no necessity to further elaborate its implementation.

Here, as an alternative, we will discuss some time-domain testing procedures that are based on measurement of other output quantities. First, as in the frequency domain, it is possible to base the test on response analysis. Based on the calculated response values and their sensitivity to certain parameters, as in multifrequency analysis, the most suitable signal and its properties that enable testing of a given fault can be sought.

In addition, the most suitable time instants in which the measurement will be performed, i.e., the moment of sampling the response, should be determined. The number of these measurements (instants) should be minimized, which will most likely affect the reduction of the testing time. Such tests are achieved by "brute force" by formulating the description of circuits with and without faults as many times as the fault list is long and repeating simulations and analyzes of the obtained results as many times as the fault list is long. Such an approach to creation of a test signal does not require special training of the test engineer (if proper simulation software is available).

Nevertheless, one has to have in mind some fundamental differences between the calculations in the frequency and time domain. First, to get the response for a single frequency in the frequency domain one needs single matrix factorization. Not much larger effort is needed to calculate all sensitivities simultaneously as described in [14] and will be discussed in the next chapter. When in time domain, to get the value of the response for a given time instant one needs to start from the origin and to proceed with numerical integration (with a time step preserving stability and assuring accuracy) until the required point on the time axis is reached. There will be many-many factorizations of the circuit matrix to reach this goal. Furthermore, calculation of sensitivities in the time domain is not common knowledge and it is especially critical when nonlinear dynamic circuits are treated [14].

Here we will try to highlight some of the existing procedures that try to offer solutions that make the test less laborious or "cheaper."

Since, as one can see, the problem of test synthesis in the time domain is much more complex, not a large number of solutions are offered. One of them [18] tries to answer all the mentioned questions, but the obtained solutions are limited to linear circuits with small parameter increments. In doing so, an analysis of the mapping of the imposed response tolerances into parameter tolerances (and vice versa) [16] is used, and then signals and moments of time are searched which map the thus obtained parameter tolerances into the time domain circuit response in the most noticeable way. The authors claim: "Any circuit that 'passes' the applied test is also guaranteed to meet its specifications."

We will consider in more detail the procedure described in [19]. It refers to the synthesis of test signals in the time domain for linear circuits where only two signals are considered, namely a sinusoid and a truncated ramp (which serves as a substitute for the step function). Sinusoidal testing is very similar to multifrequency analysis hence we will here talk about one test signal in the time domain-the truncated ramp only.

From the point of view of characterization of the response, its parameters such as steady-state value (V_{sat}), rise time τ_r (from 10 to 90% of the steady-state value), delay time τ_d (measured at 50%), and the overshoot γ (expressed as a percentage of V_{sat} of the input truncated ramp signal) are important. An illustration of all these quantities is given in Fig. 7.2.6.

Compilation of the test signal for this case, bearing in mind that it is only a truncated ramp, actually comes down to determining the slope of the ramp. The action proceeds as follows.

First, for the selected circuit and the selected set of faults, the circuit function is determined by symbolic analysis (a total of $m + 1$ function, one for the original circuit and m for the faulty circuits). Then for the known excitation of the shape:

$$V_{ramp} = (V_{sat}/T_r) \cdot [t \cdot v(t) - (t - T_r) \cdot v(t - T_r)] \tag{7.2.2}$$

where $v(t)$ is a step function, one finds $(m + 1)$ responses in a closed form as

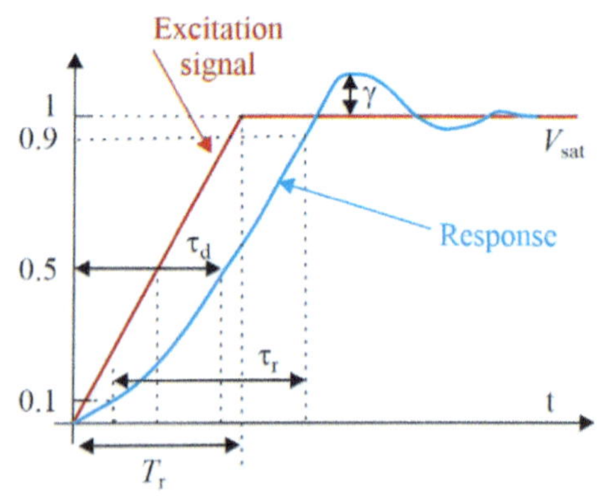

Fig. 7.2.6 Excitation in a form of a truncated ramp and the system response with parameters shown. Vertical axis normalized

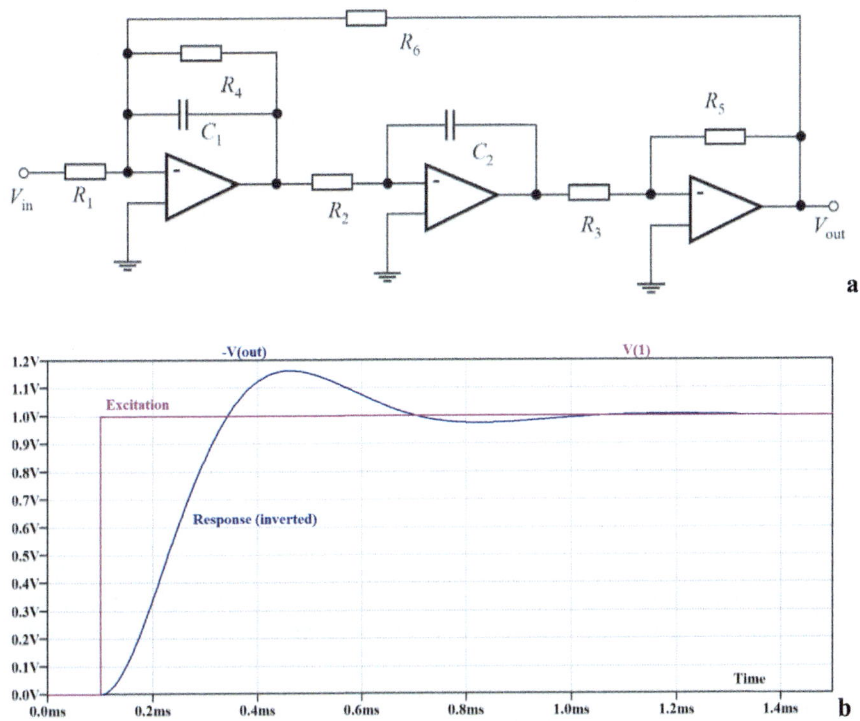

Fig. 7.2.7 **a** Second-order RC active low-pass filter ($R_1 = R_2 = R_3 = R_4 = R_5 = R_6 = 10\ \text{k}\Omega$, $C_1 = C_2 = 10$ pF) and **b** its step response for $T_r = 1$ ns

$$v_{\text{out}}(t) = \frac{1}{T_{\text{r}}}\left[\sum_{i=1}^{n}\frac{k_i}{p_i}t - \sum_{i=1}^{n}\frac{k_i}{p_i^2}\left(1 - \mathrm{e}^{-p_i t}\right)\right] \quad \text{for } t \le T_{\text{r}} \tag{7.2.3}$$

and

$$v_{\text{out}}(t) = \frac{1}{T_{\text{r}}}\left[\sum_{i=1}^{n}\frac{k_i}{p_i}T_{\text{r}} - \sum_{i=1}^{n}\frac{k_i}{p_i^2}\mathrm{e}^{-p_i t}\left(1 - \mathrm{e}^{-p_i t}\right)\right] \quad \text{for } t \ge T_{\text{r}}. \tag{7.2.4}$$

In these expressions, n is the number of poles of the circuit function, p_i are the poles, and k_i are the residues in the poles. In fact, there are $m + 1$ of these functions (responses). It should be noted here that these responses can also be obtained by numerical analysis in the time domain, but the authors considered that, given that the response for this excitation can be found in a closed form, it is more convenient to proceed by means of symbolic functions. As will be shown later in this book, the response of a linear circuit and all its replicas with a single fault, for any excitation, can be reached in the time domain without mapping to the s-domain, with extremely little effort.

The next step consists in using a special program to extract the mentioned parameters (τ_r, τ_d, γ, and V_{sat}) from the response for the original circuit and for each faulty circuit individually. By comparing the response parameters for the correct circuit and for the faulty circuit, based on the difference, the suitability of the test signal is established. If the difference is "big," the test signal is good, and if it is "small," that signal does not emphasize the fault effect, so it is not suitable for testing.

The authors illustrated this method on the example of the circuit from Fig. 7.2.7a. Here we will repeat the procedure using time domain circuit analysis based on Spice. In doing so, two test signals were used. The first had $T_r = 1$ ns (as in the original paper) and the second $T_r = 1$ ms (different from the original and accommodated to the cut-off frequency of the circuit under test). The amplitude of the input signal was $V_{sat} = 1$ V. The response properties of the fault-free circuit are shown in Tables 7.2.5 and 7.2.6 contains the deviations in the response parameters that occur when faults are introduced one by one, given the absolute value of γ, not its increment. For each fault, a circuit element value deviation of ±20% was assumed, so that individual elements appear twice in the table.

By analyzing these tables, it can be seen that even with such a simple signal, the problem of choosing its parameters is not simple. Namely, the value of the slope of the excitation signal is essential for the detectability of some faults. For example, if the delay time (which is the easiest to estimate) is taken as a measure of deviation,

Table 7.2.6 Test signals (slope of the ramp) compared

Fault	$T_r = 1$ ns				$T_r = 1$ ms			
	V_{sat}	τ_d	τ_r	γ	V_{sat}	τ_d	τ_r	γ
	Deviation in %			%	Deviation in %			%
R_1 (12 kΩ)	−16.65	14.86	34.18	−3.1	−16.66	15.91	∞	−14.18
R_1 (8 kΩ)	25.0	−14.06	−20.41	45.38	24.99	−16.76	−20.44	28.74
R_2 (12 kΩ)	0	11.56	17.08	12.8	0	−16.48	0.4390	2.662
R_2 (8 kΩ)	0	−13.23	−15.81	20.1	0	−3.977	0.2193	3.255
R_3 (12 kΩ)	0	10.73	5.77	12.8	0	3.409	0.4391	2.662
R_3 (8 kΩ)	0	−11.58	6.60	20.1	0	−3.977	−0.6598	3.255
R_4 (12 kΩ)	0	−3.315	−8.573	23.7	0	−4.261	−1.099	4.046
R_4 (8 kΩ)	0	8.253	16.42	8.10	0	4.261	1.758	1.651
R_5 (12 kΩ)	0	−10.75	−12.52	20.0	0	−2.841	0	3.221
R_5 (8 kΩ)	0	13.21	21.02	12.0	0	4.830	1.318	2.579
R_6 (12 kΩ)	20.0	−1.663	−6.60	35.4	20	−9.659	−16.92	23.19
R_6 (8 kΩ)	−20.0	0.816	17.73	−3.37	−20	18.18	∞	−17.39
C_1 (12 nF)	0	8.254	5.240	20	0	−0.5682	−1.539	3.823
C_1 (8 nF)	0	−9.1	−3.311	12	0	17.045	1.098	2.079
C_2 (12 nF)	0	9.906	17.08	12.8	0	3.977	0.4391	2.663
C_2 (8 nF)	0	−13.23	−16.47	20.2	0	−3.977	−0.2202	3.247

for element C_1, the fault being $C_1 = 8$ nF, it is incomparably more convenient to use a "slow" ramp, that is, one whose slope is 1 V/ms. Conversely, when the rise time is used as a measure of difference, to test for the same fault, it is more convenient to use a "fast" ramp whose slope is 1 V/ns.

It should be noted here that testing based on the parameters of the step function, i.e., the truncated ramp (as well as for any other signal) can be applied to nonlinear circuits with equal success. From the point of view of composing the test, the difference would be that a concurrent fault simulation would be used for a given excitation (as will be described in the following chapters, for linear and nonlinear reactive circuits), and then the response parameters would be extracted and the test signal selected in the same way.

One specific procedure for testing analog circuits in the time domain is known as the oscillation method. In this procedure, during testing, the circuit under test is converted into a new oscillating circuit [20]. Thus, by measuring the frequency of oscillations, those faults can be detected that cause a sufficiently large increase in frequency compared to a fault-free circuit. The advantage of this method would be reflected in the fact that it is not necessary to compile the test signal, and the testing time is short, considering that only a few resulting frequencies are obtained (for a fault-free or for a faulty circuit).

The application of this method, although at first glance quite simple, depends on the possibility to connect the original circuit so that it is transformed into an oscillator, and to achieve the mapping of the fault effect into the oscillation frequency. This kind of procedure is relatively difficult to systematize, the result of the systematization largely depends on the experience and creativity of the test designer. One successful application of this method to the problem of testing analog-to-digital converters is given in [20], and to the problem of analog active RC filters in [21]. We will consider this procedure in more detail later within the concept of built-in test (BIT).

Artificial neural networks (ANNs) are one of the concepts that is based on the accumulation of "knowledge." Their eventual application for testing follows from this. Namely, the neural network can be "taught" to recognize faulty circuit operation and thus be used for testing. Moreover, in more sophisticated applications, it is possible to "learn" the fault effect, that is, to "remember" the fault dictionary, so that the ANN can be used for diagnostic purposes [22].

Only one simple example of the application of neural networks for parametric fault testing based on [23] will be described here. Some more complex procedures are described in detail in [24, 25].

The concept is illustrated in Fig. 7.2.8. Shown here is the circuit under test that is excited by the signal $x(t)$. A series of output signals $y_i(t)$, $i = 1, 2, \ldots, m$, are observed, which are functions of the value of the excitation signal and the values of the parameters p_i, $i = 1, 2, \ldots, n$. The resulting output signals of the circuit are fed to the input of the ANN. Its outputs should be equal to the values of the circuit parameters p_i.

Given that deviations between the actual values of the parameters and the values of the output signals of the neural network can always be expected, the output values of the neural network are denoted with h_i, $i = 1, 2, \ldots, n$. The norm $\varepsilon = \|\mathbf{p} - \mathbf{h}\|$ is a

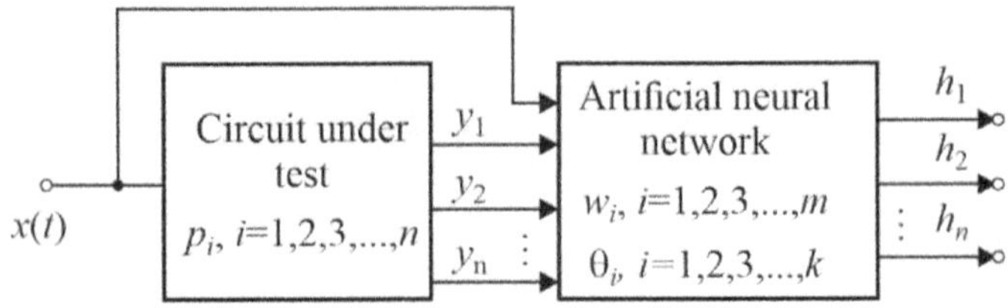

Fig. 7.2.8 A symbolic representation of using an ANN for testing

measure that tells how accurately the neural network has "learned" the values of the circuit parameters. By the way, using the terminology of automation, minimization of $\varepsilon = \|\mathbf{p} - \mathbf{h}\|$ can also be called parameter identification. The learning of the ANN is reduced to an optimization procedure that gives the values of the coefficients (so-called weights and thresholds) w_i, $i = 1, 2, \ldots, m$ and θ_i, $i = 1, 2, \ldots, m$ (as defined in Chap. 7.12) from the condition that ε is minimal.

Learning proceeds as follows. A value interval is set for each circuit parameter. This creates an n-dimensional parameter space. The boundaries of the space are determined by the expected tolerances of the parameters. Speaking in the language of neural networks, the parameter space together with the excitation signal $x(t)$ can be called the "training set" of the neural network. The value of ε is calculated by simulating the neural network as many times as there are points in the parameter space for the same set of excitation x and $\mathbf{y}$. Based on the obtained deviation, the values of the weights w and thresholds θ are corrected, until the value of ε is small enough. It is said that the neural network learning algorithm is applied, which means that the correction values Δw and $\Delta\theta$ are calculated in a special way. When the "learning" process is finished, each output h_i should show, with satisfactory accuracy, the value of p_i.

In the exploitation phase (depicted in Fig. 7.2.8), the "learned" neural network is connected to the physical circuit being tested. The coupling can be such that the neural network is physically made in the form of an electronic circuit (the so-called neuroprocessor) and directly connected to the circuit being tested. Alternatively, the measured signals are introduced into a computer in which the neural network is simulated by software. As a result of the measurement, numerical values of the circuit parameters are obtained, so the engineer who deals with the testing decides by direct comparison with the known desired values of the parameters, which parameter is the cause of the failure.

Neural networks can also be used to diagnose defects even in the case of the simultaneous presence of several defects [26]. This issue will be discussed in more details later in this book.

7.2.4.1 Selection of Testing Waveforms

Since we have familiarized ourselves with the most important concepts of testing analog electronic circuits, in the following text the waveforms of the signals that should be applied will be considered. In doing so, in some way, there will be an overlap with what has already been said, but that is with the aim of maintaining systematic presentation. It will be discussed why the waveform of the signal and the change of its parameters affect the detection of faults according to [2] and [27].

A large number of analog circuits contain nonlinear elements whose state depends on the voltage at their terminals. Some of them can be brought into cut-off and thus close the signal propagation path from the fault's site to the output. That issue could be one of the bases for reasoning about whether to choose a DC signal and with what value to calculate. Similar considerations apply when evaluating the parameters of AC signals such as frequency, amplitude, and phase. Therefore, each of these quantities (circuit variables) will be given special attention.

The significance of the value of the DC signal at the input or, in other words, the offset of the input signal, is mostly related to the influence of this value on the position of the quiescent operating point. First, the DC value of the input signal determines the parameters of the transistor (incremental) model that are valid for small signals. In this way, significant global properties of the circuit are defined, such as the nominal gain, the shape of the amplitude and phase characteristics, and the cut-off frequencies. Then, and related to that, both the DC and AC components of the supply current are determined by the value of the DC component of the input voltage. All these quantities are measurable and as such can contain information about a possible fault.

Measuring an AC signal at the supply voltage terminal is a test that more sensitively captures the fault effect than measuring a DC signal. For example, the DC power supply current of the operational amplifier will not change if the connection (permanently open in series) of the compensating capacitor in an operational amplifier is broken. Therefore, an alternating signal is necessary to test that fault.

When it comes to steady-state AC mode, all parameters of the excitation sinusoid are important for fault detection. The signal frequency, for example, is a parameter that determines the amplitude of the circuit's response for constant excitation amplitudes. Different frequency values, however, can indicate different faults in different ways. The application of a high-frequency signal can lead to a significant increase in the supply current of one part of the circuit and thus to an increase in the detectability of faults in it. Namely, with complex circuits, the supply current can be displayed as a vector (sum) of the supply currents of individual subcircuits connected in parallel. A fault in one of the subcircuits results in a change in the total supply current, but it does not have to be large enough and noticeable. When observing the power supply current of the subcircuit, however, this increase can be made noticeable and suitable for fault detection. Different values of the frequency of the excitation signal can be characteristic of different subcircuits or faults. However, the greatest importance of the frequency of the excitation signal for the detection of faults is related to the frequency dependence of the sensitivity of the output voltage to changes in circuit parameters. Fault detection using this concept will be discussed separately in this chapter.

Changing the amplitude of the excitation sinusoid will change the effective value of the circuit's supply current. When considering this, one should keep in mind the problems that arise at the boundaries of the interval. Namely, at very small amplitudes, the signal-to-noise ratio is important, and such signals should be chosen that will make the signal-to-noise ratio at the input acceptable. At the other end of the interval, large signals lead to nonlinear distortions and thus to additional signals that are measured at the output. Of course, it is most advantageous to choose the highest possible signal in order to avoid noise, without causing nonlinear distortions. This can be achieved by performing an FFT of the output signal for the chosen amplitude of the test signal and determining whether the second harmonic is acceptable. Instead, the *transfer admittance* function from the input (voltage) to the supply current (current) is most often created. In the area of the input voltage amplitudes where this quantity is constant (transfer characteristic is linear) it is allowed to set the amplitude of the input signal.

If the circuit has several input terminals, changing the phase of one signal relative to another can cause a change in the entire supply current. For example, if a circuit has two identical input stages, applying phase-separated signals to those two inputs can lead to a reduction in the overall supply current and, in this way, provide a clearer insight into the fault effect that would otherwise be masked.

To illustrate these considerations, the results given in [27] will be presented. The operational amplifier circuit of Fig. 7.2.2 will be considered meaning that the method of [27] will be applied but the numerical data will be accordingly different. For some of the faults, the most suitable test signal was investigated. The one that causes the largest deviation of the response of the faulty circuit from the response of the fault-free circuit is considered as the right choice. First, an attempt will be made to arrive at the test by an analytical procedure, and then it will be shown that even when we are unable to generate the test based on analytical considerations, we can arrive at it by simulation.

Let us first consider the determination of the DC test. The output voltage of an open-loop operational amplifier can be expressed as:

$$V_{\text{out}} = A_{\text{d}}(V_{\text{d}} + V_{\text{ofs}}) + A_{\text{c}}V_{\text{c}} \tag{7.2.5a}$$

where A_{d} is the differential gain, A_{c} is the common mode gain, V_{d} is the input differential signal, V_{ofs} is the input offset voltage, and V_{c} is the common mode signal. When a fault occurs in the operational amplifier (OA), the value of the OA model parameters will change. The difference in the output voltages of the fault-free and the faulty OA can be expressed as

$$|\Delta V_{\text{out}}| = \left|(\Delta A_{\text{d}})V_{\text{d}} + (\Delta A_{\text{s}})V_{\text{s}} + \Delta V_{\text{out,ofs}}\right| \tag{7.2.5b}$$

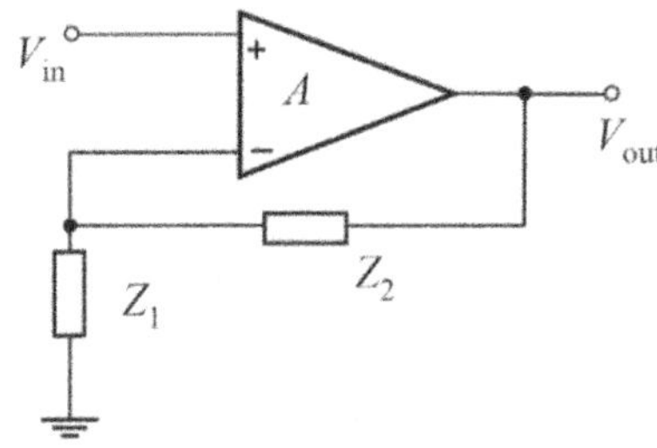

Fig. 7.2.9 Non-inverting amplifier

where ΔA_d is the increment of the differential gain, ΔA_c is the increment of the common mode gain, and $\Delta V_{out,ofs}$ is the increment in the offset output voltage of the fault-free and faulty OA. In (7.2.5) ΔA_d, ΔA_c, and $\Delta V_{out,ofs}$ can be both positive and negative.

If all possible combinations of the above three parameter increments that maximize the value of $|\Delta V_{out}|$ are considered, we find that the most suitable test signals are the following pairs of values: $(V_{d,min}; V_{c,max})$ or $(V_{d,max}; V_{c,min})$ or $(V_{d,max}; V_{c,max})$ or $(V_{d,min}; V_{s,min})$, i.e., in the vertices of the parallelepiped defined by these for values.

Generating a DC test for an operational amplifier in a real-world environment depends on the very circuit in which it is embedded. For example, the simplest situation will be considered here, which is the non-inverting amplifier shown in Fig. 7.2.9.

For this circuit, the output voltage can be shown to be

$$V_{out} = \left(1 + \frac{Z_2}{Z_1}\right) \cdot \frac{1}{1 + \frac{1}{A}\left(1 + \frac{Z_2}{Z_1}\right)} \cdot V_{in}. \tag{7.2.6a}$$

When a fault is present in the operational amplifier, the difference between the output voltages of the fault-free circuit and the faulty circuit can be represented as:

$$\Delta V_{out} = \left(1 + \frac{Z_2}{Z_1}\right)^2 \cdot \frac{A - A_f}{\left(1 + A_f + \frac{Z_2}{Z_1}\right) \cdot \left(1 + A + \frac{Z_2}{Z_1}\right)} \cdot V_{in}. \tag{7.2.6b}$$

where A is the differential gain of the fault-free, and A_f is the differential gain of the faulty OA.

In order to maximize the difference (left side of (7.2.6b)), it is necessary to supply $\pm V_{in,max}$ to the input, that is, $\pm V_{in,max}$ is a DC signal that should be used to test this circuit in the best way.

In order to test the behavior of the operational amplifier to AC signals, a sinusoidal excitation should be applied. In doing so, first of all, the right frequency should be chosen in order to maximize the difference in the response of the fault-free and the faulty circuit.

To illustrate the selection of the signal frequency, the circuit of Fig. 7.2.9 will be used again while $Z_2 = 0$ will be chosen, which, in the conditions of infinite gain of the OA, means that the resulting circuit is a unity gain amplifier. If the frequency characteristic of the operational amplifier is approximated by a function with one pole ($s_p = -\omega_p$), for the total gain we have

$$A_{jp} = \frac{A_0}{1 + A_0} \cdot \frac{1}{1 + \frac{s}{(1+A_0)\omega_p}} \tag{7.2.7a}$$

where A_0 is the differential gain at zero frequency and A_{jp} is the gain of the unity gain amplifier. The above expression can be expressed in decibels as:

$$20 \cdot \log\left|A_{jp}\right| = 20 \cdot \log\left|\frac{A_0}{1 + A_0}\right| - 20 \cdot \log\left|1 + \frac{j\omega}{(1 + A_0)\omega_p}\right| \tag{7.2.7b}$$

When there is a fault, the gain difference between the fault-free and the faulty circuit expressed in decibels is

$$\left|\Delta A_{jp}(\mathrm{dB})\right| = |C - x| \tag{7.2.8}$$

where

$$C = 20 \cdot \log\left|\frac{A_0}{1 + A_0}\right| - 20 \cdot \log\left|\frac{A_{0f}}{1 + A_{0f}}\right|;$$
$$x = 10 \cdot \log\left[\left(1 + \frac{\omega^2}{a^2}\right)/1 + \frac{\omega^2}{b^2}\right];$$
$$a = (1 + A_0) \cdot \omega_p \text{ and } b = (1 + A_{0f}) \cdot \omega_{pf},$$

and A_{0f} is the gain of the faulty operational amplifier at zero frequency.

From (7.2.8), we conclude that the largest difference in gain will be at the frequency where C and x differ the most, whether it is a maximum or a minimum. Since C does not depend on frequency, the frequency of the test sinusoid will be equal to the frequency of the extremum of x. As the derivative of x with respect to ω exhibits zeros both at the origin and at infinity, the origin and infinity become the respective points for the frequencies of minimum and maximum (7.2.8). The specific location of each frequency hinges on the relative proportion between b^2 and a^2. The frequency of the test signal will therefore be equal to the highest and the lowest frequency of the frequency range of interest.

If the phase of the output voltage is measured, a sinusoidal test signal with a different frequency value will be of interest. Namely, for the mentioned unity gain amplifier, without faults, the phase is

$$\varphi_{jp}(\omega) = -\mathrm{arctg}\frac{\omega}{(1 + A_0) \cdot \omega_p} \tag{7.2.9}$$

while the phase difference for the fault-free and faulty circuit is

$$\Delta\varphi_{jp}(\omega) = \text{arctg}\frac{\omega}{a} - \text{arctg}\frac{\omega}{b}. \tag{7.2.10}$$

The highest value of the phase difference is at the frequency $\omega = \sqrt{a \cdot b}$, so a sinusoid with that frequency is recommended as a test signal. The verification of these results was performed on the example of the OA from Fig. 7.2.2 by simulating the circuit several times, during which the analysis of the original circuit and its replicas in which faults were inserted was performed.

The results for the open-loop OA are shown in Table 7.2.7. Four faults were considered. The first and second are parametric, and the third and fourth are catastrophic or structural. Namely, a parametric fault was first taken, the origin of which is usually a design error: a change in the W/L ratio of transistor T_3 of $\pm 10\%$. Then, a change in the threshold voltage of the transistor T_8 of $\pm 30\%$ was taken, which can be attributed to the instability of the technological process. There are actually four catastrophic faults: a short circuit between the drain and the source and a short circuit between the gate and drain of transistor T_5 and an open drain and open gate of transistor T_1.

The results of the simulation are the differences in the output voltage values for the fault-free and the faulty circuit. When it comes to the first fault (where the parameter gets two values), looking at the table, we conclude that for the case where the deviation is -10%, the test signals are $V_{\text{in\_c,max}}$ and $V_{\text{in\_d,max}}$, and for the case when the deviation is $+10\%$, the test signals are $V_{\text{in\_c,min}}$ and $V_{\text{in\_d,min}}$. These results were obtained, of course, by simulation for the complete range of signals. The results of one such simulation, but for the case of an amplifier with a gain of 20 dB, are shown in Fig. 7.2.10.

The second fault, according to the simulation results shown in Table 7.2.7, requires the tests: $V_{\text{in\_c,max}}$ and $V_{\text{in\_d,min}}$ for the case when the threshold voltage change is –

Table 7.2.7 Numerical values of the difference between the output voltage of the fault-free and the faulty circuit

Faults →		T_3		T_8		T_5		T_1	
Input voltage ↓		fault W/L (%)		fault V_T (%)		fault SC		fault OC	
Vin_c	Vin_d	−10	+10	−30	+30	D-S	G-D	D	G
min	min	−0.29	4.67	1.31	−0.04	4.58	−0.28	3.91	0.00
min	max	−4.09	0.87	0.69	−3.81	0.29	−4.57	3.91	0.00
max	min	−1.03	3.74	2.86	0.33	4.09	3.56	0.29	1.27
max	max	−4.61	0.16	0.19	−3.24	−0.33	−0.20	0.29	1.27

SC short circuit and *OC* open circuit

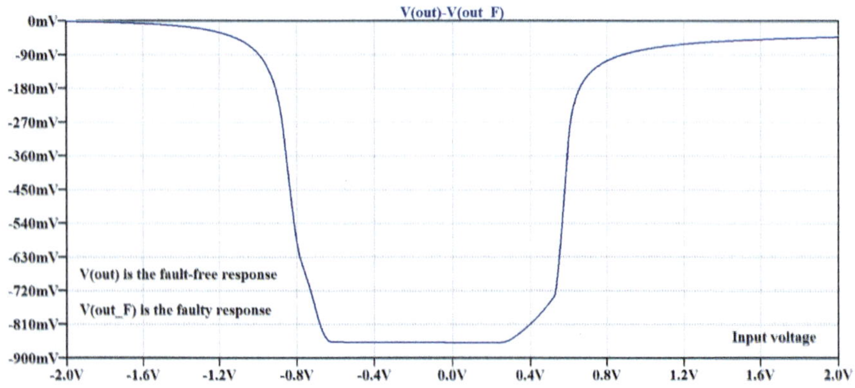

Fig. 7.2.10 Dependence of the difference between the output voltages of the fault-free and the faulty circuit with a fault in T_3 consisting of a change in W/L for -10%. The circuit of Fig. 7.2.9 with a gain of 20 dB ($Z_1 = R_1 = 10\ \text{k}\Omega$ and $Z_2 = R_2 = 100\ \text{k}\Omega$) was simulated

30% and $V_{in\_c,min}$ and $V_{in\_d,max}$ for the case when the threshold voltage increment of the transistor T_8 is +30%.

When it comes to faults of the short-circuit type, for the transistor T_5, we can see from the table that the appropriate test signals are: ($V_{in\_c,min}$, $V_{in\_d,min}$) for the case of short circuit between drain and source and ($V_{in\_c,min}$, $V_{in\_d,max}$) for the case of the short circuit between the gate and the drain.

Finally, the disconnected drain of transistor T_1 can be tested with either a pair of ($V_{in\_c,min}$, $V_{in\_d,min}$) or a pair of ($V_{in\_c,min}$, $V_{in\_d,max}$), and the disconnected gate of the same transistor can be tested with either a pair of ($V_{in\_c,max}$, $V_{in\_d,min}$) either by a pair ($V_{in\_c,max}$ and $V_{in\_d,max}$). Such results are obtained by simulation, an example of which is shown in Fig. 7.2.11.

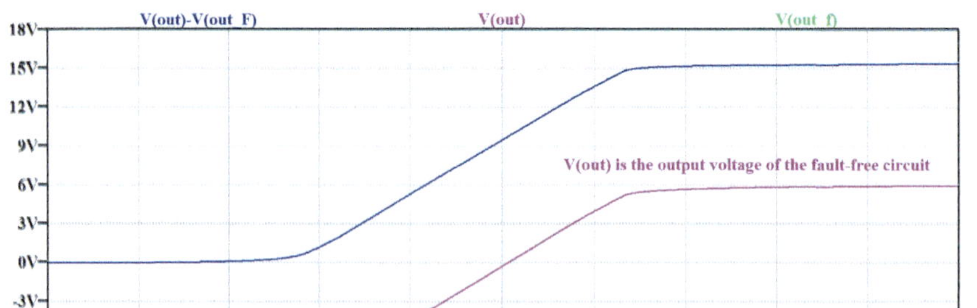

Fig. 7.2.11 (The same circuit as in Fig. 7.2.10) Dependence of the difference between the output voltages of the fault-free and the faulty circuit when the transistor T_4 has a disconnected gate

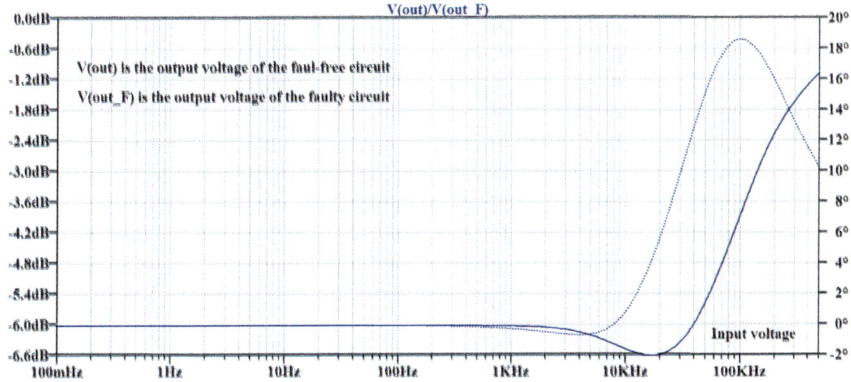

Fig. 7.2.12 (The same circuit as in Fig. 7.2.10) Dependence of the difference between the output voltages (in dB) and phases (in degrees) of the fault-free and the faulty circuit when the transistor T_7 has the *W*/*L* ratio increased for 50%

Summarizing the results of the previous analysis of DC regimes for an open-loop OA, we conclude that all test signals require extreme values (minimum or maximum) except for those related to short circuits and open circuits. They deviate because there is a distortion of the transfer characteristic of the amplifier, which becomes nonlinear.

When dealing with an amplifier with unity gain, the most suitable test signals are maximum ones.

In order to find the best frequencies that maximize the difference between the fault-free and the faulty circuit, a non-inverting amplifier with a gain of 20 dB will be considered. The simulation results for the case when a fault is installed: in T_7, *W*/*L* is increased by +50%, are shown in Fig. 7.2.12. It is easy to conclude that the best test signal from the point of view of the amplitude characteristic is the one whose frequency is the smallest, i.e., 1 Hz, and from the point of view of the phase characteristic the one whose frequency is around 46 kHz, which is very close to 100 kHz.

7.2.4.2 Selection of Test Signals and Creation of the Test Sequence

A test synthesis procedure for analog electronic circuits can be recommended based on the previous considerations. The main goal is to find the simplest set of test signals that will reveal all the assumed faults with the cheapest experiment and in the shortest time. The following procedure is recommended [27].

1. Perform a DC analysis in the circuit by varying the value of the input DC signal (offset) over a full range (from $-V_{DD}$ to $+V_{DD}$, for example). At the same time, for each point, determine the sensitivity of the output quantity (supply current

or output voltage, for example) to all possible parametric and structural faults. The sensitivity analysis can be done by the method of adjoint networks [14, 16] or by the methods of concurrent fault simulation, which will be described in the following chapters.
2. For faults that are shown not to be sensitized by the input DC signal, i.e., for which the sensitivity is low, a sensitivity analysis with respect to changes in the input signal should be performed. This means that the sensitivity of the local voltage or current related to the fault to changes in the input signal is checked. This will establish whether there is a path of sensitization at all, that is, whether the fault is detectable with a DC signal. If it turns out that there is no DC sensitization path, one should switch to another type of signal, and if there is a DC sensitization path, and the fault is not sensitized, one should go back to step 1 and complete the range of change of the input voltage or reduce the step of its change.
3. For all sensitized faults, a fault simulation should be performed with the so obtained DC signals and determine whether they are really testable. In doing so, the observability of the fault is checked. If observability is proven, the fault is testable and can be deleted from the fault list (delete step of the algorithm from Fig. 7.1.7).
4. For each DC offset value, perform an analysis in the frequency domain with multiple values of the input signal frequency. In this way, either the sensitivity of the output quantity to faults or the sensitivity of the voltage or current at the given node to changes in the input signal will be determined. This determines the signal candidates (DC component + sinusoid) that will serve to test the given fault. In all this, it should be kept in mind that the frequency analysis is performed for given offsets (DC signals), which means that for each new value of the DC component, new models of active components will be generated by the simulator.
5. Simulate faults in the time domain with the obtained signal candidates from the previous section.

That would exhaust the procedure. When evaluating this procedure, it should be kept in mind that because the DC component affects the operating points of the active components, it cannot be said that all the required test signals will be found. In other words, this procedure does not guarantee that it will be possible to test every fault. The goal of the authors was, however, to offer an algorithm that is likely to generate all test signals.

Once the list of faults that are not testable with the signals generated by this algorithm is determined (which is assumed not to be large), the simulation of the faults is approached with an exhaustive list of signals until a suitable one is found. In doing so, it is possible to change the domain so that the test signals are searched for in a different waveform using time domain simulation.

When the test is compiled according to the procedure described above or in another way, the following situation arises. A number of signals (tests) are available, each of which enables the detection of one or more faults. From the point of view of the application of this result, the only task that still needs to be solved is determining the sequence of signal application. By choosing the optimal sequence of signals, it is possible to shorten the testing time of the manufactured circuit. Namely, such a sequence of tests should be set that will enable faulty circuits to be found at the lowest price, i.e., in the shortest time.

Bearing in mind that for a large number of analog circuits that can be purchased on the market, tests are provided by the manufacturer, the only thing left for the user of the finished product is to arrange the tests obtained in this way so that the testing can be done as quickly as possible. In such a situation, determining the order of application of the tests is the only thing the test designer has to do.

The goal of editing the test sequence is to make the faulty circuit show the fault effect as early as possible. If it is already faulty and has incurred some costs, at least its testing should be cheaper. The procedure we will describe does not save on testing the fault-free circuits and it does not have to be applied if the production is so stable that the probable number of faulty products is very small.

In order to speed up the testing process, some weights or priorities can be assigned to the test signals, such as

- Tests for faults with the highest probability of failure should be applied first.
- Then the tests that take the shortest testing time should be applied, and
- then tests should be applied which can be claimed to be independent of the previous tests.

The probability of failure is directly related to the production itself, and such information must be obtained from the manufacturer. This is especially true for catastrophic faults. One example of such a dataset for a Philips technology can be found in [28]. When it comes to parametric faults, instead of the probability of failure, there is a distribution of parameter values and tolerance limits that determine the limits of the acceptance domain in the parameter space: $(\overline{p}_j, \underline{p}_j)$ for $j = 1, 2, \ldots, n$, where n is the total number of test signals, i.e., when every test signal covers only a single fault, the number of faults. In order to obtain the probability of failure, it is necessary to generate random numbers with a given distribution for each parameter p_j^k, $k = 1, 2, \ldots, m$ and compare them with the limits of the interval. The algorithm shown in Fig. 7.2.13 describes the computation of all probabilities for parametric faults. m is a large number equivalent to the number of Monte Carlo simulations.

Fig. 7.2.13 Calculation of probability of failure (v_j) for parametric faults

```
for j=1 to n;   {
        counter=0;
        for k=1 to m;            {
        generate random value of p_j^k;
        if p_j^k > p̄_j and p_j^k < p_j
        counter = counter +1;
    vj= counter/m
    }
```

Another parameter that determines the total duration of testing is the duration of a single test. Each test, for a given device under test, is assigned a duration of testing according to its complexity, which we will be denoted here τ_j for $j = 1, \ldots, n$.

In the last case (under number 3), the following is meant. If one test signal detects a certain type of faults and if after the application of that test it turns out that the corresponding fault is not present, a new test signal should be applied, which is intended to test a completely different type of fault. Or, if the previous test addresses a fault that is physically located on one part of the board (the chip), the next one might test for a similar fault that is physically on a completely different end. This increases the likelihood of revealing the failure.

In this sense, certain pairs of faults can be attributed "distance," which we will be denoted here with r. It would be a number that is all the higher if the nature of the technological causes of the faults are less correlated. A simplified example of such a "distance matrix" is given in Table 7.2.8. MOS integrated circuits with five types of faults are considered: short circuit (SC), open circuit (OC), gate breakdown (GB), increment W/L (DW/L), and threshold voltage increment (DV_T).

With all this in mind, each test can be assigned a number (price) that reflects its properties according to the above criteria. The following number will be selected here:

$$C_j^R = \frac{1}{r_{j,j-1}} \cdot \frac{\tau_j}{v_j} \quad \text{for } R = 1, 2, \ldots, R_{\max} \text{ and } j = 1, 2, \ldots, n, \tag{7.2.11}$$

where R is the ordinary number of the test combination, the number of combinations or possible sequences is $R_{\max} = n!$, and it can be taken that $r_{1,0} = 1$. C_j^R says that low cost, and thus high priority, should have a test whose duration of testing is short, the probability of a fault is high, and which is "far" from the previous fault.

We should not lose sight of the fact that C_j^R depends on the order of listing the tests. Thus, each possible order of testing gets its total cost as

$$C^R = \sum_1^n C_j \tag{7.2.12}$$

Table 7.2.8 Fault distance matrix

Distances $r \searrow$	SC	OC	GB	DW/L	DV_T
SC	1	3	3	3	3
OC	3	1	3	3	3
GB	3	3	1	2	2
DW/L	3	3	2	1	3
DV_T	3	3	2	3	1

Table 7.2.9 Values needed to calculate C^R

Test no. → Properties ↓	(1)	(2)	(3)	(4)	(5)
Typ1	OC	DW/L	DW/L	DV_T	SC
$\tau_{\text{normalized}}$	5	17	14	8	11
v	0.015	0.005	0.005	0.008	0.03
τ/v	1000/3	3400	2800	1000	1100/3

Table 7.2.10 Values of R (the quantities C are divided by 1000)

R	Order	$C_1{}^R$	$C_2{}^R$	$C_3{}^R$	$C_4{}^R$	$C_5{}^R$	C^R
1	12345	1/3	3.4/3	2.8	1/3	1.1/9	4.722
2	12354	1/3	3.4/3	2.8	1.1/9	1/3	4.722
3	12435	1/3	3.4/3	1/3	2.8/3	1.1/3	3.1
⋮	⋮	⋮	⋮	⋮	⋮	⋮	⋮
120	54321	1.1/3	1/3	2.8/3	3.4	1/3	5.367

All that remains is to arrange the test sets (sets of orders) in order with increasing C^R value.

To illustrate, an example of test sequencing for a simple circuit with five potential faults will be considered. The necessary data for implementing the above procedure for the technology described in Table 7.2.8 are listed in Table 7.2.9.

Based on this data, the content of Table 7.2.10 can be created, in which, due to the volume, only four rows are listed. The optimal order is obtained by finding the smallest number in the last column of this table.

This would complete the description of the procedure for determining the order of the test, with the need remaining to discuss its applicability. Namely, bearing in mind that the number of rows in Table 7.2.10 is, in the general case, $n!$, the realization of this algorithm becomes very expensive. For the case $n = 5$, as above, the number of possible orders is $R_{\max} = 120$. For more complex examples, however, that number grows very quickly. For $n = 10$, $R_{\max} \approx 3.6 \times 10^6$, and for $n = 20$ even $R_{\max} \approx 2.4 \times 10^{12}$.

Our ambition is to test VLSI circuits as well, and in that case we should talk about hundreds of tests. This problem is solved by avoiding the repeated calculations that appear in the mentioned table. However, those solutions will not be discussed here. One algorithm for solving this problem, which for its implementation requires knowledge from the field of dynamic programming can be found in [29].

References

1. Soma M (1996) Automatic test generation algorithms for analogue circuits. IEE Proc Circuits Dev Syst 143(6):366–373
2. Spina R, Upadhyaya S (1997) Linear circuit fault diagnosis using neuromorphic analyzers. IEEE Trans CAS II Analog Digit Signal Process 44(3):188–196
3. Levi M (1981) CMOS is most testable. In: Proceedings of the international test conference, pp 217–220
4. Milovanović D, Litovski V (1998) I_{ddq} versus V_{ddt} testing—a comparative study. In: Proceedings of the electrical engineering and electronics conference, EE'98, Gabrovo, Bulgaria, pp 131–135
5. Suparjo BS, Wilkins BR (1992) Steady state measurements. In: IEE colloquium on testing mixed signal circuits, London, p 118
6. Cole EI et al (1997) Transient power supply voltage (v_{DDT}) analysis and detecting IC defects. In: Proceedings of the international test conference, Washington D.C., pp 23–31
7. Malaiya YK, Su SYH (1982) A new fault model and testing techniques for CMOS devices. In: International test conference, pp 25–34
8. Bell IM, Spinks SJ, da Silva JM (1996) Supply current test of analogue and mixed signal circuits. IEE Proc Circuits Dev Syst 143(6):399–407
9. Sachdev M, de Gyvez JP (2007) Defect-oriented testing for nano-metric CMOS VLSI circuits. Springer, Berlin
10. Chang SJ, Lee CL, Chen JE (1997) Functional test pattern generation for CMOS operational amplifier. In: Proceedings of the 15th IEEE VLSI test symposium, Monterey, US, pp 267–272
11. Mielke JA (1996) Frequency domain testing of ADCs. IEEE Des Test Comput 13(1):64–69
12. Binns RJ, Taylor D, Pritchard TI (1996) Generating, capturing, and processing supply current signatures from analogue macros in mixed-signal ASICs. Microelectron J 27(8):723–729
13. Litovski V (2006) Fundamentals of electronics, theory, solved problems and exam questions. Akademska Misao, Belgrade, Serbia (in Serbian). http://litovski.com/
14. Litovski VB, Zwoliński M (1997) VLSI circuit simulation and optimization. Chapman and Hall, London
15. Slamani M, Kaminska B (1995) Multifrequency analysis of faults in analog circuits. IEEE Des Test Comput 12(2):70–80
16. Litovski VB (2000) Electronic circuit design, simulation, optimization, testing and physical design. Nova Jugoslavija, Vranje, Serbia (in Serbian)
17. Liu R-W (ed) (1991) Testing and diagnosis of analog circuits and systems. Van Nostrand Reinhold, New York
18. Variyam PN, Chatterjee A (1997) Test generation for comprehensive testing of linear analog circuits using transient response sampling. In: Proceedings of the IEEE/ACM international conference on CAD, ICCAD'97, San Hose, California, pp 382–385
19. Balivada A, Chen J, Abraham JA (1996) Analog testing with time response parameters. IEEE Des Test Comput 13(2):18–25
20. Arabi K, Kaminska B (1996) Oscillation-test strategy for analog and mixed-signal integrated circuits. In: Proceedings of the 14th IEEE VLSI test symposium (VTS 1996), Princeton, New Jersey, pp 476–482
21. Arabi K, Kaminska B (1999) Oscillation-test methodology for low-cost testing of active filters. IEEE Trans Instrum Meas 48(4):798–806
22. Litovski V, Andrejević M, Zwoliński M (2006) Analogue electronic circuit diagnosis based on ANNs. Microelectron Reliab 46(8):1382–1391
23. Materka A (1995) Neural network technique for parametric testing of mixed-signal circuits. Electron Lett 31(3):183–184
24. Benjamins R, Jansweijer W (1994) Toward a competence theory of diagnosis. IEEE Expert 9(5):43–52

25. Yang ZR, Zwoliński M, Chalk CD, Williams AC (2000) Applying a robust heteroscedactic neural network to analog fault detection and classification. IEEE Trans Comput Aided Des 19(1):142–151
26. Maidon Y, Jervis BW, Dutton N, Lesage S (1997) Diagnosis of multifaults in analogue circuits using multilayer perceptrons. IEE Proc Circuits Dev Syst 144(3):149–154
27. Chalk CD, Zwoliński M, Wilkins BR (1997) Test stimulus generation for steady-state analysis of analogue and mixed-signal circuits. Res J. (University of Southampton)
28. Harvey RJA, Richardson A, Kerkhoff HG (1995) Defect oriented test development based on inductive fault analysis. Collection of papers presented at the International mixed-signal testing workshop, Grenoble, France, pp 2–9
29. Huss SD, Gyurcsik RS (1991) Optimal ordering of analog integrated circuit tests to minimize test time. Proc of the 28 ACM/IEEE Design Automation Conference. Paper 30(1):494–499

7.3 Fault Simulation in Linear Electronic Circuits

Abstract A procedure is described in this chapter that allows concurrent fault simulation of linear electronic circuits. It is based on the Modified Nodal Analysis which is extensively used for circuit equation formulation. The concept of stamps is followed and proper stamps are given for all linear fault-free and faulty circuit elements including two- and four-terminal ones. A special feature of the method is its ability to model catastrophic faults. The method insists for the LU factorization of the circuit matrix to be implemented in order to reach its effectiveness. Examples are given performing fault simulation in DC regime and in time domain simulation. Finally, a short study of the efficiency of the method is given showing its advantages. Special warning is given for fault simulation of linear circuit being created as linearized models of nonlinear ones since change in some circuit elements may affect the quiescent operating point of the original nonlinear circuit and so making the fault simulation inaccurate.

7.3.1 Introduction

In this chapter, an effective method for simulating faults in linear electronic circuits will be presented. The importance of this method stems from the need for multiple circuit simulations when estimating the fault effects and verifying the test. In addition, such calculations can be important for tolerance analysis [1], yield estimation as well as fault simulation for diagnostic purposes [2]. The method has been reported several times, and the most complete versions of its implementation can be found in [3, 4].

When the value of one circuit parameter is changed, to calculate the increment value of the response (e.g., the output voltage or the output current) two analyses are required: one for the original circuit to establish the fault-free value of the response, say v_{out}, and one for the modified circuit (or circuit with a fault) whose output voltage will be marked with $\hat{v}_{out} = v_{out} + \Delta v_{out}$. When we have M parameters changed, $M + 1$ analysis will be required: one for the original circuit and M for the modified circuits. It should be kept in mind that when the value of one parameter changes, the circuit matrix also changes, so there is a need for the circuit matrix to be factorized M

V. B. Litovski, *Lecture Notes in Analog Electronics*, Lecture Notes in Electrical Engineering 1265, https://doi.org/10.1007/978-981-97-8257-4_3

+ 1 times. For large M, complex circuits, and for repeated analyses in the frequency or time domain, the fault analysis becomes expensive even for linear circuits. For example, in the simple circuit of Fig. 7.2.4 there are $M = 15$ parameters. If the goal is to determine the values of the frequencies given in the third column of Table 7.2.2, it is necessary to calculate the dependence of the sensitivity as function of the frequency in the interval from 0 to 20 kHz, say in 100 points. Therefore, the required total number of factorizations of the system matrix will be $100(M + 1) = 1600$.

In order to avoid this in [1, 5] a method based on the differential form of Tellegen's theorem [6] was applied in the frequency domain. To calculate M response increments, at a given frequency, it requires $M + 1$ adjoint network analysis [7] plus one solution of a newly created system of linear equations of order M. Of course, $M + 1$ adjoint network analysis means only one factorization of the circuit matrix and $M + 1$ back substitutions. Moreover, although its implementation seems quite complicated, the application of this method can be considered simple. However, this approach, in addition to being limited to the frequency domain, suffers from a significant additional limitation: it cannot be applied when catastrophic faults are present in the circuit.

As far as we know, no proposals have been made for a rationalized deterministic analysis of tolerances with large increments of parameters in the time domain.

It should be said that analysis of large change tolerances and simulation of soft faults are basically not different. In the case of catastrophic faults, however, the topology (graph) of the circuit changes, so it is usually necessary to describe the circuit again in order to perform a completely new analysis for each fault of this type. Procedures based on this concept are described in [2, 8].

Here we propose a simple method based on the fundamental idea of the Modified Nodal Analysis (MNA) [1, 9]. It will enable the analysis of parametric and catastrophic faults with minimal effort. Its main advantages are the following. First, it introduces the fault as a circuit element as an extension to the circuit description. Therefore, it does not require the creation of separate circuit description files for each fault individually. In this way, considerable savings are achieved in the preparation of circuit description files (which can be prone to errors), as well as savings in memory space for storing these files. Second, during the analysis, at the given time instant (or, for the given frequency value), the formulation of the equations is not repeated for each new fault. When evaluating this, it should be borne in mind that the equation formulation can constitute a significant part of the time required to realize one step of the analysis (formulation + solution). In this way, the new method offers additional savings. Third and finally, practically, only one factorization of the circuit matrix is required for the solution of the system of equations describing both the original and all modified circuits. Bearing in mind that factorization of the circuit matrix is the most time-consuming step at each time point or for each frequency value, significant savings in the total simulation time can be expected if long simulation period (broad frequency range) is considered.

In this chapter, using the example of a modified element (conductance), the method will be introduced first. Next, a procedure for modeling the increments of other lumped elements will be described. The specifics of the analysis in the frequency and time domain will be the subject of special consideration. Simulation of catastrophic faults follows. Finally, some aspects of the application of this method will be considered, as well as limitations from the point of view of circuit classes and elements to which the method is applicable. Presentations are based on [4, 10].

7.3.2 Description of the Basic Idea of the Method

Given a circuit N described by

$$\mathbf{Y} \cdot \mathbf{x} = \mathbf{y} \tag{7.3.1a}$$

where **Y** is the matrix of the system of equations formulated by MNA, **x** is the vector of unknown voltages and currents and **y** is the excitation vector. Let the conductance G, of the chosen branch, be modified so that its increment is $\Delta G = g$. For this case, the original (N) and the modified ($\widehat{\text{N}}$) circuits can be represented in Fig. 7.3.1a and b, respectively.

As can be seen, a pair of terminals was created parallel to the conductance G in the original circuit. These are open which is expressed by the zero-valued current source. At the same time in $\widehat{\text{N}}$ the zero-valued current source is replaced by the increment g of the conductance G. For the circuits of Fig. 7.3.1 the following branch equations apply:

$$i_A = 0 \tag{7.3.1b}$$

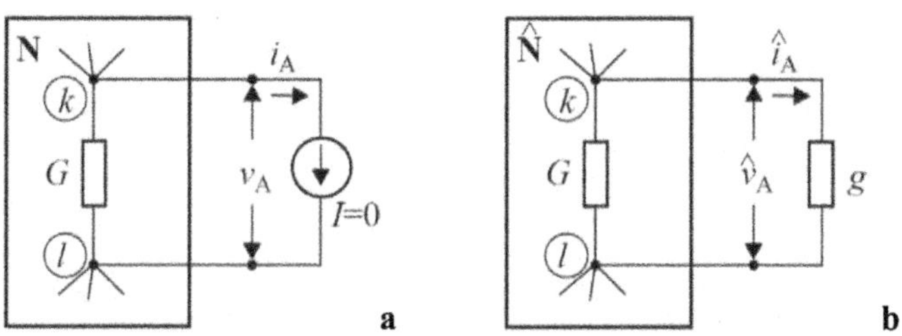

variables→ equations ↓	Δv_k	Δv_l	Δi_A (vector **A**)	RHS
k			1	
l			-1	
n+1 (vector **C**)	g	-g		-$g\cdot(v_k - v_l)$

c

Fig. 7.3.1 **a** Original circuit, **b** the modified circuit, and **c** a stamp that enables the formulation of the equations of the modified circuit. RHS stands for right-hand side of the system of equations

$$g\hat{v}_{\mathrm{A}} - \hat{i}_{\mathrm{A}} = g\hat{v}_k - g\hat{v}_l - \hat{i}_{\mathrm{A}} = 0 \tag{7.3.1c}$$

where ^ denotes a variable in the "modified" circuit. v_k and v_l denote the k-th and l-th node voltages, respectively. Expressions (7.3.1b) and (7.3.1c) are the equations of the new branches that will be used when formulating the equations using the MNA for the original and the modified circuit, respectively.

The increment of the matrix of the system of equations in the case of a single fault, can also be expressed as

$$\Delta \mathbf{Y} = g \cdot \begin{bmatrix} 0 \ldots 0 \underset{k}{1} 0 \ldots 0 \underset{l}{-1} 0 \ldots 0 \end{bmatrix}^{\mathrm{T}} = g \cdot \mathbf{A} \tag{7.3.1d}$$

where $\mathbf{A}$ is a column matrix in which all elements are equal to zero except for the kth which is equal to 1 and the lth which is equal to -1.

The corresponding system of equations describing the circuit of Fig. 7.3.1a by the MNA method is

$$\begin{bmatrix} \mathbf{Y} & \mathbf{A} \\ \mathbf{B} & -1 \end{bmatrix} \cdot \begin{bmatrix} \mathbf{x} \\ i_{\mathrm{A}} \end{bmatrix} = \begin{bmatrix} \mathbf{y} \\ 0 \end{bmatrix}, \tag{7.3.2a}$$

where $\mathbf{B}$ is a row matrix whose elements are all equal to zero. All vectors and matrices are n-dimensional, where n is the number of unknowns (voltage and current) generated by the MNA.

From now on, it will be considered that the system (7.3.2a) was solved, so it is considered that the quantities $\mathbf{x}$ and i_{A} are known. Bearing in mind that $i_{\mathrm{A}} = 0$, it becomes irrelevant where the new port is created or, what is the same, which conductance is modified: $\mathbf{x}$ remains unchanged for any modification of the original circuit of this type.

If multiple faults are present, for the increment of the system matrix one can write:

$$\Delta \mathbf{Y} = \mathbf{P}\mathbf{D}\mathbf{P}^{\mathrm{T}}. \tag{7.3.2b}$$

Here, $\mathbf{D} = \begin{bmatrix} g_1 & \cdots & 0 \\ \vdots & \ddots & \vdots \\ 0 & \cdots & g_m \end{bmatrix}$, m is the number of elements whose value is subject to change, and, $\mathbf{P} = \begin{bmatrix} \mathbf{A}_1 & \ldots & \mathbf{A}_m \end{bmatrix}$ an $n \times m$ matrix, with $\mathbf{A}_i$ defined as in (7.3.1d) for each element whose conductivity is changed individually.

Now let us write down the system of equations according to MNA for the circuit from Fig. 7.3.1b:

$$\begin{bmatrix} \mathbf{Y} & \mathbf{A} \\ \mathbf{C} & -1 \end{bmatrix} \begin{bmatrix} \hat{\mathbf{x}} \\ \hat{i}_{\mathrm{A}} \end{bmatrix} = \begin{bmatrix} \mathbf{y} \\ 0 \end{bmatrix} \tag{7.3.3}$$

The meaning of the new quantity **C** in this system is easily expressed if the numbering of the nodes and the corresponding node equations in the original circuit and in the perturbed one are the same: $\boldsymbol{C} = g \cdot \boldsymbol{A}^{\mathrm{T}}$. Of course: $\mathbf{B} + \mathbf{C} = \mathbf{C}$.

To find the voltage and current increments in the modified circuit due to the change in *G*, the following notation will be used

$$\begin{bmatrix} \hat{\mathbf{x}} \\ \hat{i}_{\mathrm{A}} \end{bmatrix} = \begin{bmatrix} \mathbf{x} \\ i_{\mathrm{A}} \end{bmatrix} + \begin{bmatrix} \Delta\mathbf{x} \\ \Delta i_{\mathrm{A}} \end{bmatrix}. \tag{7.3.4}$$

Substituting in (7.3.3) gives

$$\begin{aligned}\begin{bmatrix} \mathbf{Y} & \mathbf{A} \\ \mathbf{C} & -1 \end{bmatrix}\begin{bmatrix} \Delta\mathbf{x} \\ \Delta i_{\mathrm{A}} \end{bmatrix} &= \begin{bmatrix} \mathbf{y} \\ 0 \end{bmatrix} - \begin{bmatrix} \mathbf{Y} & \mathbf{A} \\ \mathbf{C} & -1 \end{bmatrix}\begin{bmatrix} \mathbf{x} \\ i_{\mathrm{A}} \end{bmatrix} \\ &= \begin{bmatrix} \mathbf{y} \\ 0 \end{bmatrix} - \begin{bmatrix} \mathbf{Y} & \mathbf{A} \\ \mathbf{B} & -1 \end{bmatrix}\begin{bmatrix} \mathbf{x} \\ i_{\mathrm{A}} \end{bmatrix} - \begin{bmatrix} 0 & 0 \\ \mathbf{C} & 0 \end{bmatrix}\begin{bmatrix} \mathbf{x} \\ i_{\mathrm{A}} \end{bmatrix} = \begin{bmatrix} 0 \\ -g \cdot (v_k - v_l) \end{bmatrix}.\end{aligned} \tag{7.3.5}$$

The newly created system of equations expresses the increments of voltages and currents in the modified circuit as a function of the circuit elements of the circuit with a changed topology (*C* instead of *B*) and with excitation determined by the analysis of the original circuit which was already done. *The solution of this system, therefore, is not the node voltages and branch currents, but their increments.* The final voltages and currents in the modified circuit will be obtained after the substitution in (7.3.4).

Equation (7.3.5) is also valid for the case when the number of branches whose conductivity changes is larger than one. In that case, **B** becomes a matrix ($n \times m$) with all elements equal to zero, and instead of **A**, **P** will be used.

Figure 7.3.1c represents the stamp which enables the formulation of the equations for the modified circuit in case the subject of the perturbation is the conductance of the conductive branch. This table should be considered as the basis for determining the vectors **A** and **C**. Namely, the column in the stamp below the label Δi_{A} contains nonzero elements of the vector **A**, and the last row of table gives the nonzero elements of the vector **C**.

It is of crucial importance for the application of this method to apply *LU factorization* when solving this system of equations. In that case, the following decomposition can be used

$$\mathbf{Y} = \mathbf{L}_{\mathrm{Y}} \cdot \mathbf{U}_{\mathrm{Y}} \tag{7.3.6}$$

where $\mathbf{L}_{\mathrm{Y}}$ and $\mathbf{U}_{\mathrm{Y}}$ are the lower and upper triangular matrices, respectively. Also, it is of crucial importance that the factorization of the matrices of the system (7.3.2a) and (7.3.5) be performed simultaneously. The idea is that, when considering that (7.3.6) is common to both systems, if the same pivots are used, only a slight additional effort will be required to factorize the matrix of system (7.3.5). This is the basis of the power of this method. Bearing in mind that in the simulation of faults in digital

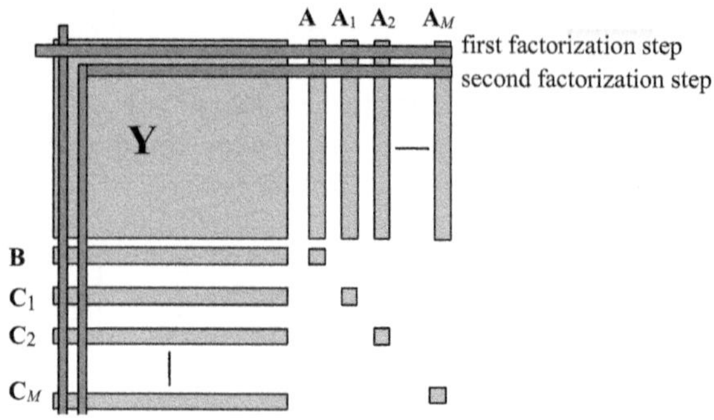

Fig. 7.3.2 Illustration of the concurrent factorization procedure. The kernel **Y** is common to all $M + 1$ extensions

circuits, the simultaneous simulation of several faults is usually called concurrent, and here we adopt the name "concurrent fault simulation.".

For the next modified conductance, the structure of the matrix **A** in (7.3.3) and (7.3.5) will change insofar as k and l will change their values. "1" and "−1" will migrate through **A**. Therefore, it will be necessary to solve (7.3.5) again for each new modification. Of course, the factorizations of all matrices will be realized simultaneously. From the point of view of computational complexity, only one factorization and $M + 1$ forward and backward substitutions are required for the fault (tolerance) analysis of circuits with M modified conductances.

In order to get an idea of the factorization procedure itself as well as its complexity Fig. 7.3.2 shows the system matrix **Y** together with the corresponding extensions. With $\mathbf{C}_1$, $\mathbf{C}_2$, …, $\mathbf{C}_M$ and with $\mathbf{A}_1$, $\mathbf{A}_2$, …, $\mathbf{A}_M$, are marked matrix-rows or matrix-columns, respectively, which refer to each fault individually. The factorization takes place so that when the corresponding column of **Y** is processed, it continues with the corresponding element in $\mathbf{C}_1$ as if it were part of a new matrix equal to **Y** expanded by $\mathbf{C}_1$. The last action gets repeated so that the role of the element of $\mathbf{C}_1$ is taken over by an element of $\mathbf{C}_2$, etc. The same applies when processing rows. Now, instead of **C**, **A** is processed.

It can be seen that, in fact, the factorizations of all $M + 1$ matrices are performed simultaneously, with the kernel being factorized once. Assuming that the circuit modification does not change the structure of **Y** and bearing in mind the pronounced sparsity of **A** and **C** (i.e., P), the number of operations for all factorizations combined can be expected to be slightly larger than the number of operations for the factorization of **Y**. The final appearance of the factored matrix together with its additions is shown in Fig. 7.3.3.

When this method is compared with the method based on adjoint networks [5], it can be said that the advantage of this method is that in this way the increments of all voltages and (required) currents in the circuit are obtained, which is not the case in [5]. An increment of only one output quantity is obtained there. We will see later that by suitable modeling of catastrophic faults, this method can be easily extended to such a case, which is contrary to the basic idea in [5]. Additionally, as will be shown shortly, this method applies equally effectively directly to the time domain.

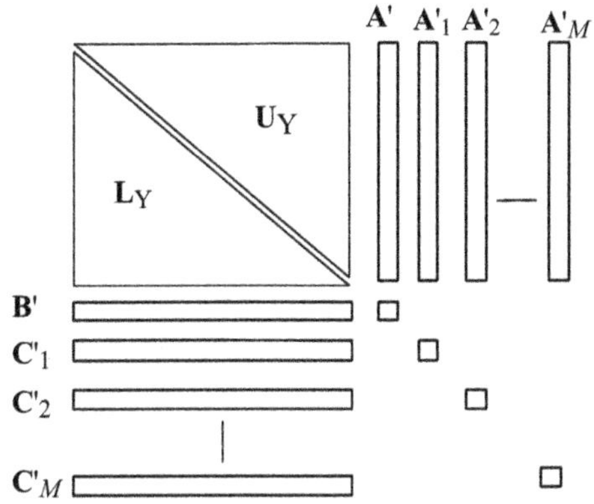

Fig. 7.3.3 Layout of the factored matrix of the original circuit and its additions (marked with ′) related to each fault individually

7.3.3 Tolerances Analysis in the Frequency and Time Domain

When considering the simulation of faults in the frequency domain, the quantities **Y**, **v**, **j**, **A**, **B**, **D**, and $i_\mathbf{A}$ in (7.3.2) and (7.3.5) become complex. If complex arithmetic is not used, the following relations will apply $\mathbf{Y} = \mathbf{Y}_\mathrm{r} + j \cdot \mathbf{Y}_\mathrm{i}$, $\mathbf{x} = \mathbf{x}_\mathrm{r} + j \cdot \mathbf{x}_\mathrm{i}$, $\mathbf{y} = \mathbf{y}_\mathrm{r} + j \cdot \mathbf{y}_\mathrm{i}$, $\mathbf{A} = \mathbf{A}_\mathrm{r} + j \cdot \mathbf{A}_\mathrm{i} = \mathbf{A}_\mathrm{r} + j \cdot \mathbf{0}$, $\mathbf{B} = \mathbf{B}_\mathrm{r} + j \cdot \mathbf{B}_\mathrm{i} = \mathbf{0} + j \cdot \mathbf{0}$, $\mathbf{D} = \mathbf{D}_\mathrm{r} + j \cdot \mathbf{D}_\mathrm{i}$, and $i_\mathrm{A} = i_\mathrm{Ar} + j \cdot i_\mathrm{Ai}$. Substituting this into (7.3.2a) gives

$$\begin{bmatrix} \mathbf{Y}_\mathrm{r} & -\mathbf{Y}_\mathrm{i} & \mathbf{A}_\mathrm{r} & -\mathbf{A}_\mathrm{i} \\ \mathbf{Y}_\mathrm{i} & \mathbf{Y}_\mathrm{r} & \mathbf{A}_\mathrm{i} & \mathbf{A}_\mathrm{r} \\ \mathbf{B}_\mathrm{r} & -\mathbf{B}_\mathrm{i} & -1 & 0 \\ \mathbf{B}_\mathrm{i} & \mathbf{B}_\mathrm{r} & 0 & -1 \end{bmatrix} \cdot \begin{bmatrix} \mathbf{x}_\mathrm{r} \\ \mathbf{x}_\mathrm{i} \\ i_\mathrm{Ar} \\ i_\mathrm{Ai} \end{bmatrix} = \begin{matrix} \mathbf{y}_\mathrm{r} \\ \mathbf{y}_\mathrm{i} \\ 0 \\ 0 \end{matrix} \tag{7.3.7}$$

and substitution in (7.3.5) gives

$$\begin{bmatrix} \mathbf{Y}_\mathrm{r} & -\mathbf{Y}_\mathrm{i} & \mathbf{A}_\mathrm{r} & -\mathbf{A}_\mathrm{i} \\ \mathbf{Y}_\mathrm{i} & \mathbf{Y}_\mathrm{r} & \mathbf{A}_\mathrm{i} & \mathbf{A}_\mathrm{r} \\ \mathbf{C}_\mathrm{r} & -\mathbf{C}_\mathrm{i} & -1 & 0 \\ \mathbf{C}_\mathrm{i} & \mathbf{C}_\mathrm{r} & 0 & -1 \end{bmatrix} \cdot \begin{bmatrix} \Delta\mathbf{x}_\mathrm{r} \\ \Delta\mathbf{x}_\mathrm{i} \\ \Delta i_\mathrm{Ar} \\ \Delta i_\mathrm{Ai} \end{bmatrix} = \begin{matrix} 0 \\ 0 \\ -g \cdot (V_{k\mathrm{r}} - V_{l\mathrm{r}}) \\ -g \cdot (V_{k\mathrm{i}} - V_{l\mathrm{i}}) \end{matrix}. \tag{7.3.8}$$

By carefully considering (7.3.7) and (7.3.8), it can be concluded that they are essentially similar to (7.3.2a) and (7.3.5), respectively, so all conclusions related to the realization of solving (7.3.5) are based on solving (7.3.2a) can be applied to the realization of the solution (7.3.8), which would be based on the solution (7.3.7). Bearing this in mind, the following text will manipulate the original versions of the equations, i.e., (7.3.2a) and (7.3.5) will be dealt with in the following text. It is important to keep in mind that when determining the frequency characteristics, all circuits (original and modified) must be analyzed at the same points on the frequency axis, that is, when calculating the frequency characteristics, the frequency increment must be the same for all circuits.

Analyzing (7.3.5) and (7.3.8), we see that to the RHS vector the property of sparsity can be attributed too. Bearing in mind the direct substitution (procedure when solving systems of linear equations by LU factorization [10]) we conclude that it is reduced to one (for DC analysis) or two (frequency domain analysis) operations. Considering the exceptional sparsity of the matrix in the frequency domain analysis as well as the large number of points on the frequency axis for which the system of equations needs to be solved, the exploitation of this property of (7.3.5) and (7.3.8) becomes significant from the point of view of the total time required to solve the system of equations. It has been shown [3] that savings can reach as much as 20% of the total computation time.

Let us now consider the modeling of modifications (increments) of other circuit elements. From the point of view of this method, they can be divided into three categories. In the first there will be elements described by a current equation. These are conductivity (G), capacitance (C), current controlled current source (CCCS), and voltage controlled current source (VCVS). The following parameters are modified in this category: conductance ($\Delta G = g$), capacitance ($\Delta C = c$), current gain ($\Delta\beta$), and transconductance (Δg_{m}).

For each of these elements the branch current can be represented as a new variable $\hat{\imath}_{\mathrm{A}}$ in the same way as it was done for the conductance. In Eq. (7.3.5), the changes will refer to changes in $\mathbf{C}$ as well as in the RHS vector, which will be expressed in a different way. For example, for a VCCS, according to Fig. 7.3.4a, the equation of the branch that creates $\mathbf{C}$ will be

$$(\Delta g_{\mathrm{m}})\widehat{V}_i - (\Delta g_{\mathrm{m}})\widehat{V}_j - \hat{\imath}_{\mathrm{A}} = 0 \tag{7.3.9}$$

where $\widehat{V}_i$ and $\widehat{V}_j$ are the node voltages that make up the controlling voltage ($\widehat{V}_i - \widehat{V}_j = \widehat{V}$) in the modified circuit. For the VCCS case, the equation equivalent to (7.3.5) will be

$$\begin{bmatrix} \mathbf{Y} & \mathbf{A} \\ \mathbf{C} & -1 \end{bmatrix} \begin{bmatrix} \Delta \mathbf{x} \\ \Delta i_{\mathrm{A}} \end{bmatrix} = \begin{bmatrix} 0 \\ -g_{\mathrm{m}} \cdot \left(V_i - V_j\right) \end{bmatrix}. \tag{7.3.10}$$

Similar expressions can be derived for CCCS (Fig. 7.3.4b) where I_{cd} is the controlling current, and for the capacitor (Fig. 7.3.4c). In doing so, one uses $(\Delta\beta)(\Delta I_{\mathrm{cd}}) - \Delta I_{\mathrm{A}} = 0$.

Figure 7.3.4d, e show the stamps for the formulation of the equations in the modified circuit. They can be used for both time and frequency domains. If, however, for the frequency domain, the equations are displayed according to (7.3.8), then the stamps are created in Fig. 7.3.4f–h. The reader is left to perform a similar transformation for the stamps from Fig. 7.3.4d, e.

The second category consists of elements described by voltage equations: resistance (R), inductance (L), current controlled voltage source (CCVS), and voltage controlled voltage source (VCVS).

a

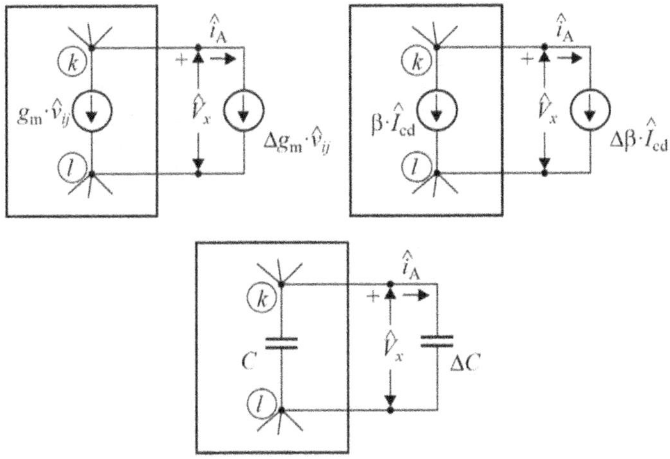

b

variables→ equations ↓	ΔV_k	ΔV_l	ΔI_A	RHS
k			1	
l			-1	
n+1	Δg_m	$-\Delta g_m$		$-\Delta g_m \cdot (V_i - V_j)$

variables→ equations ↓	ΔV_k	ΔV_l	ΔI_{cd}	ΔI_A	RHS
k				1	
l				-1	
n+1			$\Delta\beta$	-1	

Fig. 7.3.4 Circuit modification to introduce increments of **a** transconductance, **b** current gain, and **c** capacitance **d** inductance. Corresponding stamps for modified branches: **d** for transconductance and **e** for CCCS. Stamps for modified circuit for frequency domain analysis **f** conductance, **g** capacitance, and **h** inductance

c

variables→ equations ↓	$\Delta v_{k\mathrm{r}}$	$\Delta v_{k\mathrm{i}}$	$\Delta v_{l\mathrm{r}}$	$\Delta v_{l\mathrm{r}}$	Δi_{Ar}	Δi_{Ai}	RHS
k					1		
l					-1		
$n+k$						1	
$n+l$						-1	
$2n+1$	g		$-g$		-1		$-g(v_{k\mathrm{r}}-v_{l\mathrm{r}})$
$2n+2$		g		$-g$		-1	$-g(v_{k\mathrm{i}}-v_{l\mathrm{i}})$

variables→ equations ↓	$\Delta v_{k\mathrm{r}}$	$\Delta v_{k\mathrm{i}}$	$\Delta v_{l\mathrm{r}}$	$\Delta v_{l\mathrm{r}}$	Δi_{Ar}	Δi_{Ai}	RHS
k					1		
l					-1		
$n+k$						1	
$n+l$						-1	
$2n+1$		$-\omega C$		ωC	-1		$\omega \cdot \Delta C \cdot (v_{k\mathrm{i}} - v_{l\mathrm{i}})$
$2n+2$	ωC		$-\omega C$			-1	$-\omega \cdot \Delta C \cdot (v_{k\mathrm{r}} - v_{l\mathrm{r}})$

variables→ equations ↓	$\Delta v_{k\mathrm{r}}$	$\Delta v_{k\mathrm{i}}$	$\Delta v_{l\mathrm{r}}$	$\Delta v_{l\mathrm{r}}$	Δi_{Ar}	Δi_{Ai}	RHS
k					1		
l					-1		
$n+k$						1	
$n+l$						-1	
$2n+1$		$-\omega L$		ωL	-1		$\omega \cdot (\Delta L) \cdot i_{L\mathrm{i}}$
$2n+2$	ωL		$-\omega L$			-1	$-(\Delta L) \cdot i_{L\mathrm{r}}$

Fig. 7.3.4 (continued)

In the case of the second category of circuits, the following parameters are subject to modification: resistance ($\Delta R = r$), inductance ($\Delta L = l$), transresistance (Δr_{m}), and voltage gain ($\Delta\mu$). The problem with introducing increments is the following. If the parameter of the element described by the voltage equation is changed, the effect can be displayed in two ways. First is to add an increment to the existing element value and thereby create the conditions for repeating the analysis. This means that we apply the classical method, which we found to be computationally inefficient. Another possibility is to display the increment by adding a new element in series with the original one. In that case, the VCVS would have a series connection of two VCVS: $\mu \cdot V$ and $(\Delta\mu) \cdot V$. A new node would be created between them, thus creating a new matrix **Y**. The procedure that is a subject of this section would not be applicable. We do not want to destroy the matrix **Y**, so, by our method, the original voltage source is first transformed into a current source. In this way, it will be possible for the increment to be expressed by a current element that is connected in parallel with the newly created port, as was also the case for the conductance and the VCCS.

When the equations of the original circuit are formulated using the MNA all these elements create a new equation in the original system. This has the same effect as if a gyrator were already installed in the circuit, which is loaded with an element described by a current equation. That means that the introduction of the gyrator did not change the matrix of the circuit **Y** but represents only a schematic interpretation of the new equation introduced by the MNA for ideal voltage sources (Yes, the reader must familiarize her/himself with the MNA.).

As the first example, the VCVS shown in Fig. 7.3.5 will be discussed. Figure 7.3.5a represents the original unmodified circuit before the transformation (before introduction of the gyrator), and Fig. 7.3.5b shows the same circuit with the addition of a gyrator. It should be noted that the rightmost source is now VCCS and that μ now plays the role of transconductance. The numerical value of μ is, of course, unchanged. The equation of the branch related to VCVS in the original description by MNA is

$$v_k - v_j = \mu \cdot v = \mu \cdot (v_i - v_j). \tag{7.3.11}$$

In this expression, v_k and v_l are the node voltages in the original circuit measured at the VCVS itself, and $v = v_i - v_j$ is the controlling voltage as before. It is important to note again that (7.3.11) is part of the system of equations describing the original circuit, thus contributing to the matrix **Y**.

The advantages afforded by the representation of the original circuit given in Fig. 7.3.5b can be understood if Fig. 7.3.5c is considered. Namely, in order to introduce an increment ($\Delta\mu \cdot V$), in the circuit of Fig. 7.3.5c, a new port was created parallel to the VCCS, which was shown in Fig. 7.3.5b denoted by $\mu \cdot V$. The equation defining **C** now has the form

$$\Delta\mu \cdot (\hat{v}_i - \hat{v}_j) - \hat{\imath}_{\mathrm{A}} = 0. \tag{7.3.12}$$

which is formally identical to the one used to model the fault of the transconductance.

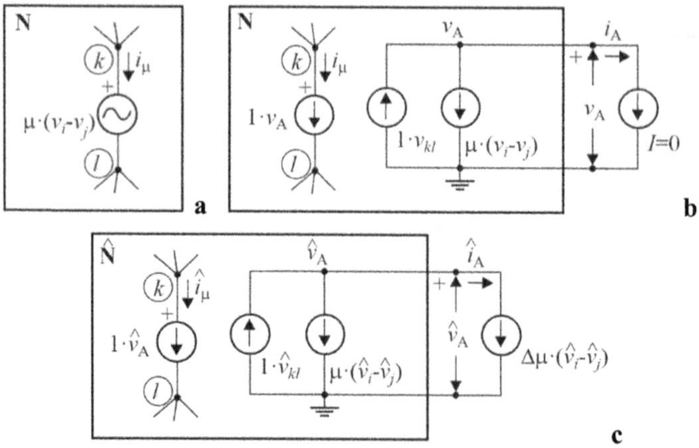

Fig. 7.3.5 **a** Original VCVS, **b** original VCVS modeled by gyrator (adapted for MNA), and **c** modified VCVS. **d** Stamp for the original VCVS and **e** stamp for the modified VCVS

Fig. 7.3.5 (continued)

variables→ equations↓	v_k	v_l	v_i	v_j	$i_\mu = v_A$	RHS
k					1	
l					-1	
new equation	1	-1	-μ	μ		

variables→ equations↓	Δv_i	Δv_j	Δi_A	RHS
new equation			1	
n+1	Δμ	-Δμ	-1	$-(\Delta\mu)\cdot(v_i - v_j)$

Similar considerations apply to other elements belonging to the same category. For the modified circuit corresponding to the resistor only the schematic will be shown here. Figure 7.3.6a represents the original resistor. Figure 7.3.6b shows the introduction of the gyrator, i.e., the new equation in the original description. The rightmost element is a conductance which has as many Siemens as R has Ohms. Figure 7.3.6c represents the modified circuit. Now ΔG also has as many Siemens as ΔR in Ohms. The rest of Fig. 7.3.6 refers to the automation of equation formulation.

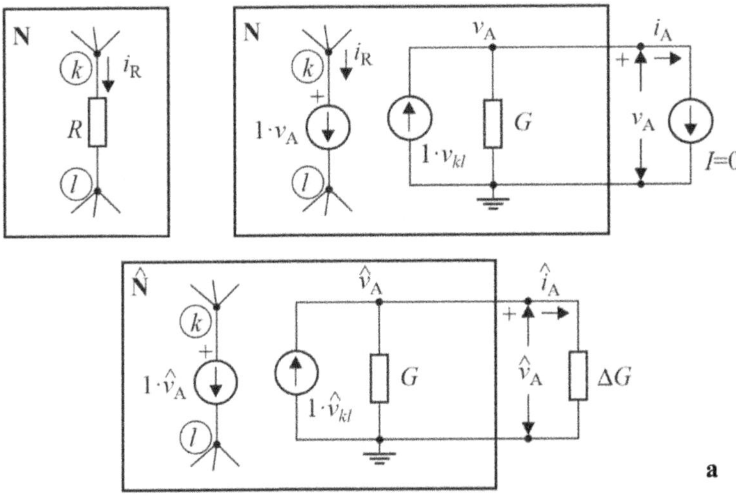

a

variables→ equations↓	v_k	v_l	v_A	RHS
k			1	
l			-1	
new equation	-1	1	*R*	

variables→ equations↓	Δv_A	Δi_A	RHS
new equation		1	
$n+1$	ΔR	-1	$-(\Delta R)\cdot v_A$

b

Fig. 7.3.6 **a** Original resistor, **b** original resistor modeled by a gyrator, and **c** modified resistor. **d** Stamp for the original resistor and **e** stamp for the modified branch. *G* is a conductance that has as many Siemens as *R* has Ohms

It should be noted that the stamps developed in Figs. 7.3.5 and 7.3.6 apply to both the time and frequency domain.

A similar transformation is shown in Fig. 7.3.7. It refers to inductance. Now, instead of conductivity, a capacitor appears. A suitable stamp to use in the frequency domain can be obtained from Fig. 7.3.6d, e if R is replaced by sL and ΔR by $s \cdot \Delta L$. The final form of this stamp (for frequency domain analysis) is given in Fig. 7.3.4h.

The third category of elements that will be covered here will include bilateral four-poles whose transmission is controlled by one parameter. There are four such four-poles: transformer (T), gyrator (Gy), negative impedance converter with voltage inversion (VNIC), and negative impedance converter with current inversion (INIC). It should be emphasized that these are abstract elements introduced by circuit theory

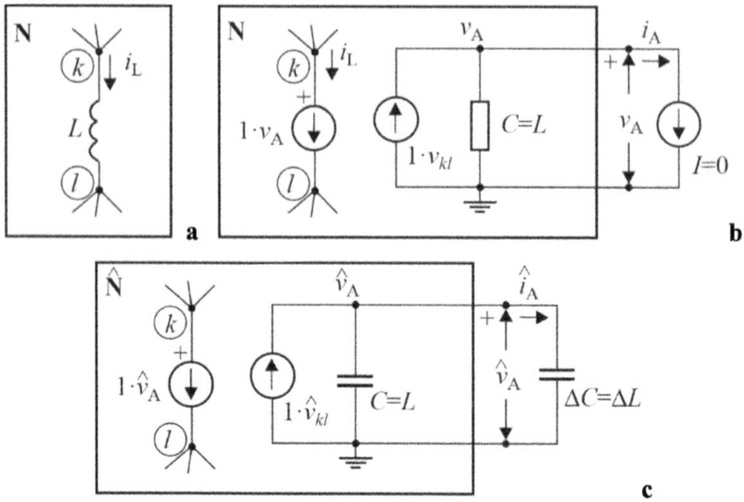

Fig. 7.3.7 **a** Original coil, **b** original coil modeled with a gyrator, and **c** modified coil. C is a capacitance that has as many Farads as L has Henrys

which have a nature similar to controlled sources but are bilateral. Namely, the controlled source is only a model of the phenomena that occur in the circuit. The importance of these considerations comes to the fore especially when processing transformers. The four-pole that will be discussed now will be called a transformer, but it should not be confused with a real transformer. Finally, the equations presented here will apply to instantaneous values.

Due to bilaterality, the transmission coefficient of these four-poles can be seen from both ports and therefore its influence should be taken twice. The basic idea in generating the fault model of these elements starts from the fact that they can be viewed as two coupled controlled sources, each transmitting the signal unilaterally. This means that the previous theory that was developed for controlled sources can be directly used here. Therefore, relatively less attention will be paid to these elements here.

The most commonly used element from this category is the transformer. That is why it will be discussed in this section. As a consequence of the necessity to introduce the terminal currents as new variables, this type of element will introduce two new equations in the MNA. Accordingly, the system corresponding to system (7.3.3) will now have two more equations than the number of nodes in the circuit.

In order to apply the concept that has been discussed so far for the other elements, the transformer is modeled with two gyrators as in Fig. 7.3.8. Figure 7.3.8a represents the original circuit of the transformer with the appropriate notation, and Fig. 7.3.8b represents a model of a transformer in which two gyrators are used. The contribution of the transformer to the system of equations of type (7.3.2) for the unmodified circuit can be extracted from the following

$$\begin{aligned} i_1 &= \frac{-1}{n} \cdot i_2 \\ v_1 &= n \cdot v_2. \end{aligned} \tag{7.3.13}$$

The corresponding stamp is shown in Fig. 7.3.8d. Let us repeat once again that this element introduces two new currents and by itself generates two new variables in the original system of equations written using MNA.

Introducing modification of parameter n is simple if two controlled sources are created as shown in Fig. 7.3.8c. Of course, for this case the matrices **A** and **C** have two columns or two rows, respectively, but that is not important considering that the basic idea is preserved. The matrix **C** is created according to the following pair of equations:

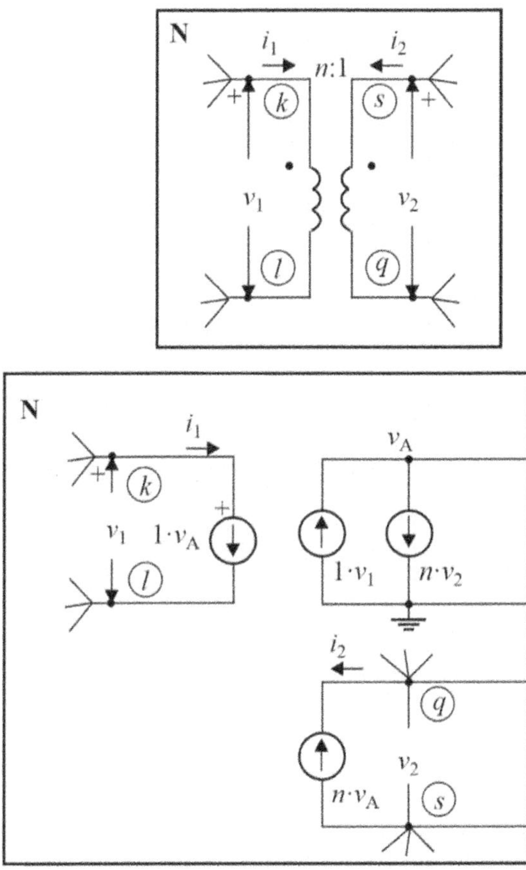

Fig. 7.3.8 Transformer. **a** Original version, **b** modeled via gyrator. Transformer. **c** Added sources for modeling tolerances, **d** stamp for the original circuit, and **e** stamp for the modified circuit

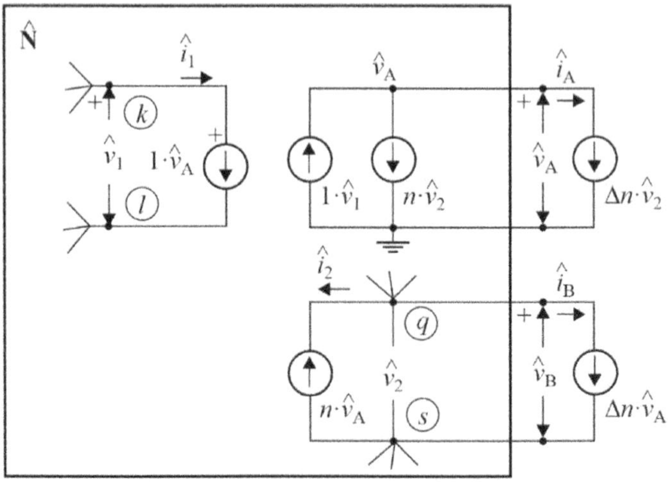

variables→ equations↓	v_k	v_l	v_s	v_q	i_1	i_2	RHS
k					1		
l					-1		
s						1	
q						-1	
new equation 1	1	-1	-*n*	*n*			
new equation 2					*n*	1	

variables→ equations↓	Δv_s	Δv_q	Δi_1	Δi_A	Δi_B	RHS
s				1		
q					1	
new equation 1					-1	
n+1	-Δ*n*	Δ*n*		-1		$(\Delta n)\cdot(v_s - v_q)$
n+2			Δ*n*		-1	$-(\Delta n)\cdot i_1$

Fig. 7.3.8 (continued)

$$\begin{aligned} \Delta n \cdot \hat{v}_2 - \hat{i}_A &= 0 \\ \Delta n \cdot \hat{v}_x + \hat{i}_B &= 0 \end{aligned} \tag{7.3.14}$$

Therefore, the system corresponding to (7.3.5) will now have $n + 2$ equations. The corresponding stamp is shown in Fig. 7.3.8e.

It is left to the reader, based on this example, to derive the transformations and corresponding stamps for the negative impedance converters and the gyrator.

Applying the above concepts to time domain analysis is straightforward. Namely, no conditions were set when creating the system (7.3.2) and (7.3.5). Therefore, all the conclusions drawn so far remain.

From the point of view of practical implementation, the following considerations are of interest. So far, two reactive elements have been introduced: the capacitor and the coil. If a gyrator is used to formulate the equations in the case of an inductive branch as in Fig. 7.3.7, it remains to resolve the influence of the equation describing the capacitive branch in the time domain. To this end, without any loss of generality, Euler's rule for numerical discretization will be used. If so, the capacitive branch described in the time domain by

$$i_{\mathrm{C}} = C\frac{\mathrm{d}v_{\mathrm{C}}}{\mathrm{d}t}, \tag{7.3.15}$$

will be discretized as follows:

$$i_{\mathrm{C}}^{n+1} = G_{\mathrm{C}}^{n} \cdot v_{\mathrm{C}}^{n+1} + i_{\mathrm{Cs}}^{n} \tag{7.3.16}$$

where degrees n and $n+1$ denote the previous and next time instants, respectively, $G_{\mathrm{C}}^{n} = C/h^{n}$, h^{n} is the time step: $h^{n} = t^{n+1} - t^{n}$, and $i_{\mathrm{Cs}}^{n} = -C \cdot v_{\mathrm{C}}^{n}/h^{n}$. It should be remembered that all quantities marked with "n" are known. Namely, for $n = 0$, v_{C}^{0} is the boundary condition that must be set in order to start the simulation (for both the fault-free and the faulty circuit).

Equation (7.3.16) hides the entire difference between the time and frequency domains. Here as well as in the frequency domain, the capacitive branch gives one conductivity, but here it also gives a current source that affects the RHS vector of the system of equations. That is the difference. Now the system of equations equivalent to the system (7.3.2) will be

$$\begin{bmatrix} \mathbf{Y}^{n} & \mathbf{A} \\ \mathbf{B} & 1 \end{bmatrix} \begin{bmatrix} \mathbf{x}^{n+1} \\ i_{\mathrm{A}}^{n+1} \end{bmatrix} = \begin{bmatrix} \mathbf{y}^{n+1} - \mathbf{j}_{\mathrm{Cs}}^{n} \\ 0 \end{bmatrix}, \tag{7.3.17}$$

where $(j_{\mathrm{Cs}}^{n})_{k}$, is the kth element of the vector $\boldsymbol{j}_{\mathrm{Cs}}^{n}$, and represents the sum of all i_{Cs}^{n} leaving the kth node in the original circuit N. $\boldsymbol{y}^{n+1}$ is the value of $\mathbf{y}$ at $t = t^{n+1}$.

If equal discretization steps are used for the original and modified circuit, introducing the capacitor modification will lead to the same matrix $\mathbf{Y}^{n}$ but to a changed vector $\boldsymbol{j}_{\mathrm{Cs}}^{n}$. The new value of that vector will be marked with $\hat{\boldsymbol{j}}_{\mathrm{Cs}}^{n}$. Now, for the modified circuit in which the capacitor has changed its value by ΔC, it can be written

$$\begin{bmatrix} \mathbf{Y}^{n} & \mathbf{A} \\ \mathbf{C}^{n} & -1 \end{bmatrix} \begin{bmatrix} \hat{\mathbf{x}}^{n+1} \\ \hat{i}_{\mathrm{A}}^{n+1} \end{bmatrix} = \begin{bmatrix} \mathbf{y}^{n+1} - \hat{\mathbf{j}}_{\mathrm{Cs}}^{n} \\ -i_{\Delta\mathrm{Cs}}^{n} \end{bmatrix}, \tag{7.3.18}$$

where $\mathbf{C}^{n} = G_{\Delta C}^{n}\mathbf{A}^{\mathrm{T}}$ is under the condition that, as before, we kept the ordinal number of the node and the ordinal number of the equation that describes that node equal. $G_{\Delta C}^{n} = (\Delta C)/h^{n}$ and $i_{\Delta\mathrm{Cs}}^{n} = -G_{\Delta C}^{n}\hat{v}_{\mathrm{C}}^{n}$.

variables→ equations ↓	Δv_k^{n+1}	Δv_l^{n+1}	Δi_A^{n+1}	RHS
k			1	$i_{Cs}^n - \hat{i}_{Cs}^n$
l			-1	$-(i_{Cs}^n - \hat{i}_{Cs}^n)$
N+1	G_C^n	$-G_C^n$	-1	$G_{\Delta C}^n(\hat{v}_k^n - \hat{v}_l^n) - G_{\Delta C}^n(v_k^{n+1} - v_l^{n+1})$

Fig. 7.3.9 Stamp for modified capacitor branch in time domain simulation

By introducing $\hat{x}^{n+1} = x^{n+1} + \Delta x^{n+1}$ and $\hat{i}_A^{n+1} = i_A^{n+1} + \Delta i_A^{n+1} = 0 + \Delta i_A^{n+1}$, we create the final form of a linear system of equations that enables the simultaneous calculation of voltage and current increments in the time domain:

$$\begin{bmatrix} \mathbf{Y}^n & \mathbf{A} \\ \mathbf{C}^n & -1 \end{bmatrix} \begin{bmatrix} \Delta \mathbf{x}^{n+1} \\ \Delta i_A^{n+1} \end{bmatrix} = \begin{bmatrix} \mathbf{j}_{Cs}^n - \hat{\mathbf{j}}_{Cs}^n \\ G_{\Delta C}^n\left(\hat{v}_k^n - \hat{v}_l^n\right) - G_{\Delta C}^n\left(v_k^{n+1} - v_l^{n+1}\right) \end{bmatrix}. \tag{7.3.19}$$

According to this, for the capacitor connected to the circuit as in Fig. 7.3.4c, the stamp valid for a circuit with a modified capacitance value is given in Fig. 7.3.9.

The inductance stamp will not be discussed in detail here, because after introducing the gyrator as in Fig. 7.3.7b, the branch leads to the capacitor just treated.

Coupled inductances as a specific element for analysis in the time domain (in the frequency domain they can be viewed as a four-pole, i.e., an element with two controlled sources) will be considered separately. They are described by the following system of equations.

$$\begin{aligned} v_1 &= L_1 \frac{di_1}{dt} + M \frac{di_2}{dt} \\ v_2 &= L_2 \frac{di_2}{dt} + M \frac{di_1}{dt} \end{aligned} \tag{7.3.20}$$

The schematic diagram showing this model is given in Fig. 7.3.10a. In order to simplify the representation, symbols for differentiation operators were used. It can be seen that the coupled inductances are modeled with two CCVSs, which requires the introduction of two new currents when formulating the equations for the original circuit. This fact is shown in Fig. 7.3.10b where gyrators are introduced to convert CCVSs into CCCSs for the modified circuit. The corresponding stamp for the original circuit is shown in Fig. 7.3.10c. In order to preserve transparency, the symbolic representation of differentiation has been retained. Note, Fig. 7.3.10b represents the modification of the original circuit for the case when M is incremented only. The corresponding stamp is shown in Fig. 7.3.10d. Similar to before, in order to preserve clarity, the symbolic representation of differentiation is retained.

This would complete the presentations on the new procedure for efficient calculation of linear circuit tolerances. Although a detailed analysis of the efficiency will be given later, we note here once again that when evaluating the efficiency of this

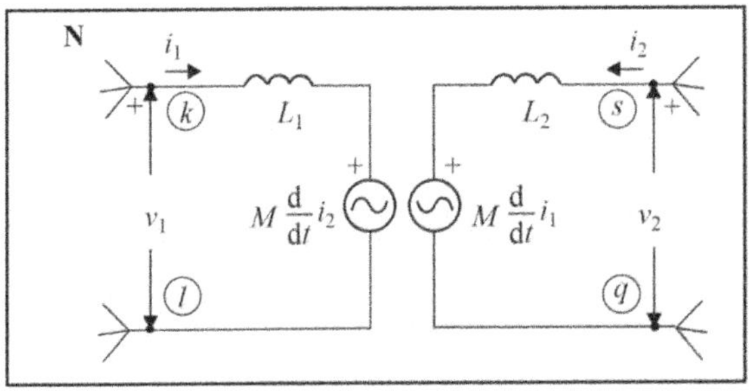

a

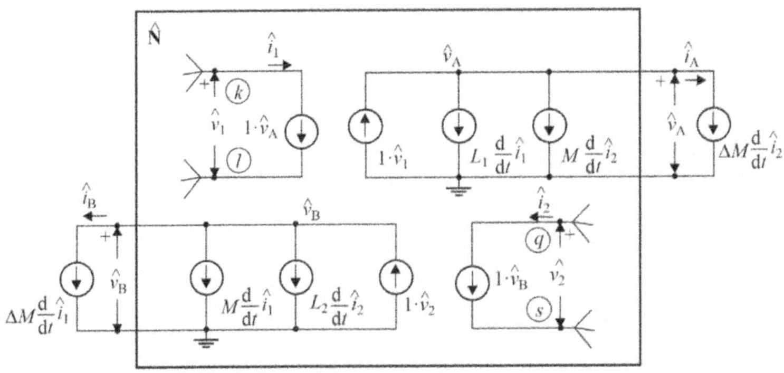

b

variables→ equations ↓	v_k	v_l	v_s	v_q	i_1=v_A	i_2=v_B	RHS
k					1		
l					-1		
s						1	
q						-1	
new equation 1	1	-1			$-L_1\frac{d}{dt}$	$-M\frac{d}{dt}$	
new equation 2			1	-1	$-M\frac{d}{dt}$	$-L_2\frac{d}{dt}$	

c

variables→ equations ↓	Δi_1	Δi_2	Δi_A	Δi_B	RHS
new equation 1			1		
new equation 2				1	
n+1		$-M\frac{d}{dt}$	1		$-M\frac{d}{dt}i_2$
n+2	$-M\frac{d}{dt}$			1	$-M\frac{d}{dt}i_1$

d

Fig. 7.3.10 **a** Coupled inductances and **b** modified circuit after introducing gyrators to transform CCVSs into CCCSs and enabling introducing the increment of M. **c** Stamp for coupled inductances in the original circuit. **d** Stamp for coupled inductances in the modified circuit

method of fault simulation in the time domain, one should keep in mind that the only alternative to it is the $M + 1$ analysis of new circuits resulting from modifications.

7.3.4 Simulation of Catastrophic Faults

In this section, an extension of the procedure described above will be presented that allows the same concepts to be applied to the case of catastrophic faults. It is about the modeling of short and open circuits. When introducing the new method, the procedure for introducing (modeling) faults will be described first, and then the simulation procedure, which is basically identical to the one described in [11].

A short circuit between two nodes results in a reduction in the number of nodes, and thus a change in the matrix **Y** in the original formulation using MNA. In order to avoid this, introducing a short-circuit type fault between two nodes is achieved by using a zero-value voltage source as shown in Fig. 7.3.11a. The equation that creates the vector **C** now has the form

$$\hat{v}_A = \hat{v}_a - \hat{v}_b = 0. \tag{7.3.21}$$

The corresponding stamp that applies to the formulation of the equations in the modified circuit is given in Fig. 7.3.11c. It looks very simply, and it is. This concludes the consideration of introducing the short-circuit model and simulating this very frequent fault.

Modeling of the open-circuit fault, if one still insists on the need to preserve the unmodified circuit matrix **Y**, leads to special difficulties. The reason for this is that

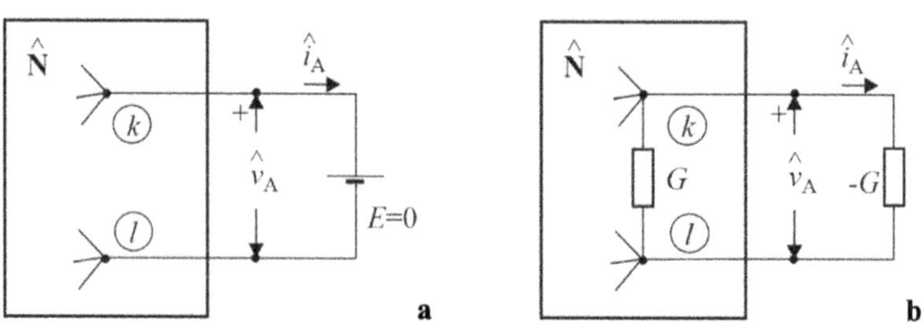

variables→ equations ↓	Δv_k	Δv_l	Δi_A	RHS
k			1	
l			-1	
n+1	1	-1		v_k-v_l

c

Fig. 7.3.11 **a** A short circuit modeled by a zero-value voltage source, **b** an open circuit realized by parallel connection of G and $-G$, and **c** the corresponding stamp for the modified circuit given under (**a**)

a new node is created when a break occurs. It must be introduced before modeling (introducing) the fault without disturbing the voltages and currents in the original circuit. In the following text, a description of the procedure that solves this problem will be given, but before that, some special cases that offer easier specific solutions will be considered.

When it is necessary to model the broken connection which is in series with a conductance G, the open circuit can be modeled by introducing, in parallel, a negative admittance whose value is $\Delta G = g = -G$. In this way, the total admittance in the parallel connection becomes equal to zero, that is, the resistance becomes infinite. The method of introducing negative conductance is shown in Fig. 7.3.11b. Similar concepts can be used in capacitive and inductive branches where, for the inductance, as before, the gyrator should be introduced first.

Controlled current sources are easily "interrupted" if a current source of equal value with the opposite sign is connected in parallel. Voltage sources, however, will again, like the coil, require a gyrator. An ideal independent current source can be "turned off" by connecting an equal current source but of the opposite sign, etc.

There are situations, however, when these ideas cannot be applied. Such an example is a break in the node itself, for example, where k conductors that meet in the node are divided into two groups. This kind of fault is not rare in real circuits if it is considered that a node, which is lumped in the sense of circuit theory, actually abstracts a series of connections. Some of those links may be broken. Figure 7.3.12a shows one connection (node) which is shown only as one point in the circuit diagram. Figure 7.3.12b, however, indicates some places within the "node" that may be interrupted. In addition, the break at the point of coupling of complex elements (four-poles, for example), must also be modeled separately.

For such cases, but also in any other case, the general idea, shown in Fig. 7.3.13, can be used. Figure 7.3.13a shows the "connection" to be broken. Figure 7.3.13b represents the model of the same connection obtained by introducing a zero-valued voltage source. One of the nodes, k or l, is now new for the original circuit. This solved the basic problem, which is the introduction of a new node, without disturbing the voltages and currents in the original circuit.

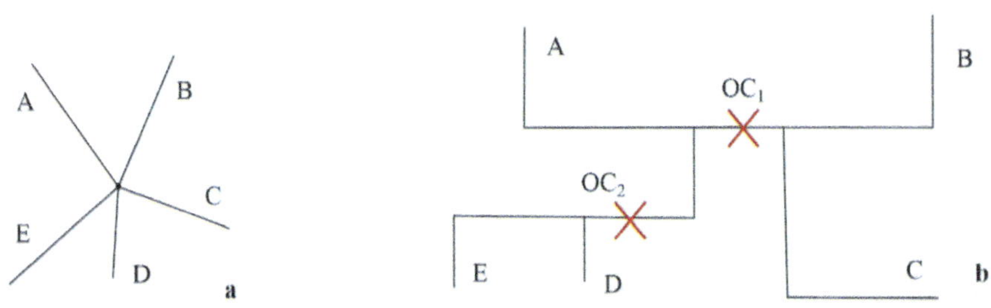

Fig. 7.3.12 **a** A node in symbolic form with branches connected to it and **b** a "node" on a printed circuit board with possible break points within a single circuit-theoretical node

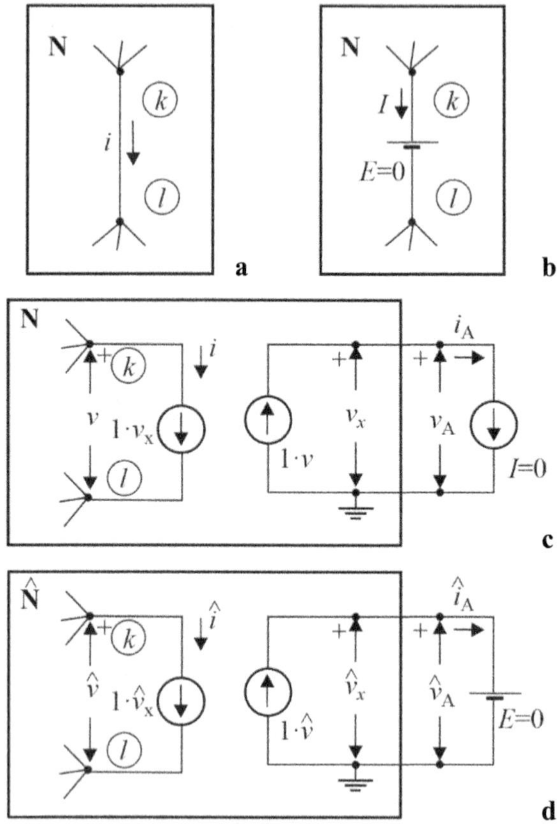

variables→ equations ↓	*vk*	*vl*	*i*	RHS
k			1	
l			-1	
new equation	1	-1		

e

variables→ equations ↓	Δi	Δi_A	RHS
new equation		1	
n+1	1		v_k-v_l

f

Fig. 7.3.13 **a** Short circuit, **b** short circuit modeled by an ideal zero-value voltage source, **c** short circuit modeled by a gyrator loaded with a zero-value current source—used in the original circuit **d** open circuit modeled by a gyrator loaded with a zero-value voltage source—used in the faulty circuit. **e** Stamp for original circuit and **f** stamp for modified circuit

In order to enable the introduction of faults, the newly introduced zero-valued voltage source will be modeled by a gyrator loaded by a zero-value current source, as in Fig. 7.3.13c. A zero-value current source is in fact open circuit. This increases the number of nodes in the original circuit, compared to the case of a correct circuit, by two, but in turn conditions are created that the fault can be treated as a circuit element.

Before continuing, it should be emphasized here that the introduction of the gyrator to model the short circuit at the point of the fault that should occur, which in fact represents the open circuit, is in fact the first change in the topology of the original, unmodified, circuit that was introduced for fault simulation using a new method. Namely, all gyrators that have been introduced so far for the purpose of modeling resistors, coils, voltage sources, and transformers are created per se always when using MNA, so they do not represent a change due to fault simulation. At the same time, it can be observed that, in fact, the place of interruption for the introduction of open circuit can be declared as an element: gyrator loaded with open circuit.

Let us now consider the introduction of the fault itself. Open circuit at the breaking point is achieved if the newly introduced gyrator is loaded by a zero-valued voltage source. It is shown in Fig. 7.3.13d. Figure 7.3.13e represents the stamp that enables the formulation of equations in the original circuit, and Fig. 7.3.13f is given an addition that refers to the modified circuit.

This completes the description of short- and open-circuit modeling, that is, the procedure for introducing catastrophic faults. From the point of view of simulation, the new method enables the treatment of catastrophic faults exactly the same as the parametric ones.

7.3.5 Applicability Considerations and Limitations

In this section, two examples will first be shown as an illustration of the way equations are formulated and how the methods work. Then, an overview of the properties of the new method will be given, and at the end, an insight into the possible limitations of its application will be given.

As an illustration of the method, the example of the circuit from Fig. 7.3.14a will be considered first. Three faults will be introduced. The fault F_1 represents an increase in G_4 for $\Delta G_4 = g$. The fault F_2 represents a break in the wire inside node 2 as shown in the figure, and the fault F_3 is a short circuit between node 2 and ground. Based on that, according to the theory that was presented earlier, the circuit of Fig. 7.3.14b was created. It enables the successive introduction of faults. As can be seen, simulation F_2 (break of a wire within a node) imposes the introduction of two new nodes. The circuit of Fig. 7.3.14b plays the role of the original circuit and the system of equations corresponding to system (7.3.2) for this circuit is

$$\mathbf{Y} = \begin{bmatrix} G_1 + G_2 & -G_1 & 0 & -G_2 & 0 \\ -G_1 & G_1 + G_4 & -G_4 & 0 & -1 \\ 0 & -G_4 & G_3 + G_4 + G_5 & -G_3 & 0 \\ -G_2 & 0 & -G_3 & G_2 + G_3 & 1 \\ 0 & -1 & 0 & 1 & 0 \end{bmatrix}, \tag{7.3.22a}$$

$$\mathbf{v} = \begin{bmatrix} v_1 \; v_2 \; v_3 \; v_4 \; v_5 \end{bmatrix}^{\mathrm{T}}, \text{ and } \mathbf{j} = \begin{bmatrix} I \; 0 \; 0 \; 0 \; 0 \end{bmatrix}^{\mathrm{T}}. \tag{7.3.22b}$$

After the formulation of the system (7.3.22), one can approach the fault simulation. When the fault F_1 is introduced, a system is created that is equivalent to (7.3.5):

$$\begin{bmatrix} \boldsymbol{Y} & \boldsymbol{A} \\ \boldsymbol{C} & -1 \end{bmatrix} \begin{bmatrix} \Delta \boldsymbol{v} \\ \Delta i_{\mathrm{A}} \end{bmatrix} = \begin{bmatrix} 0 \\ -g(v_2 - v_3) \end{bmatrix} \tag{7.3.23}$$

where for the first fault, it is: $\boldsymbol{A}_1^{\mathrm{T}} = \begin{bmatrix} 0 \; 1 \; -1 \; 0 \; 0 \end{bmatrix}$ and $\boldsymbol{C}_1 = \begin{bmatrix} 0 \; g \; -g \; 0 \; 0 \end{bmatrix}$.

For faults F_2 and F_3, instead of (7.3.5), a system of equations should be solved in the form:

$$\begin{bmatrix} \mathbf{Y} & \mathbf{A} \\ \mathbf{C} & 0 \end{bmatrix} \begin{bmatrix} \Delta \mathbf{v} \\ \Delta i_{\mathrm{A}} \end{bmatrix} = \begin{bmatrix} \mathbf{0} \\ -v_x \end{bmatrix} \tag{7.3.24}$$

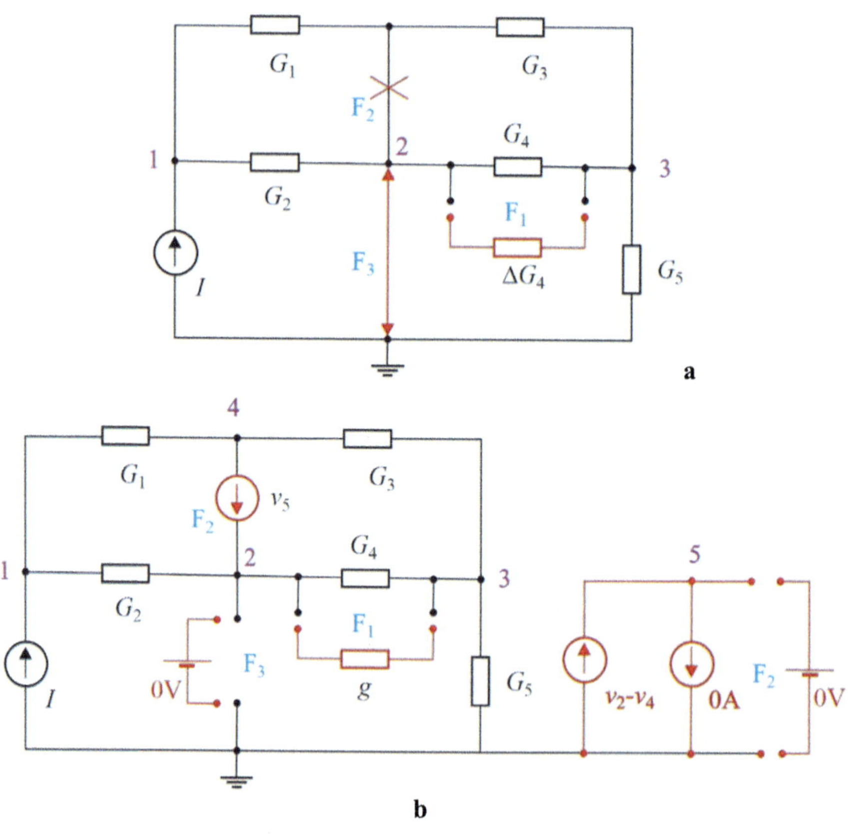

Fig. 7.3.14 **a** Original circuit with faults shown and **b** adapted original circuit with indicated elements that will be successively introduced to simulate faults

where for F_2: $x = 5$, $\boldsymbol{A}_2^{\mathrm{T}} = \begin{bmatrix} 0\,0\,0\,0\,1 \end{bmatrix}$ and $\boldsymbol{C}_2 = \begin{bmatrix} 0\,0\,0\,0\,1 \end{bmatrix}$; and for F_3: $x = 2$, $\boldsymbol{A}_3^{\mathrm{T}} = \begin{bmatrix} 0\,1\,0\,0\,0 \end{bmatrix}$ and $\boldsymbol{C}_3 = \begin{bmatrix} 0\,1\,0\,0\,0 \end{bmatrix}$.

This completes the description of circuit transformation and equation formulation for the first example. If selected $G_1 = (1/8)\ \Omega^{-1}$, $G_2 = (1/4)\ \Omega^{-1}$, $G_3 = (1/2)\ \Omega^{-1}$, $G_4 = (1/3)\ \Omega^{-1}$, $\Delta G_4 = (1/2)\ \Omega^{-1}$, and $I = 1$ A, the system matrix becomes:

$$
\begin{array}{cccccc}
 & \mathbf{Y} & \mathbf{A}_0 & \mathbf{A}_1 & \mathbf{A}_2 & \mathbf{A}_3 \\
\mathbf{Y} & \begin{bmatrix} 3/8 & -1/8 & 0 & -1/4 & 0 \\ -1/8 & 11/24 & -1/3 & 0 & -1 \\ 0 & -1/3 & 11/6 & -1/2 & 0 \\ -1/4 & 0 & -1/2 & 3/4 & 1 \\ 0 & -1 & 0 & 1 & 0 \end{bmatrix} & \begin{bmatrix} 0 \\ 1 \\ -1 \\ 0 \\ 0 \end{bmatrix} & \begin{bmatrix} 0 \\ 1 \\ -1 \\ 0 \\ 1 \end{bmatrix} & \begin{bmatrix} 0 \\ 0 \\ 0 \\ 0 \\ 1 \end{bmatrix} & \begin{bmatrix} 0 \\ 1 \\ 0 \\ 0 \\ 0 \end{bmatrix} \\
\mathbf{B} & [0\ \ 0\ \ 0\ \ 0\ \ 0] & [1] & & & \\
\mathbf{C}_1 & [0\ \ 1/2\ \ -1/2\ \ 0\ \ 0] & & [-1] & & \\
\mathbf{C}_2 & [0\ \ 0\ \ 0\ \ 0\ \ 1] & & & [0] & \\
\mathbf{C}_3 & [0\ \ 1\ \ 0\ \ 0\ \ 0] & & & & [0]
\end{array}
\tag{7.3.25}
$$

We can be easily recognized the structure that is explained in Fig. 7.3.2. Here A_0 is the last column in the original matrix. Four RHS vectors are associated with these four matrices:

$$
\mathrm{RHS}_0 = \begin{bmatrix} 1 \\ 0 \\ 0 \\ 0 \\ 0 \end{bmatrix}, \mathrm{RHS}_1 = \begin{bmatrix} 0 \\ 0 \\ 0 \\ 0 \\ -\frac{1}{2}(v_2 - v_3) \end{bmatrix}, \mathrm{RHS}_2 = \begin{bmatrix} 0 \\ 0 \\ 0 \\ 0 \\ -v_5 \end{bmatrix}, \text{ and } \mathrm{RHS}_3 = \begin{bmatrix} 0 \\ 0 \\ 0 \\ 0 \\ -v_2 \end{bmatrix}.
$$

After the factorization of the system of equations shown above, the following factorized matrix is created according to Fig. 7.3.3:

$$
\begin{array}{cc}
 & \mathbf{U}_{\mathrm{Y}} \\
\mathbf{L}_{\mathrm{Y}} & \begin{bmatrix} 0.375\backslash 1 & -0.333 & 0.000 & -0.666 & 0.000 \\ -0.125 & 0.4166\backslash 1 & -0.800 & -0.200 & -2.4 \\ 0.000 & -0.333 & 1.566\backslash 1 & -0.362 & -0.511 \\ -0.25 & -0.083 & -0.566 & 0.362\backslash 1 & 1.41 \\ 0.000 & -1.00 & -0.80 & 0.511 & -3.53\backslash 1 \end{bmatrix}
\end{array}
$$

$$
\begin{array}{cccc}
\mathbf{A}_0' & \mathbf{A}_1' & \mathbf{A}_2' & \mathbf{A}_3' \\
\begin{bmatrix} 0.000 \\ 2.40 \\ -0.128 \\ 0.353 \\ -0.600 \end{bmatrix} & \begin{bmatrix} 0.000 \\ 2.40 \\ -0.128 \\ 0.353 \\ -0.600 \end{bmatrix} & \begin{bmatrix} 0.000 \\ 0.000 \\ 0.000 \\ 0.000 \\ -0.283 \end{bmatrix} & \begin{bmatrix} 0.000 \\ 2.40 \\ 0.511 \\ 1.353 \\ -0.600 \end{bmatrix}
\end{array}
$$

$$\begin{array}{ll} \mathbf{B}' & [0 \quad 0 \quad 0 \quad 0 \quad 0] \\ \mathbf{C}'_1 & [0.00\ 0.500\ -0.10\ 0.640\ 1.089] \\ \mathbf{C}'_2 & [0 \quad 0 \quad 0 \quad 0 \quad 1] \\ \mathbf{C}'_3 & [0 \quad 1 \quad 0.800\ 0.489\ 2.12] \end{array}$$

[1\1]

[−1.6\1]

[0.283\1]

[−2.2\1]

with $\mathbf{L}_Y$ and $\mathbf{U}_Y$, the LU factors of the matrix $\mathbf{Y}$ are marked here, and with the prime, "′", the columns and rows of the matrix that were created after the LU factorization are marked. After solving, one gets:

$$\begin{bmatrix} v_1 \\ v_2 \\ v_3 \\ v_4 \\ v_5 \\ i_A \end{bmatrix} = \begin{bmatrix} 4.86666 \\ 2.2 \\ 1.0 \\ 2.2 \\ 0.066666 \\ 0.0 \end{bmatrix}, \begin{bmatrix} \Delta v \\ \Delta i_A \end{bmatrix}_1 = \begin{bmatrix} -0.45 \\ -0.45 \\ 0.0 \\ -0.45 \\ 0.225 \\ 0.375 \end{bmatrix}, \begin{bmatrix} \Delta v \\ \Delta i_A \end{bmatrix}_2 = \begin{bmatrix} 0.0157 \\ -0.1412 \\ 0.0 \\ 0.09142 \\ -0.06666 \\ -0.2353 \end{bmatrix},$$

$$\text{and} \begin{bmatrix} \Delta v \\ \Delta i_A \end{bmatrix}_3 = \begin{bmatrix} -2.2 \\ -2.2 \\ -1.0 \\ -2.2 \\ 0.600 \\ 1.00 \end{bmatrix}.$$

Based on this, for the node voltages in the faulty circuit we get

$$\begin{bmatrix} \hat{v}_1 \\ \hat{v}_2 \\ \hat{v}_3 \\ \hat{v}_4 \end{bmatrix}_1 = \begin{bmatrix} 4.41666 \\ 1.75 \\ 1 \\ 1.75 \end{bmatrix}, \begin{bmatrix} \hat{v}_1 \\ \hat{v}_2 \\ \hat{v}_3 \\ \hat{v}_4 \end{bmatrix}_2 = \begin{bmatrix} 4.85098 \\ 2.058824 \\ 1 \\ 2.294117 \end{bmatrix}, \text{and} \begin{bmatrix} \hat{v}_1 \\ \hat{v}_2 \\ \hat{v}_3 \\ \hat{v}_4 \end{bmatrix}_3 = \begin{bmatrix} 2.66666 \\ 0.0 \\ 0.0 \\ 0.0 \end{bmatrix}$$

This completes the fault analysis in the circuit shown in Fig. 7.3.12a.

As the second example, the circuit from Fig. 7.3.15a will be taken which will be simulated in the time domain. The current excitation is a step function of amplitude 1 A. The values of the circuit elements are $G_1 = 1\ \Omega^{-1}$, $G_2 = 1\ \Omega^{-1}$, $G_3 = 0.1\ \Omega^{-1}$, $C_1 = 1$ F, $L = 1$ H, $n = 5$, and $C_2 = 0.1$ F. As shown in the figure the faults considered are:

- F_1 is the first fault that represents a short circuit of the transformer primary.
- F_2 is the second fault that represents the break of the secondary winding of the transformer; and

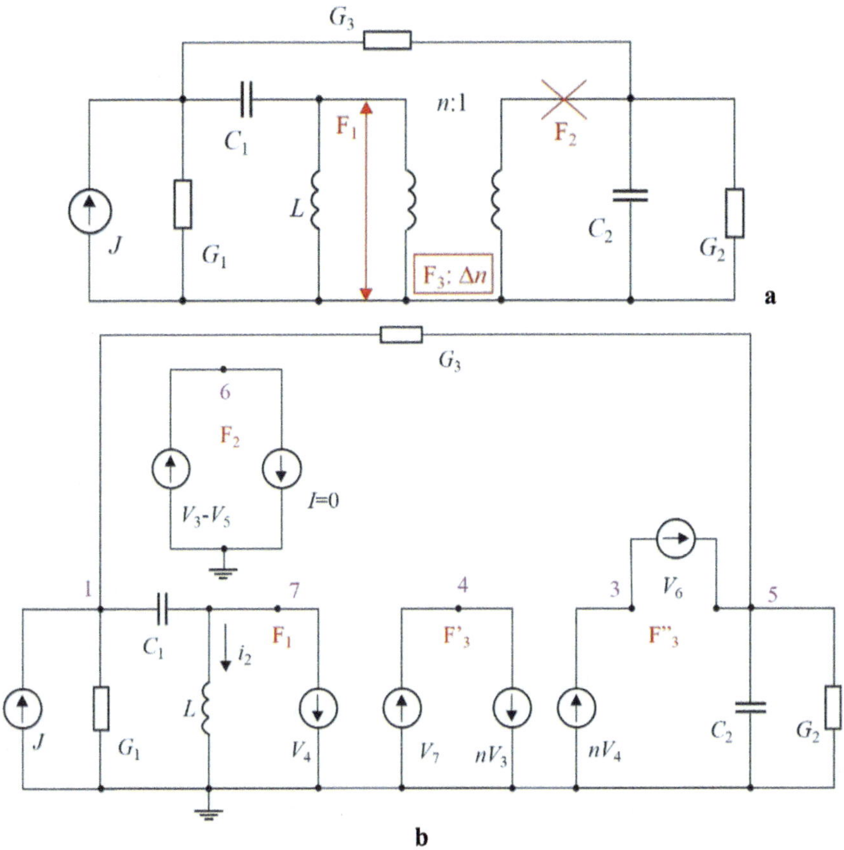

Fig. 7.3.15 **a** Original circuit with marked faults and **b** original circuit with changes that allow formulation using MNA and introduction of fault models

- F_3 is the third fault in which the turn ratio of the transformer is increased by $\Delta n = 4$.

According to the above set of faults, the original circuit was rearranged as shown in Fig. 7.3.15b. The node numbering shown in the figure was generated after pivoting, which is aimed to avoid a zero-valued matrix element on the main diagonal. The ports are shown that were created at the places where the elements serving for the insertion of faults will be connected. For F_1 it will be a zero-valued ideal voltage source which will be connected between node 7 and ground, for F_2 it will again be an ideal zero-valued voltage source but now connected between node 6 and ground, and for F_3 it will be two new ports to which two VCCSs will be connected ($\Delta n \cdot \widehat{V}_3$ at F_3' and $\Delta n \cdot \widehat{V}_4$ at F_3').

A one-step Euler formula with a constant step $h = 0.007$ was used in the simulation. The **Y** matrix was generated only once (for each instant) for the circuit of

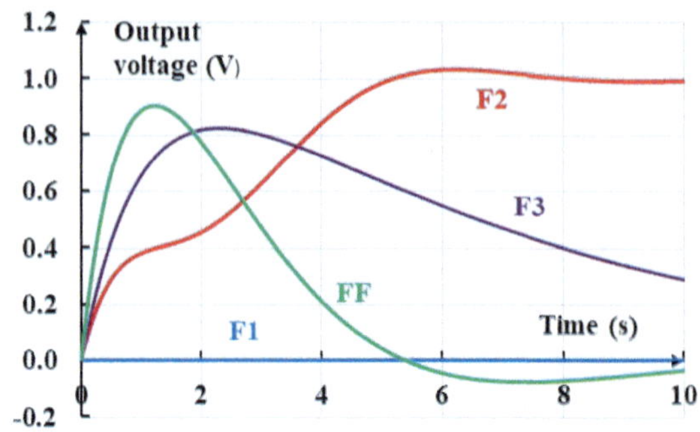

Fig. 7.3.16 Responses of a fault-free (FF) and the faulty circuits

Fig. 7.3.15b, and three pairs of **A**–**C** matrices and three different RHS vectors were generated for each fault individually. It should be kept in mind that if a constant step is used or if the step from the previous and the next moment do not differ, the circuit matrix can be kept in the factorized form and thus save in computation time.

The simulation results are shown in Fig. 7.3.16. The aim of these results, of course, was to verify the simulation method. However, they can be used to evaluate the step function as a test signal. In addition, considering the efficiency of the method, in order to find the appropriate test signal, it is possible to repeat the simulation with different excitations, at a low cost.

We can now summarize the properties of the proposed method. As mentioned in the introduction, the first saving is achieved when describing the faulty circuits. Here, faults can be viewed as circuit elements and can be listed in the description of the fault-free circuit along with other elements. By expanding the command system of the input language, the simulator would be enabled to recognize whether it is expected to simulate only the fault-free circuit or both the fault-free and the faulty circuit.

In the equation formulation phase, the following can be considered. First, one can intuitively accept the concept that if the faults are in fact circuit elements, stamps can be developed for them similar to those developed for all other circuit elements according to MNA as suggested in [1, 8] (The reader may partly get familiarized with the MNA and the proper equation formulation method based on stamps using https://books.google.rs/books?id=Ca9DJs9-SPYC&printsec=copyright&redir_esc=y#v=onepage&q&f=false). Thus, the formulation of the equations can be automated in the same way as it was done for all other elements of the circuit. In addition, it is not necessary to repeat the entire equation formulation process for each fault. For individual faults only the corresponding **A**, **C** and RHS vector should be created and stored. When one considers the multiple repetition of the formulation of the equation for the frequency and time domain, it can be concluded that considerable savings are also achieved in this way. It should be kept in mind that for most faults, vectors **A**, **C**, as well as the RHS vector, do not depend on frequency or time, so they do not need to be regenerated. Compared with possible other methods, it should be expected that during the simulation of catastrophic faults, due to a change in the circuit structure, there will be a need to renumber the system of equations, which may also require time. In the new method, renumeration is done only once.

The reader should note the interwoven structure of the algorithm meaning that the equation formulation and equation solution methods are interdependent. That, of course, is not a specificity of this method since similar situations are seen frequently in circuit analysis algorithms.

We will first consider the complexity of the calculation in the phase of solving the system of equations on the example of simulating faults in DC regime. The overall complexity of the procedure can be divided into two parts. First, one factorization of the matrix of order n is required followed by recalculation (M times) of $(n + 1)$ columns and rows. When looking at the complexity of these actions, one should keep in mind Figs. 7.3.2 and 7.3.3. The total effort for the factorization is higher than when factoring a matrix of order n, and less than when factorizing a matrix of order $(n + M)$. Bearing all that in mind, it can be said that the volume of work on one factorization with the new method is θ times larger than for the factorization of the original circuit. We estimate the value of θ here to be less than 1.5, considering the exceptional sparsity of the vectors **A** and **C**.

Second, $M + 1$ forward and backward substitutions are required for a circuit containing M faults.

Let us now consider the complexity of the calculation when the simulation is performed in the time domain. Since the circuit equations are discretized, the problem of analysis in the time domain is reduced to multiple analyses of DC regimes, and all the advantages obtained during DC analysis remain here as well but are applied to one step of discretization. It should be borne in mind that the step is usually variable and that its value is adjusted to the response. If a faulty circuit is considered as a changed circuit and its response will be changed, then a different discretization flow can be expected for each variant individually. The concurrent (simultaneous) simulation introduced here insists on equal discretization steps for the original and the faulty circuits. Therefore, it will be necessary at every time instant to determine the most suitable value of the discretization step for all circuits (original and faulty circuits) and to choose the smallest one as the most suitable.

The problem is similar for simulation in the frequency domain. Namely, if the fault simulation is performed in a selective circuit (which is analogous to a circuit with a stiff system of differential equations), the frequency increment assumes the same role as the time step in the simulation in the time domain. Changing the circuit element can change the skirt selectivity of the amplitude characteristic and thus require a smaller (or larger) increment in frequency. Or, it can move the location of the bandwidth on the frequency axis and thus change the frequency intervals where the frequency increments were small, etc. As a result of these considerations, we can say that even in frequency domain analysis when using the new method, since all circuits are considered at equal frequencies, the smallest increment in frequency determines the progress of the simulation for all, the original and faulty circuits. To simplify, one may use the same increment in frequency for all simulations which, however, may slow down some of them being less selective.

In the following text, only the time domain will be considered.

Let the original circuit be simulated in N time steps.

When the classic fault simulation method is used, the i-th modified circuit is analyzed in $N + \Delta N_i$ time steps. It should always be kept in mind that we are talking about complex linear circuits and that only one fault is considered at a time. This means that in the worst case only two local time constants will be changed in the faulty circuit compared to the fault-free one. As a consequence, ΔN_i (which can be both positive and negative) should be significantly less than N. In other words, it can be expected that the number of points on the time axis will not be significantly changed.

When the classical method is applied, $M + 1$ simulation is performed (of the original and M faulty circuits) so that the total number of factorizations of the matrix (linear circuit) is

$$\mathrm{NN} = \sum_{0}^{M}(N + \Delta N_i) = (M + 1) + \sum_{1}^{M} \Delta N_i. \tag{7.3.26}$$

In the most favorable (best) case, all increments ΔN_i would be negative, so the minimum value of NN is given by

$$\mathrm{NN}_{\min} = (M + 1) \cdot N - \sum_{1}^{M} |\Delta N_i|. \tag{7.3.27}$$

The highest value of NN is reached by changing the sign in front of the sum in (7.3.27). The mean, and most likely value, is $\mathrm{NN}_{\text{average}} = (M + 1)N$.

When the new method is applied, all $(M + 1)$ simulations take place simultaneously so that, per instant, it is calculated with one factorization that is θ times more complex than the usual one. The time step, however, is adjusted according to the fastest edge relative to the slowest simulation. It indicates the worst possible situation. It occurs when each fault affects the simulation in such a way that, in this case, the worst case, i.e., the largest number of simulations, is given by

$$\mathrm{NN}'_{\max} = \theta \cdot N + \sum_{1}^{M} |\Delta N_i|. \tag{7.3.28}$$

Bearing in mind that the initial assumption is that $N \gg |\Delta N|$, we conclude that also

$$\mathrm{NN}_{\min} \gg \mathrm{NN}'_{\max} \tag{7.3.29a}$$

or

$$(M + 1 - \theta) \cdot N \gg 2 \sum_{1}^{M} |\Delta N_i|, \tag{7.3.29b}$$

which means that significant savings in simulation time can be achieved by applying the new method. We should not forget that the most favorable situation for the classic method is compared with the most unfavorable for the new method. A comparison of the mean values of NN would give an even more favorable picture of the advantages of the new method. This would complete the consideration of the effectiveness of the new method from the point of view of speed of work.

Sometimes it is of interest to simulate the effects of multiple faults acting simultaneously in a circuit. Applying the new method to such cases is as simple as applying it to single faults. If, for example, two short circuits are to be simulated, two new zero-valued voltage sources will be introduced. As a consequence the matrices **A** and **C** will become of order $n \times 2$. This is of course also applicable to the frequency domain.

Something that is particularly specific to the time domain is the intermittent fault. It is a phenomenon that appears and disappears during the operation of the circuit. When the device is sealed, for example, a printed circuit board heats up and any cold solder joint on it becomes exposed. At lower dissipation or when the cover is removed, the board cools down and there are no fault effects. The nature of the simulation procedure proposed here lends itself particularly well to the simulation of such faults. Namely, the presence of the element that models the fault does not affect the circuit matrix, so even when it is not present, the circuit matrix remains the same. Simply, when there is no fault the corresponding matrix additions (**A** and **C**) are not processed. Thus, if $t_1 < t < t_2$ is the time interval in which the fault is active, the simulation will proceed as follows. In the interval $0 < t < t_1$, a normal simulation of an ordinary circuit is performed. At the end of that interval, the simulation of the circuit with faults begins, with the last simulation results from the faulty circuit being taken as boundary conditions for the simulation of the new circuit. At the time instant t_2, the opposite is done.

Let us now consider some limitations of this method. This, as well as all other methods for simulating tolerances and faults in analog circuits with large parameter increments, when it comes to simulating "linear electronic circuits" should be applied with utmost care. By linear electronic circuits we mean linear circuits that were created when linear models of electronic components (which are otherwise nonlinear) are substituted into the electrical circuit diagram. The parameters of those models depend on the DC operating conditions in the original nonlinear circuits. A fault in the linear circuit is also a fault in the original one, so it can change the operating conditions in it. This can result in changing a large number of parameters simultaneously. For the sake of illustration, the following simple example will be considered.

Figure 7.3.17 shows a part of the CE amplifier circuit that contains a bipolar transistor. If the frequency domain is observed, we can analyze this circuit without problems. A change (fault) in the value of C will not affect the DC regime and thus, neither the parameters of the transistor model, so the simulation of the fault can be done without problems. A significant change in R, however, will affect not only the response of the linear circuit but also the operating conditions in the original (nonlinear) circuit, so that the quiescent operating point of the transistor and the parameters in its linear model used for frequency (time) analysis. This means that

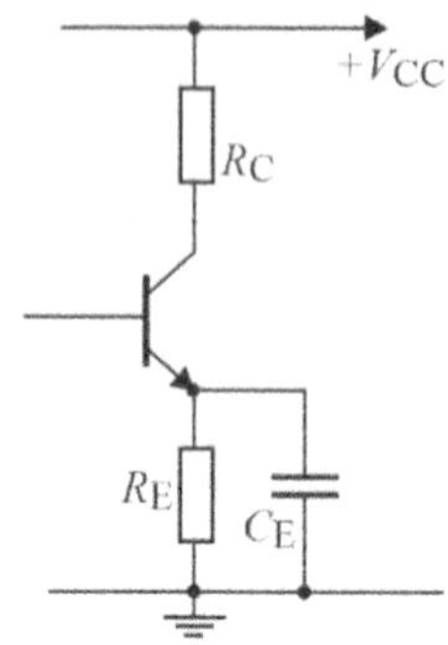

Fig. 7.3.17 Illustration of the influence of DC regimes on the parameters of the linear model

the inherent nonlinearity of this circuit requires the simulation of nonlinear reactive circuit faults to be applied in such cases.

References

1. Litovski VB, Zwoliński M (1996) VLSI simulation and optimization. Chapman and Hall, London. https://books.google.rs/books?id=Ca9DJs9-SPYC&printsec=copyright&redir_esc=y#v=onepage&q&f=false
2. Bell IM, Spinks SJ (1995) Analogue fault simulation for the structural approach to analogue and mixed-signal IC testing. Collection of papers presented at the International mixed signal testing workshop, Grenoble, France, pp 10–14
3. Ilić T, Litovski VB, Litovski IV, Stojilković SM (2000) Computationally efficient large-change statistical analysis of linear electronic circuits. Microelectron Reliab 40(6):1023–1038
4. Litovski V, Litovski I, Zwoliński M (2004) Concurrent analogue fault simulation, the equation formulation aspect. Int J Circuit Theor Appl 32(6):467–507
5. Gadenz RN, Rezai-Fakhr MG, Temes GC (1973) A method for the computation of large tolerance effects. IEEE Trans Circuits Syst CAS 20(6):704–708
6. Tellegen BDH (1952) A general network theorem with applications. Philips Res Rep 7:259–269
7. Director SW, Rohrer RA (1969) The generalized adjoint network and network sensitivities. IEEE Trans Circuits Theory CT 16(3):318–323
8. Paralias E, Rueda A, Huertas JL (1995) An on-line testing approach for pipelined A/D converters. Collection of papers presented at the International mixed signal testing workshop, Grenoble, France, pp 44–49
9. Ho CW, Ruehli AI, Brennan PA (1975) The modified nodal approach to network analysis. IEEE Trans Circuits Syst CAS 22(6):504–509
10. Litovski VB, Zwoliński M, Litovski IV (1996) Large-change-sensitivity computation in linear electronic circuits. In: Proceedings of the 18th international seminar on semiconductor and hybrid technologies, Sozopol, Bulgaria, pp 268–274
11. Litovski VB (2000) Electronic circuit design, simulation, optimization, testing and physical design. Nova Jugoslavija, Vranje, Serbia (in Serbian)

7.4 Fault Simulation in Nonlinear Analog Circuits

Abstract Fault simulation of nonlinear analog circuits is usually done as DC or time domain simulation. In that, the crucial difference with linear circuits is in the property that due to the nonlinearity of the circuit elements the fault, even a soft (parametric) one, may disturb the DC regime so creating difference between the responses of the fault-free and the faulty circuit and among the faulty circuits representing different faults at any time instant. In this chapter, being aware of this property, algorithms are described that attempt to reduce the time needed for repetitive simulation of the faulty circuits based on similarity of the circuit structure and similarity of the responses in some period of time.

7.4.1 Fault Simulation in Nonlinear Analog Circuits

When we say nonlinear analog circuits, we mean nonlinear reactive circuits with nonlinear resistive and reactive elements. A fault in such circuits, whether parametric or structural, can have complex consequences. In the most favorable case, the consequence of the fault is a change in the value of the signal level in the circuit. It often happens, however, that the fault affects the way the circuit works to such an extent that it completely changes its function. A short circuit between the electrodes of a transistor, for example, can completely change the operation of the circuit, and at the same time, change the simulation mode of the faulty circuit.

Bearing this in mind, the fault simulation in nonlinear reactive circuits becomes a specific problem. And here, as the first idea for simulating faults, the classic concept would be imposed, which is to create as many new circuits as there are faults and simulate each circuit separately with a specific list of signals for each fault [1], which is illustrated in Fig. 7.4.1.

The number of faults in an analog circuit, however, if both parametric and structural faults are counted, is practically unlimited. For a circuit with n nodes, for example, only faults representing short circuit between connections can be $n(n-1)/2$. That is why the list of faults can be very long. In addition, the number of types of test signals can be large as well. Ultimately, each simulation in the time domain

V. B. Litovski, *Lecture Notes in Analog Electronics*, Lecture Notes in Electrical Engineering 1265, https://doi.org/10.1007/978-981-97-8257-4_4

```
Prepare the fault list;
until all faults from the list are exhausted do{
  Select the next fault from the list;
  until all test signals are exhausted do{
      Compile the circuit description for the selected fault;
      Compose an excitation for the selected fault;
      Simulate the faulty circuit;
      }
  for the rest of elements from the fault list {
      Delete from the fault list those elements (faults) that have been
simulated;
  }
}
```

Fig. 7.4.1 Exhaustive simulation of faults in a nonlinear analog circuit

is an endeavor in itself. Hence the need to rationalize this algorithm so that, at least, possible repetition is eliminated.

The first step in this sense is to rationalize the generation of a new circuit description. When dealing with structural or catastrophic faults, for this purpose, the short- or open-circuit model introduced in the previous chapter can be used. Thus, most of the circuit description (netlist) is repeated, and at the end of the description only some lines related to the short-circuit model are added. At the same time, the notation (names) of the nodes remains the same as in the original circuit, which makes it easier to monitor the fault effects.

When working with parametric faults, two cases can be encounter. In the first, it is about changing the parameter value of one circuit element. This situation is most suitable for circuits with discrete components or integrated circuits where the fault is actually the origin of a designer's error (wrong dimension, for example). Here too, it is possible to rationalize the description (netlist) of the circuit in a way very similar to introducing a structural fault. In the case of integrated circuits, however, it is a very common situation that some of the model parameters of all (or most) components are changed at the same time for reasons that have a common cause, such as concentration in the substrate, for example. In such a case, the description of the circuit must be repeated in its entirety, and no particular savings can be made at that stage.

The biggest savings, however, can be achieved in the simulation phase. Here, too, it is assumed that the fault-free and the faulty circuit are slightly different, so it is expected that their responses are similar. To make it easier to see, let us look at the circuit diagram of the CMOS operational amplifier from Fig. 7.2.2a. The current flowing through the current source T_5 is controlled only by the transistor T_8 and the biasing resistors T_9–T_{10}–T_{11} and is independent on the behavior of other parts of the circuit. Therefore, if a short-circuit type fault would appear in the circuit, which does not refer to the nodes where the power supply and ground are connected, as well as to the nodes where the mentioned transistors (T_5, T_8, T_9, T_{10}, and T_{11}) are connected, the current of the constant current source would be unchanged. Thus, the contribution of these five transistors to the system of equations describing the circuit at each instant and each iteration remains unchanged after the fault occurs. This is where the idea of concurrent fault simulation lies, which was proposed in [2]. The goal of concurrent fault simulation is to avoid the computational redundancy by avoiding the calculation of the values of the elements of the linearized models of those transistors in the faulty circuit that are known not to change their contribution to the circuit. Here one is not to forget the exceptional numerical effort needed to create a linearized dynamic model of a transistor in every iteration and every time instant during time domain analysis.

To better explain the last statement, before considering the concurrent fault simulation in more detail, we will return to the basic characteristics of the simulation of nonlinear reactive circuits in the time domain as described in [3]. Simply put, the following two steps need to be taken:

1. At each time instant t^n (n being the time instants counter), where the numerical integration step is h^{n-1}, a system of nonlinear equations is solved using Newton's method. To do that linearized models of the nonlinear elements must be regenerated in every iteration.
2. Then if, after convergence, the solution satisfies the specified error, a new value of the integration step h^n is proposed for the instant $t^{n+1} = t^n + h^n$. Otherwise, the solution is rejected and, after reducing the integration step, the search for a new solution is started.

This explanation is expressed in more detail in Fig. 7.4.2 where the algorithm for time domain simulation of nonlinear dynamic circuits is depicted [3].

Within the algorithm given below $\mathbf{v}$ and $\mathbf{i}$ stand for the vectors of the unknown voltages and currents, respectively. As initial values of the first iteration at the instant $t = h$ the following is used: ${v_1}^{1,0}$ and ${v_2}^{1,0}$. The first number in the degree is related to the time instant counter and the other to the iteration counter.

It is important to keep in mind that the cost of calculation mainly consists of two components: calculating the parameter values of linearized and discretized component models and solving the system of linear equations having sparse matrices. The duration of calculating the values of the model parameters is proportional to the number of elements, and as shown in [4], the time required to solve a system of equations, if methods for handling sparse matrices are applied, is proportional to $k^{1.5}$, where k is the order of the system equation (usually the number of nodes in the circuit). Therefore, the total computation cost can be expressed as [5]:

$$C = l \cdot N \cdot (R_1 \cdot k^{1.5} + R_2 \cdot d), \tag{7.4.1}$$

where l is the average number of iterations per simulation time instant, N is the total number of simulation steps (instants) including those when convergence or acceptable accuracy was not achieved, and d is the number of nonlinear components in the circuit. R_1 and R_2 are constants, where R_1 represents a price related to the solution of the system of sparse linear equations, and R_2 represents a price related to

```
perform steady state analysis and generate node voltages (v^0) and selected
        branch currents (i^0) so that:
  x^0=[(v^0)^T (i^0)^T]^T;
  Coose h;
  n=0;
  /*time loop*/
  while (t<T) {
    m=0;
    predict x^{n+1,0};
    /*iterative loop*/
  until convergerce reached do {
        formulate discretized model of the reactive elements;
        generate linearized models of all nonlinear elements;
        formulate the system of linear equation describing the circuit for
        t^{n+1} and for iteration m+1;
        solve the system for x^{n+1,m+1};
        m++;
    }
    t=t+h;
    n++;
  }
```

Fig. 7.4.2 Algorithm of nonlinear reactive circuits analysis in the time domain

the equation formulation, that is, to the calculation of parameter values of linearized and discretized models of the nonlinear elements. It was experimentally shown [6] that R_2 is between 5 and 50 times larger than R_1, which depends on the complexity of the circuit and speaks of the importance of avoiding repeating the formulation of equations when simulating a faulty circuit.

Thus, a concurrent fault simulation would consist of using some of the calculations performed during the analysis of the fault-free circuit when analyzing the faulty circuit. The step **simulate** of the algorithm of Fig. 7.4.2 would be realized as described in [3], and the **generate** step from this algorithm would be realized as in Fig. 7.4.3.

In this algorithm, when formulating the equations, the matrix of the system is filled with the contributions of the linearized branches of each nonlinear element of the circuit individually. At the same time, for each component, for the previous moment, the values of the potentials of the terminals in the fault-free and the faulty circuits are compared. If it turns out that they are equal, it is possible to save time by not recalculating the contributions of the nonlinear element to the system of equations of the circuit but taking them from the analysis of the fault-free circuit for that same time instant.

The savings that can be achieved in this way can be considerable, especially if complex models of the components that make up the circuit are used. At the same time, as claimed in [7], calculations related to determining the contribution of the model of nonlinear elements to the system of linearized circuit equations in

```
do {
for each active component {
    if (the faulty circuit is analyzed & the component terminal voltages in
            the fault-free and in the faulty circuits do not differ)
        copy the linearized element model from the fault-free circuit
            equations to the faulty circuit equations;
    else
        generate a linearized branch of the nonlinear element in the faulty
            circuit;
    }
formulate the branch equations;
solve the circuit equations;
}
while (convergence not reached && maximum number of iterations not
    reached)
```

Fig. 7.4.3 Realization of generation of linearized models during concurrent simulation of nonlinear circuits

each iteration make up to 60% of all calculations in that iteration, which intuitively confirms the data from [6].

In order to be able to apply this method, as in the case of linear circuits that we discussed in the previous chapter, the fault-free circuit and its modified copies must be simulated simultaneously, that is, in parallel. This means that the systems of equations are formulated for the same time instant. The time instants counter n for all circuits should be the same. The iteration counter m for each circuit individually is different. Thus, if a divergence occurs in one of the circuits, the entire simulation must be stopped.

Similarly, since each modified copy is a problem in itself, at the beginning, it is necessary to carry out a DC simulation for each one individually in order to determine the boundary conditions in the circuit. Experience has shown that attempts to use latency at this stage have encountered serious convergence problems.

Finally, the question remains how close should the node potentials in the fault-free and in the faulty circuit be so that they can be considered equal and to avoid the evaluation of the elements of the linearized models of the nonlinear components? It is believed [2] that "close enough" is what is considered the criterion for stopping convergence in the Newton–Raphson iterative procedure [4].

Summarizing the description of this procedure, based on the results shown in [7], the time savings that may be achieved are about 20%.

Further accelerations of the procedure of simulating faults in analog circuits can be achieved if the above procedure is enriched with the following [8, 9]:

(a) reduction of the time of solving the system of linear equations in each iteration,
(b) prediction of the initial value of the response (prediction of the state) in the next circuit based on the results of the analysis of the previous one,

(c) arrangement of the sequence of faults so that the previous and next circuits are "slightly" different, and
(d) grouping of faults so that they enable the formation of groups of faulty circuits that can be simulated with the same time step. At the same time, this grouping must be dynamically adapted to the response, which means that in each step the grouping will be repeated according to the predicted value of the integration step.

In the following text, we will briefly describe the basic ideas behind these proposals.

First, to the savings that can be achieved in the equation formulation phase, we can also add savings in the solution phase. In this sense suppose that in the $(m+1)$th iteration of the faulty circuit analysis, at the time t^n, the following system of equations is needed to be solved:

$$\mathbf{G}_{\mathrm{f}}^{m}\mathbf{V}_{\mathrm{f}}^{m+1} = \mathbf{I}_{\mathrm{f}}^{m}, \tag{7.4.2a}$$

where the index f indicates that it is a faulty circuit, and the time counter n is omitted. Other notations are consistent with those used in [3].

The difference between $\mathbf{G}_{\mathrm{f}}^{m}$ and the matrix of the system of equations of the fault-free circuit $\mathbf{G}$ arises due to the differences in the partial derivatives (conductances) of the characteristics of the nonlinear elements that determine the matrix elements. Thus, a nonlinear element is said to be visible if the value of the derivative differs from the derivative in the fault-free circuit by a value larger than a preselected threshold. The visibility problem is a slightly more complex and is discussed in detail in [7]. If the differences in conductivity are denoted with δ_i^m, $i = 1, \ldots, j$, where j is the number of visible components, the difference matrix will be

$$\Delta\mathbf{G}_{\mathrm{f}}^{m} = \mathbf{G}_{\mathrm{f}}^{m} - \mathbf{G} = \mathbf{P}_{\mathrm{f}}\mathbf{D}_{\mathrm{f}}^{m}\mathbf{P}_{\mathrm{f}}^{\mathrm{T}}, \tag{7.4.2b}$$

where $\mathbf{D}_{\mathrm{f}}^{m} = \begin{bmatrix} \delta_1^m & \cdots & 0 \\ \vdots & \ddots & \vdots \\ 0 & \cdots & \delta_j^m \end{bmatrix} \in R^{j\times j}$ and $\mathbf{P}_{\mathrm{f}} = \begin{bmatrix} \mathbf{A}_1 & \mathbf{A}_2 & \cdots & \mathbf{A}_j \end{bmatrix}$ while $\mathbf{P}_{\mathrm{f}} \in R^{n\times j}$ and $\mathbf{A}_i$, $i = 1, 2, \ldots, j$, are the column matrices defined as in (7.3.2). They will have all elements equal to zero except for the kth which is equal to 1 and the lth which is equal to -1, with δ_i^m assumed to be connected between nodes k and l.

For example, for the circuit of Fig. 7.4.4, the system matrix of the fault-free circuit is given by:

$$\mathbf{G} = \begin{bmatrix} G_1 & -G_1 & 0 & 0 \\ -G_1 & G_1 + G_2 + G_{\mathrm{D1}} & -G_{\mathrm{D1}} & 0 \\ 0 & -G_{\mathrm{D1}} & G_2 + G_3 + G_{\mathrm{D1}} & -G_{\mathrm{D2}} \\ 0 & 0 & -G_{\mathrm{D2}} & G_{\mathrm{D2}} + G_4 \end{bmatrix}.$$

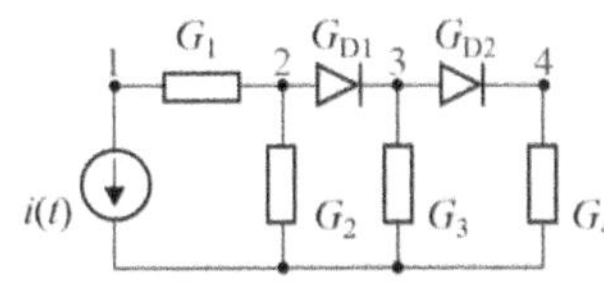

Fig. 7.4.4 Illustration of concurrent fault simulation in nonlinear circuits

G_{D1} and G_{D2} are the conductance values of diodes D_1 and D_2 at the time instant t^{n+1} and the end of the iterative process of fault-free circuit analysis.

When both diodes are visible, it is valid:

$$\mathbf{G} = \begin{bmatrix} G_1 & -G_1 & 0 & 0 \\ -G_1 & G_1 + G_2 + G_{D1} + \delta_1^m & -G_{D1} - \delta_1^m & 0 \\ 0 & -G_{D1} - \delta_1^m & G_2 + G_3 + G_{D1} + \delta_1^m + \delta_2^m & -G_{D2} - \delta_2^m \\ 0 & 0 & -G_{D2} - \delta_2^m & G_{D2} + G_4 + \delta_2^m \end{bmatrix}.$$

and the increment is

$$\Delta\mathbf{G}_f^m = \begin{bmatrix} 0 & 0 & 0 & 0 \\ 0 & \delta_1^m & -\delta_1^m & 0 \\ 0 & -\delta_1^m & \delta_1^m + \delta_2^m & -\delta_2^m \\ 0 & 0 & -\delta_2^m & \delta_2^m \end{bmatrix}.$$

Therefore, one gets $\mathbf{D}_f^m = \begin{bmatrix} \delta_1^m & 0 \\ 0 & \delta_2^m \end{bmatrix}$ and $\mathbf{P}_f = \begin{bmatrix} \mathbf{A}_1 & \mathbf{A}_2 \end{bmatrix} = \begin{bmatrix} 0 & 0 \\ 1 & 0 \\ -1 & 1 \\ 0 & -1 \end{bmatrix}$.

In that way the following system is created

$$\left(\mathbf{G} + \mathbf{P}_f\mathbf{D}_f^m\mathbf{P}_f^T\right) \cdot \mathbf{V}_f^{m+1} = \mathbf{I}_f^m, \tag{7.4.3a}$$

which describes a faulty circuit at the same time instant as the fault-free circuit.

Applying Householder's formula [10] from (7.4.3a) one gets

$$\mathbf{V}_f^{m+1} = \left\{1 - \mathbf{G}^{-1}\mathbf{P}_f \cdot \left[(\mathbf{D}_f^m)^{-1} + \mathbf{P}_f^T\mathbf{G}^{-1}\mathbf{P}_f\right]^{-1}\mathbf{P}_f^T\right\} \cdot \mathbf{G}^{-1}\mathbf{I}_f^m. \tag{7.4.3b}$$

The realization of this calculation takes place as follows. First, based on the known results of the factorization of $\mathbf{G}$ (from the analysis of the fault-free circuit) using forward and backward substitutions, $\mathbf{E}_f^m = \mathbf{G}^{-1}\mathbf{I}_f^m$ is determined.

Then the vector $\mathbf{F}_f^m = \mathbf{P}_f^T\mathbf{E}_f^m$ and *the reduced fault matrix* are formed:

$$\mathbf{R}_f^m = (\mathbf{D}_f^m)^{-1} + \mathbf{P}_f^T\mathbf{G}^{-1}\mathbf{P}_f. \tag{7.4.4}$$

This creates conditions for the system of Eq. (7.4.1) to be transformed into another system of equations whose order is reduced to *j*:

$$\mathbf{R}_{\mathrm{f}}^{m}\mathbf{H}_{\mathrm{f}}^{m} = \mathbf{F}_{\mathrm{f}}^{m}, \tag{7.4.5}$$

where $\mathbf{R}_{\mathrm{f}}^{m} \in R^{j \times j}$ and $\mathbf{H}_{\mathrm{f}}^{m}, \mathbf{F}_{\mathrm{f}}^{m} \in R^{j}$.

When this system is solved for $\mathbf{H}^{m}$, it is necessary to transform the result again into the form (7.4.2), for which it is necessary, by re-use of forward and backward substitution, to calculate:

$$\mathbf{Z}_{\mathrm{f}}^{m} = \mathbf{G}^{-1}\mathbf{P}_{\mathrm{f}}\mathbf{H}_{\mathrm{f}}^{m}. \tag{7.4.6}$$

Finally, the solution for the $(m + 1)$ th iteration is:

$$\mathbf{V}_{\mathrm{f}}^{m+1} = \mathbf{E}_{\mathrm{f}}^{m} - \mathbf{Z}_{\mathrm{f}}^{m}. \tag{7.4.7}$$

Therefore, instead of solving the linearized system of Eq. (7.4.1), it is first transformed into the lower-order system (7.4.5) which is solved and then transformed back to the original form. As a result, it can be said that savings are made at this stage by considering only the visible components, because a significantly smaller matrix is factorized in each iteration and because the transformation process described above can, sometimes, be used for multiple consecutive faults if they do not introduce structural differences.

The next saving proposed in [8] refers to the prediction of the initial value of the iterative process in the next circuit based on the final value in the previous one. Since the "states" of successive circuits differ slightly (which is achieved by arranging the order of faults), the "state" of the previous circuit is an excellent initial solution for the next one. It is possible to use it as such: by putting $\mathbf{V}_{\mathrm{f}}^{n+1,0}\big|_{\text{next circuit}} = \mathbf{V}_{\mathrm{f}}^{n+1,\infty}\big|_{\text{previous circuit}}$, or to use some interpolation algorithm. In the latter case, the following results can be used: $\mathbf{V}_{\mathrm{f}}^{n-1,\infty}\big|_{\text{next circuit}}$, $\mathbf{V}_{\mathrm{f}}^{n,\infty}\big|_{\text{next circuit}}$, and $\mathbf{V}_{\mathrm{f}}^{n,\infty}\big|_{\text{previous circuit}}$. If linear interpolation is used, the last two will be used, and in case of parabolic interpolation, all three will be needed.

Let us also consider the sequence of faults. The authors of [8] recommend that a new order of circuits should be determined at each time instant based on some measure of closeness of the numerical values of the responses. In doing so, the measure that was used is not disclosed, and it is hinted that it is a weighted sum of the values of the previous responses.

An approach to concurrent fault simulation in nonlinear reactive circuits that also applies circuit grouping based on numerical integration step size was proposed in [5]. It introduces the following two new features.

The given set of faults is divided into groups (subsets) dynamically in each time step of the simulation. Faults that, based on the error calculated for the previous time instant, should use equal or almost equal discretization steps are grouped and the same (smallest) step is chosen for the corresponding circuits. This makes it possible to perform some calculations for all the circuits in the group only once and then use them when formulating the equations for each circuit from the group. Each group, of course, has its own discretization step, which means that the progression is determined by the one that requires the largest step. In other words, it can happen that for a group with a small discretization step, several simulation steps are performed until the group with the largest step is reached. As time progresses, the faults are regrouped so that the condition of similar faults is constantly fulfilled.

All calculations at t^{n+1} for all circuits from the group are performed simultaneously. The goal of this is to minimize the number of calculations so that as many activities as possible will be performed together for all circuits from the group, which is clear based on the presentation of the procedures for solving the system of equations for similar circuits, which was discussed earlier.

Based on this, the algorithm of concurrent fault simulation would take the form as shown in Fig. 7.4.5, where the emphasis is now on grouping and time progression.

In this procedure, initially, the fault-free circuit and all the faults are in one group, and DC and the first step of the transient analysis are performed. To begin with, the same time step is chosen for all circuits. Then, the conditions for grouping are implemented. Namely, for each of the responses, an error is determined, and based on it, corresponding new step values are generated, which now serve as the basic criteria for grouping or regrouping. After the simulation step, the time advances only if the response error is acceptable. When the simulation is finished (one step) for one group, it goes to the next until the list of groups is exhausted.

For the picture to be more complete, it should always be kept in mind that one step of the simulation means an iterative procedure of analyzing the group of circuits

Given: The set ρ of faults $F=\{f_i: i=1, 2, \ldots, \rho\}$ with excitation signals defined in the duration from $t=0$ to $t=t_{stop}$;

Let:

S is a set of groups $\{G_j: j=1, 2, \ldots, l\}$, where l is variable, where each group represents a subset of the set F so that $G_k \cap G_m = \varnothing$, for every $k, m = 1, 2, \ldots, l$ and $k \neq m$.

c an index indicating the current group of defects being simulated,

t_j the current istant of simulation of the group G_j,

h_j the current value of the discretization step for the group G_j;

$G_c := F$, $t_c := 0$, $S := \varnothing$;

while $t_c < t_{stop}$ {

Perform one step of concurrent fault simulation for the circuits $\{f_r : f_r \in G_c\}$ using the time step h_c;

Based on error determine h_r and t_r for every $f_r \in G_c$;

S:= by regrouping obtained $(S \cup \{\{f_r\} : f_r \in G_c\}\}$;

Choose $G_c \in S$ so that $t_c = \text{minimum}\{t_j: G_c \in S\}$;

$S = S - \{G_c\}$;

}

return *Fault dictionary*

Fig. 7.4.5 General flow of the algorithm of concurrent fault simulation with their grouping

in the way that was described earlier here, and the final solution of which is given by (7.4.7).

By comparing this approach to fault simulation with successive (repeated) simulation for each fault using a standard simulator, it shows that large accelerations are achieved. For one filter circuit that was simulated in the time domain, for example, where 360 defects were included, an acceleration of nine times was achieved. For some simpler circuits, the accelerations are even higher. For the μA741 operational amplifier, the speedup achieved is 2.5 times, if 120 faults are simulated.

References

1. Bell IM, Spinks SJ (1995) Analogue fault simulation for the structural approach to analogue and mixed-signal IC testing. Collection of papers presented at the International mixed signal testing workshop, Grenoble, France, pp 10–14
2. Zwoliński M, Brown A, Chulk C (1997) Concurrent analogue fault simulation. In: Proceedings of the 3rd IEEE international mixed-signal testing workshop, pp 42–47
3. Litovski VB, Zwoliński M (1997) VLSI circuit simulation and optimization. Chapman and Hall, London
4. Litovski VB (2000) Electronic circuit design, simulation, optimization, testing and physical design. Nova Jugoslavija, Vranje, Serbia (in Serbian)
5. Hou J, Chatterjee A (2003) Concurrent transient fault simulation for analog circuits. IEEE Trans CAD Integr Circuits Syst 22(10):1385–1398
6. Newton AR, Sangiovanni-Vincentelli A (1983) Relaxation-based circuit simulation. IEEE Trans Electron Device ED 30:1184–1207
7. Yang ZR, Zwoliński M (1999) Fast, robust DC and transient fault simulation for nonlinear analogue circuits. In: Proceedings of the design, automation, and test Europe conference DATE'99, pp 244–248
8. Hou J, Chatterjee A (1998) CONCERT: a concurrent fault simulator for analog circuits. In: Proceedings of the 4th IEEE/ACM international mixed-signal testing workshop, pp 3–8
9. Kilic Y, Zwoliński M (2000) Concurrent transient fault simulation of nonlinear analog circuits. In: Proceedings of the 6th international mixed-signal testing workshop, Montpellier, France
10. Householder AS (1957) A survey of some closed methods for inverting matrices. SIAM J Appl Math 5(3):155–169

7.5 Testing Combinational Logic

Abstract In this chapter a glimpse of the subject to testing combinational circuit will be elaborated. It was not our intention to cover the complete subject of digital testing since it is out of the scope of this book. The idea is here, to prepare the reader for understanding the concepts of design for testability which will be elaborated later. Having in mind, however, that our aspirations are toward testing of mixed-signal (or hybrid) circuits, the explanation given here will be rather helpful to deal with the problem as a whole. Electronic testing devices, testing procedures, and test signal generation for digital circuits differ significantly from analog circuit testing. Basically, the difference lies in the diversity of the signals that appear. Namely, with digital circuits, faulty operation is established in only one way. The signal at the measuring point should have the opposite value to the value it would have if the circuit were fault-free. Hence the test signal generation strategy. It should ensure that condition. When we say test, when testing digital circuits, we mean one digital word, the bits of which determine the states of the input terminals. For example, for the gate of Fig. 7.1.6, the input word contains three bits *ABC*. A maximum of eight tests can be compiled for this gate. In order to review the procedures for generating test signals (TSG, from test signal generation) for digital circuits, we will first consider the basic principles of testing. Next, we will show the procedures for testing combinational circuits. Special attention will be paid to automatic test generation. Based on these considerations, the basic concepts of design for testability and self-testing will be presented in the next chapters. The presentations that follow are mainly based on existing textbooks or monographs (Zwoliński in Digital system design with VHDL. Pearsons, 2003 [1]; Roth in Computer logic, testing and verification. Pitman, London, 1980 [2]; Abramovici et al. in Digital system testing and testable design. Computer Science Press, New York, 1990 [3]; Rice in Computer-aided design, testing, and packaging. IEEE Computer Society Press, Piscataway, 1982 [4]; Bleeker and van den Eijnden in Boundary-scan test, a practical approach. Kluwer Academic Publishers, Dordrecht, 1993 [5]; Cortner in Digital test engineering. Wiley, New York, 1987 [6]; Wilkins in Testing digital circuits. Van Nostrand Reinhold (UK), UK, 1986 [7]; Liu in Testing and diagnosis of analog circuits and systems. Van Nostrand Reinhold, New York, 1991 [8]; Williams in VLSI testing. North-Holland, Amsterdam, 1986

V. B. Litovski, *Lecture Notes in Analog Electronics*, Lecture Notes in Electrical Engineering 1265, https://doi.org/10.1007/978-981-97-8257-4_5

[9]), standards (Williams in IEEE standard test access port and boundary-scan architecture. IEEE, New York, 1993 [10]; Williams in P1149.4 mixed-signal test bus standard. Working group meetings, minutes, and associate materials. IEEE, New York, 1995 [11]; Williams in Supplement to IEEE Std. 1149.1-1990, IEEE standard test access port and boundary-scan architecture. IEEE, New York, 1995 [12]; Williams in P1149.4 mixed-signal test bus. IEEE Standard Project. IEEE, New York, 1995 [13]; Williams in International technology roadmap for semiconductors. 2003 Edition modeling and simulation. https://www.semiconductors.org/resources/2003-international-technology-roadmap-for-semiconductors-itrs/, 2003 [14]; Williams in Testing of electronics. https://www.jtag.com/testing-of-electronics-tutorial-jtag/, 2023 [15]), as well as on original works (Roth in Diagnosis of automata failures: a calculus and a method. IBM J 278–291, 1966 [16]; Cha et al. in 9-V Algorithm for test pattern generation of combinational digital circuits. IEEE Trans Comput C 27(3):193–200, 1978 [17]; Goel in An implicit enumeration algorithm to generate tests for combinational logic circuits. IEEE Trans Comput C 30(3):215–221, 1981 [18]; Fujiwara and Shimono in On the acceleration of test generation algorithms. IEEE Trans Comput C 32(12):1137–1144, 1983 [19]).

7.5.1 Principles of Combinational Circuit Testing

The function "**compose** a test for the selected fault" in the algorithm shown in Fig. 7.1.7 is in fact the main subject of discussion in this section. In order to see the problem theoretically, let us consider the circuit from Fig. 7.5.1. It represents a combinational circuit with n inputs $\mathbf{X} = [X_1\ X_2\ X_3\ \ldots\ X_i\ \ldots\ X_n]$ and m outputs $\mathbf{F} = [F_1\ F_2\ F_3\ \ldots\ F_j\ \ldots\ F_m]$. A $G/0$ (G-at-0 or G-stuck at-0) fault test should be generated. Of course, all variables here take logic (discrete) values which are referred to as states.

The state in G can be expressed as a function of $\mathbf{X}$, as $G(\mathbf{X})$, and $\mathbf{F}$ can be expressed as a series of functions of G and $\mathbf{X}$ as $F_j = F_j(G, \mathbf{X})$. The problem of generating a test for $G/0$ can be represented as the task of solving the following system of Boolean equations:

$$G(\mathbf{X}) = 1 \qquad (7.5.1a)$$

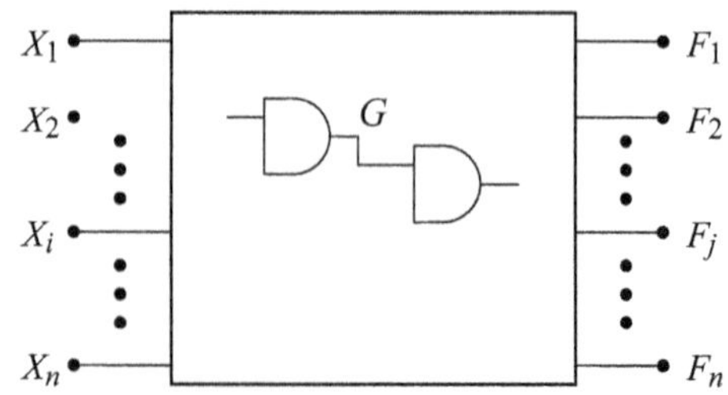

Fig. 7.5.1 Combinational circuit used to formulate the problem of generating a test signal for a given fault

$$F_j(1, \mathbf{X}) \oplus F_j(0, \mathbf{X}) = 1, \tag{7.5.1b}$$

at least for one value of j. Of course X_i can only take the values 0 and 1. The first equation talks about the detectability requirement. Namely, the states of the input variables that produce a state of G opposite to the one that causes the fault are sought (the fault sets $G/0$, and the right-hand side is equal to a logic 1). The second equation relates to observability. Those states of the input variables are sought which ensure that at least one output changes when the fault is active.

Solving Boolean equations, however, is not a task similar to solving ordinary algebraic equations. It is, in fact, a search of an n-dimensional space, in which each coordinate can only take on values of 0 and 1, with the aim of finding those points that satisfy the above set of equations. Each found value of $\mathbf{X}$ that satisfies (7.5.1) is also called a *test word or a test vector*. In the general case, a k-dimensional subspace ($k \leq n$) will be found in which all points will satisfy the above equations. In that case, each of the points in the subspace will be a test (test word) for a given fault. In other words, there may be several solutions of the above equations. Of course, it is possible for that subspace to be empty. In such a case, there is no test.

There are several procedures for searching n-dimensional spaces. The specificity of the search related to testing consists in the fact that by carefully monitoring the propagation of the signal and the fault effect, the overall effort to find a solution can be effectively reduced. This is the base for synthesis algorithms for generating the test vector $\mathbf{X}$.

To that end, we will start with the concept of *path sensitization*. Namely, the basic problem of generation test signal (TSG) is ensuring that a fault that exists at the input of a logic element produces an effect at the output of that element. As an illustration of this problem Fig. 7.5.2a will be used. It is showing an element with four inputs A, B, C, and D and one output Z. Its function can be expressed as:

$$Z = f(A, B, C, D). \tag{7.5.2}$$

From a testing point of view, however, it is more convenient to look at the gate in another way. One should consider the relationship between the output and the

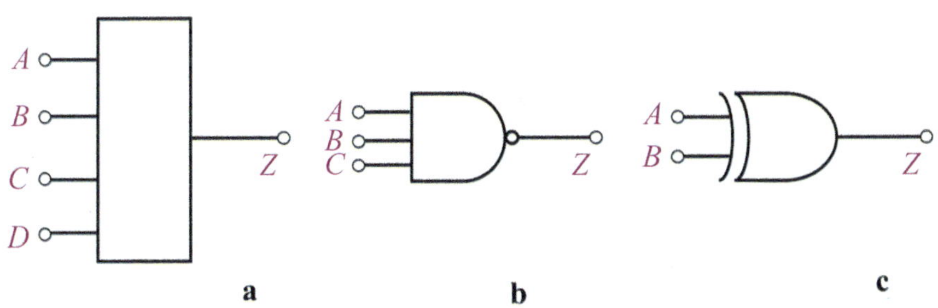

Fig. 7.5.2 Path sensitization **a** for general logic element **b** for three-input **NAND** gate, and **c** for **EXOR** circuit

corresponding input port, say A. Such a relationship could be transformed into various forms by a suitable choice of the values of the other three variables, B, C, and D. Among these forms, the most interesting, from the point of view of testing, are those in which achieves that:

- Z becomes dependent on A only and
- Z becomes completely independent of A.

For example, in Fig. 7.5.2b we have:

$Z = \overline{A}$ if $B = C = 1$ and
$Z = 1$ if $B = 0$ or $C = 0$ or $B = C = 0$.

In the first case, B and C together act as "permission" inputs and establish a "sensitized path" from input to output (from A to Z). In the second case, there are several ways to disable the operation of the gate (to brake) and to provide a fixed value of the output signal. These two procedures are the basis of TSG.

It should be noted that there is a logic circuit whose operation cannot be disabled. The Exclusive-**OR** (**EXOR**) gate, depicted in Fig. 7.5.2c, always provides a sensitized path since:

$Z = D$ if $E = 0$ and
$Z = \overline{D}$ if $E = 1$.

Now that we have developed the concepts of path sensitization and disabling signal propagation, we can proceed to test development for the simplest combinational circuits. As a first illustration, the three-input **NOR** gate shown in Fig. 7.5.3 will be used. Assuming that A, B, and C are the primary (system) inputs, and Z is the primary (system) output, the algorithm shown in Fig. 7.1.7 will be implemented.

As a first step, a list of faults will be prepared. If we stick to the permanent state model, the fault list will contain a list in which each node is fixed to one of the two states: 0 or 1. To recall, if the signal X is fixed to zero or one, we mark it as X-s-0 or X-s-1, respectively, or abbreviated $X/0$ or $X/1$, respectively. Based on that, the exhaustive list of faults will be: $\{A/0\ A/1\ B/0\ B/1\ C/0\ C/1\ Z/0\ Z/1\}$.

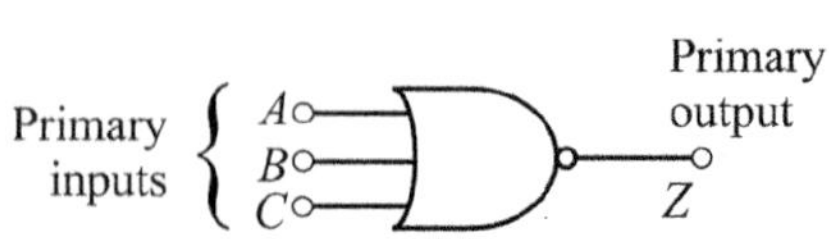

Fig. 7.5.3 Three-input **NOR** gate

Now we can start writing the test. When writing any test there are two independent requirements that must be met:

- If we intend to test whether *A*/0, we must establish such conditions for a fault-free circuit where $A = 1$. Otherwise, there would be no difference between the behavior of the fault-free and the faulty circuit.
- If the presence of a fault is to be observable, a change into an incorrect value at the location of the fault should cause a change in the primary output. In other words, a sensitized path from the fault site to the exit should be established.

For the case of a **NOR** gate, if either *B* or *C* were at logic 1, the output signal would be equal to zero regardless of the value of *A*. The fault, in this case, would not be observable from the output. We would say that the fault is blocked. If, however, $B = C = 0$ is set, we enable the transmission of the signal through the gate, so the effect of the fault is transmitted, that is, the value of *Z* is completely determined by the signal *A*.

Based on this, we conclude that the test for *A*/0 is composed of $A = 1$, $B = 0$, and $C = 0$. When this test is applied to the input of the fault-free gate, the output signal is $Z = 1$. It should be kept in mind that knowing the correct value of the output signal is information of key importance. If it were unknown, we would not be able to determine whether the gate passed or "failed" the test.

The signal (test) that has been assembled can be written in abbreviated form as 100/0, where on the left side of the forward slash are the states of the input signals that will be used as a test, and on the right side are the states of the output signal for a given test when the gate is faulty. Alternatively, the tag $A\overline{BC}/\overline{Z}$ can be used. To follow the algorithm from Fig. 7.1.7, one should now determine the *coverage* of the fault by the created test.

A fault is covered by a given test if the test allows the change represented by the given fault to be communicated to the primary output. Coverage can be established as follows:

- Establish the correct values in all nodes of the circuit and
- For each node individually determine whether a change in the correct value would cause a change in the primary output. This is best done by doing it for the output nodes first and then moving backwards to the input nodes.

For the example we are considering, after this analysis, we conclude:

1. *Z*/1 is covered (by test 100/0).
2. *A*/0 is covered (it was the fault we developed the test for).
3. Since we chose $A = 1$, the transfer from *B* and *C* to the output is blocked so *B*/1 and *C*/1 are not covered by this test.

In this way, the first cycle of TSG was completed. Having now eliminated *Z*/1 and *A*/0 from the fault list, we proceed to repeat the entire procedure for the next fault which may be *A*/1 (selected in order). Repeating the procedure completely, we get:

1. $A = 0$ in order to achieve the fault sensitization condition,
2. $B = C = 0$ to enable signal transfer through the gate,
3. accordingly, the test is 000/1, and
4. the following faults are covered: *Z*/0, *A*/1, *B*/1 i *C*/1.

Only faults *B*/0 and *C*/0 remain. The following tests can be generated for them: 010/0 and 001/0, respectively.

In this way, the procedure of generating the test for a complete set of faults in the gate of Fig. 7.5.3 was exercised. A total of four tests (or test words) are sufficient to check the presence of all faults. The reader should not be confused by the fact that one test covers multiple faults. When accepting the results of the test generation, it should be kept in mind that the goal is testing, not diagnostics. Therefore, it should be determined whether the circuit is working properly or not, not what is the cause of the malfunction. The above tests will establish whether the circuit is working or not, and it will not be known exactly (except in cases where the test covers a single fault), which cause (fault) is present. Conversely, if the circuit is shown to work correctly when tested with all four tests, it can be claimed that no faults are present in the circuit.

Testing a single logic gate where each terminal can be accessed directly is not a big problem. A somewhat more realistic problem is shown in Fig. 7.5.4. The essential difference compared to the circuit from Fig. 7.5.3 is that internal nodes *E* and *F* are introduced, which are neither primary outputs nor primary inputs. This property basically does not change the test generation method but makes it more complex because the state establishment in the internal nodes is now achieved indirectly through the primary inputs.

Consider, for example, the test for *A*/0 in the circuit of Fig. 7.5.4. The following are the steps in the test word generation process:

1. $A = 1$ to establish the fault sensitization condition.
2. $B = 1$ to allow transfer of the fault effect via the gate "1" to node *E*,
3. In order to transfer the fault effect through gate "3," that is, to complete the sensitized path to the primary output (node *Z*), $F = 0$ is required. With this we have created an example of the need to generate a given state in an internal node of the circuit. In order to satisfy this need, it is necessary to work backwards toward the input terminal of the circuit in order to establish the appropriate state

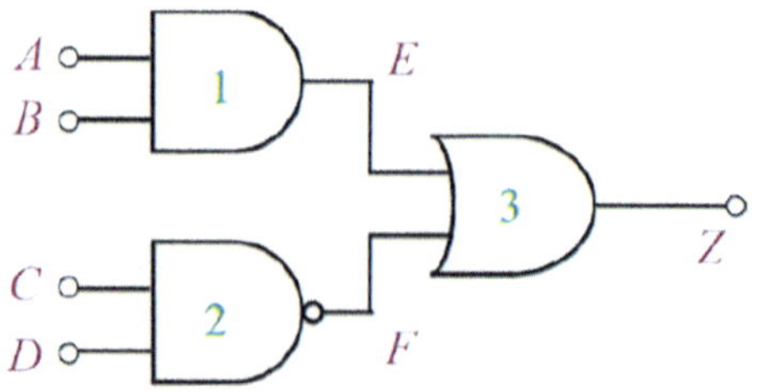

Fig. 7.5.4 Circuit with internal nodes

of the signal at the input. This need to alternatively perform the analysis in the direct and inverse direction is an important feature of TSG. For the circuit of Fig. 7.5.4 $F = 0$ if $C = D = 1$ is necessary.

So the test word for *A*/0 is 1111/1.

Based on the analysis that has been done so far for the last two examples, it could be concluded that several faults may be tested by a single test. This implies, however, that there are multiple tests that detect the effect of a single fault. As an illustration of this statement, consider the fault *F*/1 in the circuit of Fig. 7.5.4. This requires $F = 0$ which is obtained if $C = D = 1$. In order to activate the gate "3," we need $E = 0$. Since *E* is the output signal of an **AND** gate, it can be held at 0 if either of its two controlling signals is held at 0. Therefore there are three test words covering *F*/1: 0011/0, 0111/0 and 1011/0.

Matters are further complicated if the circuit contains reconvergent nodes as is the case with the circuit of Fig. 7.5.5. Suppose we apply the 111/1 test and need to consider the effect of the *B*/0 fault. The changes caused by this fault are shown in the figure, where the symbol $1 \rightarrow 0$ shows that the value one, in this case, becomes zero due to the existing fault. We observe that due to the branching (divergence) at node *B*, the fault affects two circuits ("1" and "2"), and due to the values assigned to nodes *A* and *C*, the fault effect spreads through both circuits and thus affects both inputs of the gate "3" where reconvergence occurs. Since gate "3" is a **NAND** gate, changing the input signals from 00 to 11 will cause the output to change from 1 to 0. This means that gate "3" will transmit the fault effect. It should be noted that from state 00 at the input of gate "3," the output cannot be changed by individual changes in the input signals. Combinations of 01 or 10 would not change the output of gate "3." However, due to reconvergence, there is a simultaneous change of both signals, so the fault effect spreads and the test is successful. This mechanism is known as dual path sensitization.

If gate "3" were an **OR** (or **NOR**) gate as is the case with Fig. 7.5.6 the situation would be only slightly different. Namely, in this case, individual changes in the input signals of gate "3" would cause a change in the output signal.

In the example shown in Fig. 7.5.6, both input ports change their state at the same time so that the output state changes in this case as well. Therefore, in both examples, the effects that arise from the simultaneous change of the state of both inputs of the logic circuit where we have reconvergence help each other. We say that there is a *positive reconvergence*.

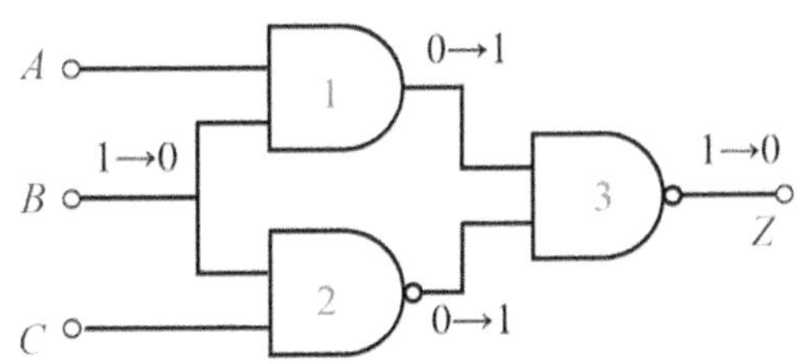

Fig. 7.5.5 Positive reconvergence

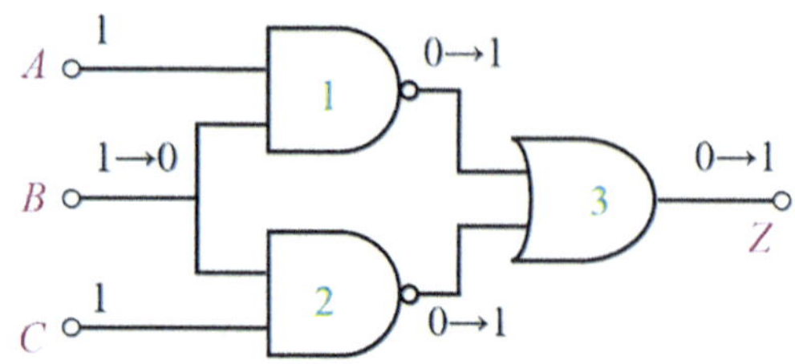

Fig. 7.5.6 Positive reconvergence where each path is separately sensitive

The situation is different in the example illustrated in Fig. 7.5.7. In this circuit, we apply the 100/1 test to test the *B*/1 fault. It can be seen that at the input of gate "3" there is a change from 10 to 01. As a consequence, the **NAND** gate controlling *Z* does not change the state at the output, even though the change 10 → 11 (if only one input was sensed) would cause a change *Z*. The effect that the transfer of the fault effect through one path is offset by the transfer of the fault effect through another path is called *negative reconvergence*. It should be noted that all three circuits discussed in connection with reconvergence have the same structure (topology). This leads us to the conclusion that despite the fact that reconvergence is easily recognized from the topology of the circuit, it is not easy to conclude whether the reconvergence is positive or negative without specific analysis.

The implications of reconvergence on the testing process depend on the circuit being considered as well as the very fault. In the circuit of Fig. 7.5.6, when testing *B*/0 with the 111/0 test, both propagation paths generate the same effect, so the reconvergence itself needs not be emphasized. In the circuit of Fig. 7.5.5 one should be a little more careful. Namely, it can happen that the test designer overlooks the joint action of both signals and draws conclusions based on individual actions. A test composed in this way would be declared unsuccessful, which, in fact, is not the case. Difficulties introduced in this way can be considered minor when the test is compiled manually. When automatically generating a test using a computer, these details should be considered in the form of a separate part of the algorithm.

During the generation of a test for a given fault, it sometimes happens that the procedure leads to a conflict situation. The problem can be illustrated by the circuit of Fig. 7.5.8 in which, among other things, we ask for a test for *D*/0. Applying the standard procedure results in the following:

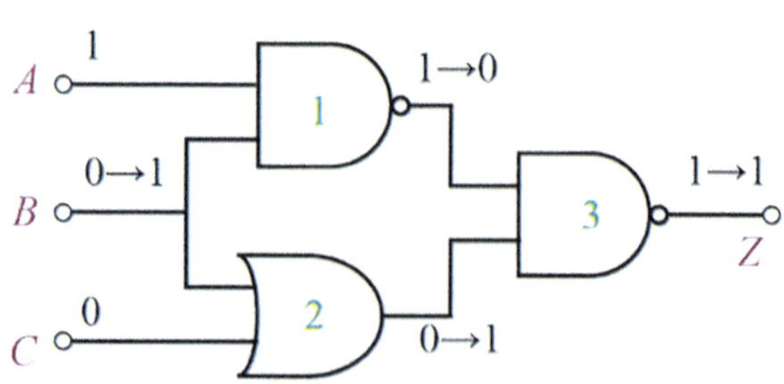

Fig. 7.5.7 Negative reconvergence

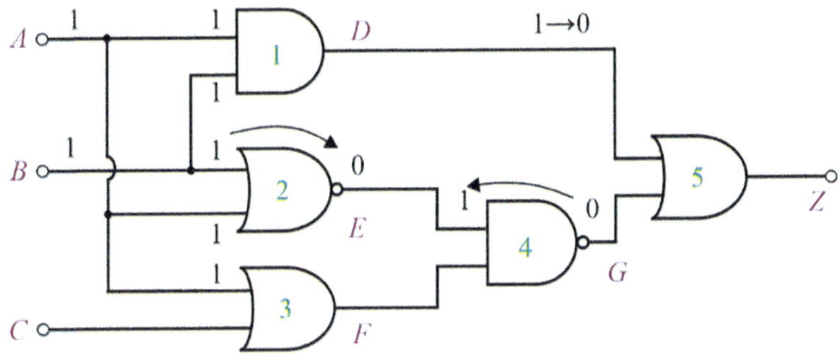

Fig. 7.5.8 Circuit with an undetectable defect. The arrows show that the signal E is needed to be forced into both 0 and 1 at the same time

1. In order to discover $D/0$, we need $D = 1$. The only way to establish this condition is to set $A = B = 1$.
2. In order to ensure observability, i.e., in order to transfer the fault effect through the "5" gate, $G = 0$ is required.
3. Looking backward, to ensure $G = 0$, the input signals of gate "4" must be at logic 1, which means that $E = F = 1$ will be required.
4. The input signals of gate "2," A and B, are already set to logic 1 so that E is already known: $E = 0$.

A conflict arose. In order to detect a given fault, it is required that the given signal (or signals) have two different values at the same time. Not a single arbitrary choice was made. All set values are mandatory. This means that the problem cannot be solved. This fault is *undetectable* and therefore *untestable.*

The existence of undetectable faults is a consequence of one of the two possible limitations imposed on the logic gate by the structure (topology) of the circuit. It is possible that:

(a) the input signals (word) of the logic gate cannot be set in one or more of the 2^n possible states (words) thus preventing detectability, or
(b) whenever the logic element is set in the desired state, other paths of signal propagation make observability impossible, that is, they make the state of the primary output independent of the state at the output of the observed (faulty) logic element.

Whichever of these situations is present in the circuit, it almost without exception arose as a result of *redundancy* in the circuit. To demonstrate this for the circuit of Fig. 7.5.8 we determine the function of the circuit. It is:

$$Z = A + B + \overline{C} \tag{7.5.3}$$

and the output of gate "4" is described by the same expression:

$$G = A + B + \overline{C}. \tag{7.5.4}$$

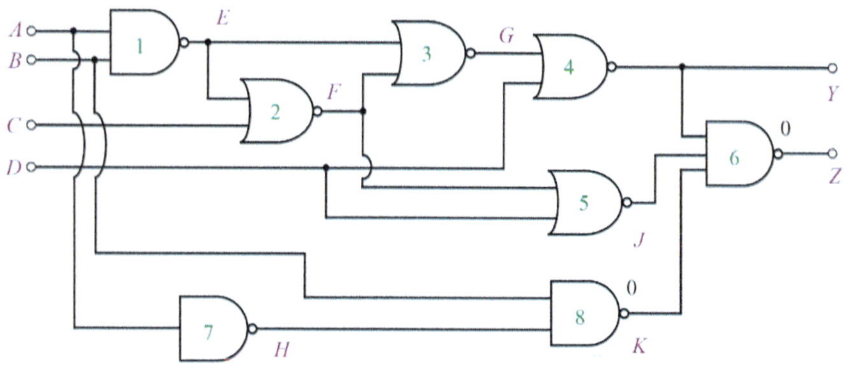

Fig. 7.5.9 Circuit with redundancy

Therefore, the gates "1" and "5" are unnecessary. If they were removed the undetectable fault would disappear.

In the example we have considered, redundant hardware is easily detected. However, this is not always the case. To demonstrate this, consider the generation of the test for the circuit of Fig. 7.5.9 for the $H/1$ fault. Applying the usual procedure leads to the following:

1. Establishing the conditions for the correct state $H = 0$ requires $A = 1$.
2. In order to transfer the fault effect through "8" $B = 1$ is required. Hence $E = 0$.
3. In order to transfer the fault effect through "6" $Y = J = 1$ is required. Now, to generate $Y = 1$ requires $G = D = 0$ and to generate $J = 1$ requires $F = D = 0$.
4. We came to a conflict. Namely, it is required that the output of gate "3," i.e., the signal G to be at logical zero, while at the same time both inputs of gate "3," the signals E and F, are also at logical 0, which cannot be satisfied. Therefore, the $H/1$ fault is undetectable.

Although the redundancy in this case is not obvious, this circuit illustrates how problems can arise in test generation if proper minimization was not applied during circuit synthesis. To illustrate this, we will find the circuit function:

$$Y = \overline{D}\left(\overline{A} + \overline{B} + \overline{C}\right) = \overline{DABC} = \overline{D + ABC}$$
$$Z = B + D. \tag{7.5.5}$$

These two functions can be obtained using only three logic gates as shown in Fig. 7.5.10. This circuit does not hide any undetectable faults.

We can detect undetectable faults in two ways. One of them is to try to compile the test as it has been done so far. This procedure, although always successful, suffers from inefficiency. Namely, if for the reason of shortening the time in the program for generating the test signal a limitation is imposed upon the time needed to find a test for a single fault, it may happen that the fact that the test was not found is not attributed to undetectability but to the complexity of the task of generating the test.

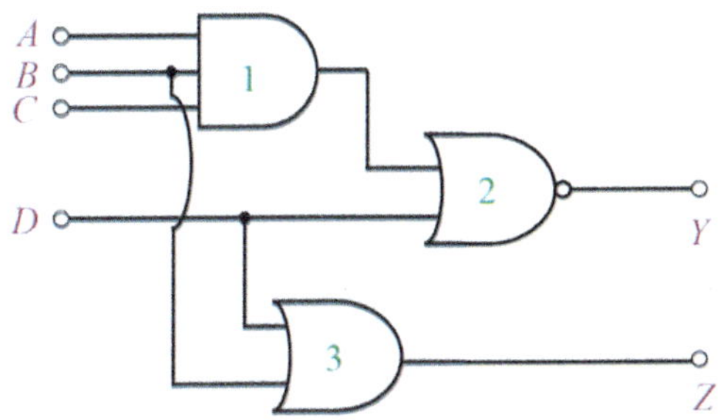

Fig. 7.5.10 Minimized circuit with the same function as that of Fig. 7.5.9

Bearing in mind, however, that undetectability is most often related to redundancy, establishing undetectability could be very useful. Thus, by removing redundancy, we can reduce the overall chip area, consumption, and in some situations, circuit delay. Removing redundancy in its own way contributes to increasing the production yield of a given technology. Namely, a logically undetectable fault, being a consequence of redundancy, is not I_{DDQ} undetectable. Thus, a redundant circuit with a fault that does not affect the function (hence, usable) will be rejected by the failure returned by the I_{DDQ} test.

Bearing this in mind, attempts are made to establish redundancy in the test generation phase as early as possible and thus achieve both goals: shortening the test generation time and eliminating redundancy [20]. We will not discuss these procedures in detail here.

Considering the properties of the last two circuits, we can say something briefly about the behavior of digital circuit designers over time. In the early stages of electronics development, when the price of electronic devices was mainly determined by the number of transistors to be installed, logic minimization was of great importance. With the advent of integrated circuits, the design price became the most significant part of the chip price, so savings in the design phase (shortening the design time by avoiding minimization) yielded more than savings in the number of transistors. With the increase in the complexity of integrated circuits, however, the cost of testing becomes extremely significant, so recently special attention has been paid to minimization in order to simplify the circuit not only to reduce the number of transistors but also, as we have seen, to enable the application of efficient methods for automatic test generation.

Of course, it should not be considered that non-testability is exclusively related to redundancy. To demonstrate this, first, let us analyze the circuit from Fig. 7.5.11a. The bus F is shown, which is excited by two tristate drivers whose input signals are A and B, and the control signals are complementary (orthogonal) and are generated from C. A pull-up resistor R is also connected to the bus. The truth table of this circuit is shown in Fig. 7.5.11b. In order to detect the fault $D/1$, $C = 1$ is required. From the last column of Fig. 7.5.11b we see, however, that the output F'' is either equal to the response F or indeterminate, which cannot be used to infer the correctness of the circuit. Therefore, $D/1$ is undetectable. At the same time, the circuit from Fig. 7.5.11a is not redundant, that is, it cannot be further simplified. There are also

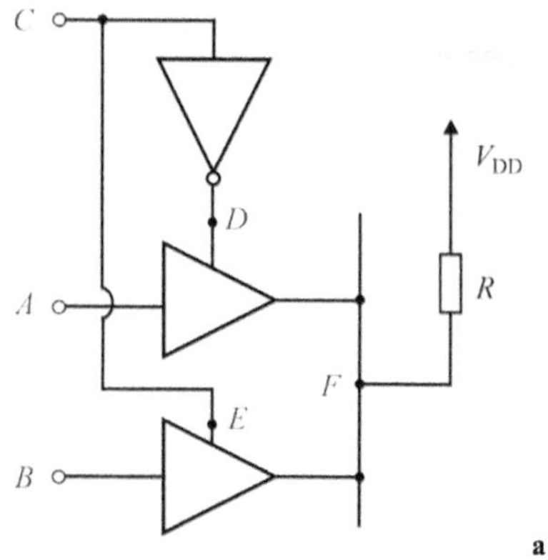

A	B	C	D	E	F	D/1	F'
0	0	0	1	0	0	1	0
0	1	0	1	0	0	1	0
1	0	0	1	0	1	1	1
1	1	0	1	0	1	1	1
0	0	1	0	1	0	1	0
0	1	1	0	1	1	1	X
1	0	1	0	1	0	1	X
1	1	1	0	1	1	1	1

b

Fig. 7.5.11 Competition on the bus with a pull-up resistor. **a** Schematic and **b** truth table

other situations, although very rare, when undetectability is not related to redundancy [21]. Similar (dual) considerations apply if a so-called pull-down resistor from bus to ground is connected.

It is left to the reader to expand the truth table of the circuit shown in Fig. 7.5.11b with fault *D*/0 and to establish its detectability.

On the other hand, it should not be considered that logic redundancy in itself is the result of bad design. There are at least two situations that can lead a designer to deliberately introduce redundancy. The first occurs when one wishes to use gates with low fan-in (for example, only two-input **NAND** gates) and to minimize the number of inverters with the restriction of using only one type of logic gate. A well-known example of such a situation is the implementation of the **EXOR** function using two-input **NAND** gates as in Fig. 7.5.12. It arises when the **EXOR** operation is decomposed as follows:

$$A\overline{B} + B\overline{A} = A\left(\overline{A} + \overline{B}\right) + B\left(\overline{A} + \overline{B}\right). \tag{7.5.6}$$

It can be seen that redundant summands were introduced, which enabled implementation with four logic gates, instead of five (including inverters), which would be required for the implementation of the original function. In this example, however, there was no possibility of appearing of non-testable faults. This means that despite the fact that, usually, redundancy is the cause of untestability, it does not always have to lead to it.

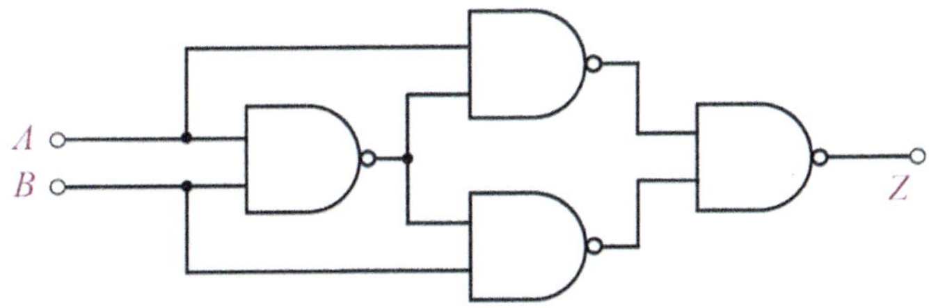

Fig. 7.5.12 Realization of the **EXOR** function with two-input **NAND** gates

The second situation arises when the designer intentionally introduces redundancy to avoid short pulses (hazards) that arise due to unequal signal propagation times through different paths. The simplest example of this type is shown in Fig. 7.5.13a, which represents the realization of the function shown by the Karnaugh map in Fig. 7.5.13b. In this circuit, the minimal realization function requires only gates "1," "3," "4" and "5." Gate "2" provides a bypass that prevents the short pulse that would occur if the input word were to change from 111 to 110.

If we try to generate a test for $D/0$ the following situation will arises.

1. For $D = 1, A = B = 1$ is necessary.
2. In order to transfer the fault effect through gate "5," $E = F = 0$ is required.
3. Gates "3" and "4" should have at least one input at logic 0 so that $E = F = 0$. Since A and B have already been determined to be at logic 1, it remains for the other inputs in "3" and "4" to be 0. These two signals are mutually inverted, which means that the previous condition cannot be fulfilled, so the fault is untestable (undetectable).

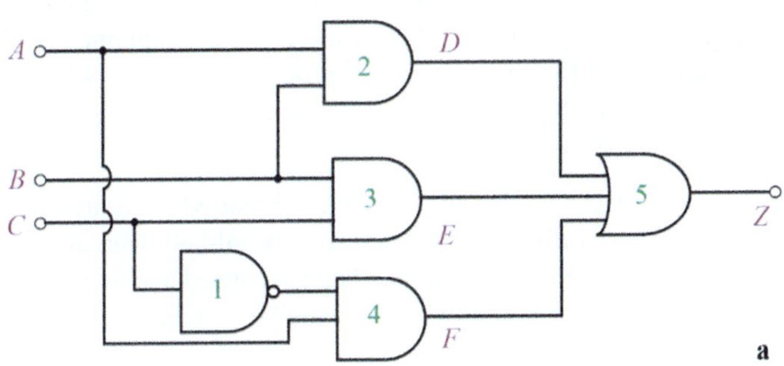

a

$AB\rightarrow$ $C\downarrow$	00	01	11	10
0			1	1
1		1	1	

b

Fig. 7.5.13 Intentionally introduced redundancy. **a** Circuit with the race problem eliminated and **b** Karnaugh map

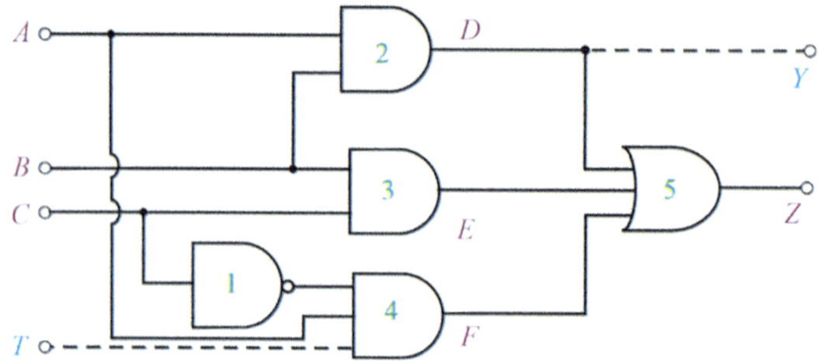

Fig. 7.5.14 Modified circuit from Fig. 7.5.13 to increase testability

Therefore, we are faced with a circuit that is not fully testable due to redundancy that is deliberately introduced for legitimate reasons. In such situations, we can act in one of the following ways:

a. Accept the fact that this circuit is not 100% testable and hope that a fault that is not testable will not appear during the production and exploitation of the circuit.
b. Remove the redundant element and thus the non-testable fault and hope that no hazard will occur.
c. Modify the circuit to make the fault observable. This can be done in different ways, and the simplest would be to *introduce one additional primary output*. In this way, as shown in Fig. 7.5.14, the terminal marked *Y* was created which is connected to gate "2."
d. Modify the circuit in order to provide additional control of signal transfer on existing paths. There are several ways to implement this idea. One of them is illustrated in Fig. 7.5.14. Namely, a *primary input was added* to gate "4," marked with *T*. During normal operation of the circuit, $T = 1$ is maintained, so the circuit behaves the same as before. For testing purposes, $T = 0$ is set, which forces the output terminal of gate "4" to logic 0 regardless of the values of *A* and *C*, which was necessary for testing *D*/0.

By introducing the last two options, in fact, we started a discussion about the concept of design for testability (DFT). Namely, it is unlikely that any of these two ideas will be applied if the circuit has already been produced and the testing problem is established at the time of exploitation. It can even be said that such an intervention is, most often, impossible.

Therefore, if all available options are to be applied (some of which will be discussed briefly later), testing should be considered in the earliest stages of design to identify potential problems and consider possible solutions before the project is completed and delivered to production.

Before we consider the procedures for automation of the test generation for combinational circuits, we will briefly talk about faults that cannot be classified in the permanent state category, that is, faults that cannot be modeled with *X*-s-0 or *X*-s-1. One of these faults is the so-called short circuit or bridging fault.

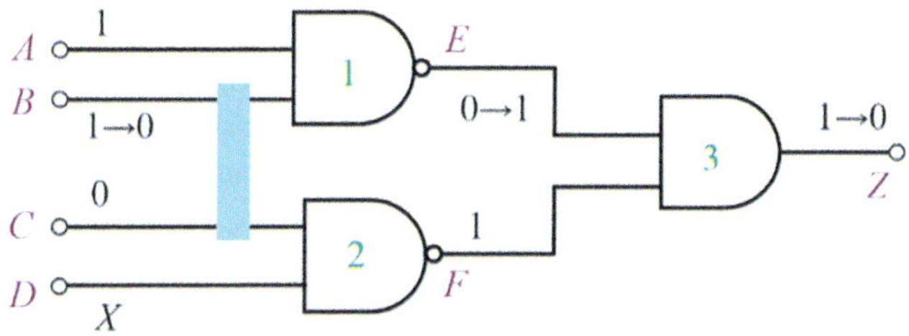

Fig. 7.5.15 Short-circuit testing

It is an unintentionally established connection between two nodes in the circuit. The cause of such a fault is a drop of metal used for connection (solder in printed circuit boards, metal in integrated circuits). The difference in the fault effect compared to faults that cause a permanent state is that now both (short-circuited) nodes can be in both states, 0 or 1. So, if in a fault-free circuit both nodes were in a (same) state of logic 0 or in the (same) state of a logic 1, the fault effect does not exist in the logical sense of the word, while electrically, a change in the values of the currents flowing through the short-circuited points can occur. However, in cases where the states of the two nodes, in a fault-free circuit, would be different, there is a need to "resolve" the conflict and establish the state in the "common" node. The result will greatly depend on the technology in which the circuit will be manufactured. In standard TTL, for example, logic 1 will dominate, in ECL logic 1 will dominate, and in CMOS technology the short-circuit condition will depend on the dimensions (resistance) of the transistors that form the paths to the power supply and to ground.

As an illustration of the need to include a short circuit in the list of faults, let us consider the circuit in Fig. 7.5.15. The usual procedure is to determine that all faults of the permanent state type can be tested with the following set of test words:

0110/0; 1001/0; 0111/1; 1110/1

These tests, however, cannot detect a short circuit between B and C for the simple reason that in each of the above tests, $B = C$ already exists.

The procedure for generating a test for this type of fault is basically the same as for generating a test for a permanent condition. First, we establish the condition of sensitization. In this case it is $B \neq C$. This can be achieved in two ways: either by setting $B = 1$ and $C = 0$ or by setting $B = 0$ and $C = 1$. If, for the sake of an example, we assume that the circuit is realized in TTL technology (One would need much more information to make an example using CMOS), then in both cases the result of the short circuit will be logic 0. The rest of the actions in test generation are:

1. If we choose $B = 1$ and $C = 0$ the effect will be that B will be pulled down toward zero.
2. In order to propagate the fault effect through "1," $A = 1$ is required.
3. In order for the fault effect to be transferred through gate "3," $F = 1$ is required. This condition is already met by $C = 0$, so the value of D is irrelevant. Possible tests are therefore 1101/1 and 1100/1.

If the procedure is repeated with $B = 0$ and $C = 1$, the role of D is taken by A, so the tests are 1011/1 0011/1. All together it would be: 110X/1 and X011/1 (where X stands for both 1 and 0).

Let us now consider a short circuit between E and F. In order to be able to test it, it is necessary to establish the condition that E and F are different ($EF = 01$ or $EF = 10$). In that case, Z would be equal to logic 1. However, whichever of these two tests is applied, the short circuit will result in $EF = 00$, which again makes $Z = 1$. So this short circuit is undetectable.

In addition to this, when classifying possible short circuits in the list of faults, we encounter another difficulty. Short circuits between each pair of nodes cannot be entered in the fault list. A circuit with a thousand nodes has about half a million pairs of nodes. On the other hand, there is generally no possibility of a short circuit between two nodes unless they are adjacent on a board (printed or silicon). At the same time, according to Fig. 7.3.12, the node consists of a series of strips that can be short-circuited with an adjacent strip. Information about the position of the strips (connection segments), however, is not available at the stage of the logical and electrical synthesis of the circuit. This information is only available after designing the topology of the circuit. If one insists on exhaustive modeling of such faults, as can be seen, the test will be generated (partially) even after topology synthesis. This means that it is possible to redesign the circuit to improve testability or to reduce the probability of a short circuit. If, on the other hand, there is no insistence on the exhaustive inclusion of short circuits in the fault list, often, only short circuits between adjacent gate (cells) terminals are modeled. Of course, due to the high probability of the occurrence of such faults, special attention should be paid to them, which we will do later.

Faults of the open-circuit type, undeniably, result in the interruption of signal transfer. This means that a new circuit topology and therefore a new function are created. From the point of view of testing, however, the interruption of signal transfer simultaneously means that testing is impossible. On the other hand, breaking the connection leading to the input of the active element (transistor) will cause a change in the function of that element. Generally, this kind of interruption can be translated into a fault of the permanent state type. Generating a new function of a logic gate that has such a fault is a serious task in fault modeling. That is considered to be out of the scope of this book.

7.5.1.1 Automation of Test Generation for Combinational Circuits

As we have seen in the previous section, the test for a given fault can generally be generated based on a deterministic procedure based on systematic analysis and setting conditions that enable detectability and observability. In this section, an attempt will be made to present the principles for the automatic generation of combinational circuit tests. For the sake of completeness, a stochastic procedure will also be given. The very algorithms for test generation are out of the scope of this book.

First, the notion of a *minimal test set* will be established, and then the concept of D-notation will be introduced. Based on this, we hope, the reader will become capable to understand the algorithms for automatic test generation given in literature exclusively devoted to digital testing.

In order to analyze the behavior of combinational circuits from the point of view of the fault effect, it is useful to first evaluate the coverage of faults with possible tests. Such an assessment will be made here on the example of the circuit from Fig. 7.5.16. Given that there are three primary inputs, there are eight possible test vectors. They will be labeled t_0 to t_7, where the index is equal to the decimal equivalent of the value of the binary number ABC.

The circuit has seven nodes, and therefore, 14 possible faults of the permanent state type. The coverage of faults by a given test can be expressed in a convenient way by a table in which the "+" sign indicates that the fault is covered by the test. The result is shown in Table 7.5.1, which is usually called the *fault matrix*. It is generated by multiple simulation of the faulty circuit. Note the logic simulation including faulty circuit simulation is beyond the scope of this book.

A minimum test set can be established by searching this table. It is the smallest set that contains a sufficient number of test words to be able to test every fault. It is generated as follows:

1. Each column containing only one plus represents a fault covered by one test word only. Therefore, if we want that fault to be tested, the corresponding test must

Fig. 7.5.16 Logic circuit for an example of generating a fault coverage matrix

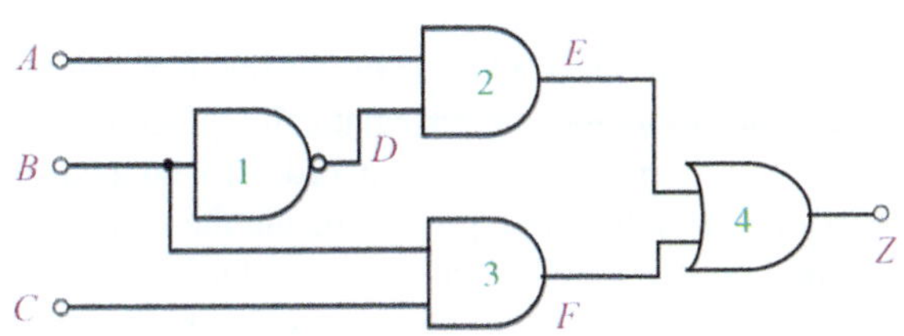

Table 7.5.1 Fault coverage matrix with tests for the circuit of Fig. 7.5.16

	A/0	A/1	B/0	B/1	C/0	C/1	D/0	D/1	E/0	E/1	F/0	F/1	Z/0	Z/1
t_0		+								+		+		+
t_1		+		+						+		+		+
t_2						+				+		+		+
t_3			+		+						+		+	
t_4	+			+			+		+				+	
t_5	+						+		+				+	
t_6			+			+		+		+		+		+
t_7					+						+		+	

be included in the minimum test set. Such tests are called *essential*. In the given example, the t_6 test is essential because it is the only one that covers the fault *D*/ 1.

2. When we have listed all essential tests, all faults covered by those tests are deleted from the table. Now it is necessary to find the smallest number of additional tests (again, we should not forget that by testing we are checking whether the circuit works correctly, not which one of the faults is active) that will cover the other faults. In this example, t_6, which we found to be the only essential test, also covers *B*/0, *C*/1, *D*/1, *E*/1, *F*/1, and *Z*/1. The rest of the faults can be covered in several ways. In order to find them, it is possible to determine one of the possible sets through a careful search, and it is also possible to derive a Boolean expression that contains all possible tests. The rest of the tests can be extracted from such an expression, which together with the essential one, make up the minimum test. For the example we are considering, the following Boolean expression is generated from Table 7.5.1: $(t_4 + t_5)(t_0 + t_1)(t_1 + t_4)(t_3 + t_7)(t_3 + t_4 + t_5 + t_7)$.

   This expression is created by visiting each column in Table 7.5.1 if it has not already been visited. The first pair in brackets $(t_4 + t_5)$ refers to the first column, the second to the second column, the third to the fourth column, etc. Minimization this expression yields: $t_4(t_0 + t_1)(t_3 + t_7) + t_1t_5(t_3 + t_7)$ or $t_3t_4(t_0 + t_1) + t_4t_7(t_0 + t_1) + t_1t_5(t_3 + t_7)$ etc.

Based on this, we conclude that in addition to the essential test t_6, only three more tests are needed: $t_3t_4(t_0 + t_1)$ or $t_4t_7(t_0 + t_1)$ or $t_1t_5(t_3 + t_7)$. There are six possible combinations (ways) of how those three tests can be chosen. Accordingly, the solution would be a test that would successively bring four test words to the input. For example, t_6, t_3, t_4, and t_0.

The presented procedure is exhaustive but tiresome and is therefore not recommended as a method for systematically generating a minimal test set for complex circuits. It is presented here in order to understand the basic mechanisms that determine the generation procedure and the structure of the minimal test set.

The first step at TSG was to form a fault list. For the circuit of Fig. 7.5.16, each node is assumed to be individually fixed to both states. This resulted in a fault matrix with 14 columns. For complex circuits, the number of columns (and the number of rows) becomes too large and leads to extremely long processing times, so for the sake of simplification, a procedure known as *fault collapsing* is applied.

In order to demonstrate the two principles underlying this procedure, two situations that arise when considering fault coverage will be observed. These situations are shown in Table 7.5.2. It contains part of the fault matrix of a hypothetical circuit, which refers to the faults *P*, *Q*, *R*, and *S*, which are tested only with the tests t_i, t_j, t_k, and t_l.

Let us first consider faults *P* and *Q*. We notice that the distribution of pluses for both tests is identical. Any test that covers fault *P* simultaneously detects fault *Q* and vice versa. *P* and *Q* are faults that are *unrecognizable*. This is usually denoted as $P \leftrightarrow Q$. Since every test that covers *P* also covers *Q*, there is no need for both faults to appear in the table.

Table 7.5.2 Fault coverage matrix allowing collapsing

	P	Q	R	S
t_i	+	+		
t_j			+	
t_k	+	+	+	+
t_l			+	+

In Table 7.5.1 there are three groups of faults that are unrecognizable. They are $\{A/0, D/0, E/0\}$, $\{E/1, F/1, Z/1\}$, and $\{C/0, F/0\}$. Based on this, the list of faults can be reduced by keeping only one representative of the group (say the first one) and ignoring the others.

The other situation in the fault matrix that allows reduction of the number of faults is shown by the faults *R* and *S* in Table 7.5.2. In this case, the set of pluses related to fault *S* is a subset of the set of pluses related to fault *R*. To test fault *S*, we can use either t_k or t_l. Whichever one we use, however, will also cover the fault *R*. The fault *S* is said to *dominate* the fault *R*. This is denoted by $S \rightarrow R$. Based on *dominance*, fault *R* can be omitted from the fault matrix. The following should be considered. First, it is an asymmetric operation. Therefore, only *R* is excluded. Second, the fault that is omitted has a larger number of pluses, although, intuitively, this seems wrong.

In Table 7.5.1 there are several examples of dominance: $A/0 \rightarrow Z/0$; $A/1 \rightarrow E/1$; $C/0 \rightarrow Z/0$; $C/1 \rightarrow E/1$; $D/1 \rightarrow B/0, C/1, E/1$.

When faults that are dominated as well as faults that are undetectable are excluded, the list of faults is reduced to only five elements *A*/0, *A*/1, *B*/1, *C*/0, and *D*/1. It should be remembered that this result inspires more optimism than it is realistic. Namely, in practice, in the best case, by applying these ideas, merging faults can cut the list of faults in half (at most).

Applying the idea of fault collapsing has so far required reasoning based on the fault matrix. Unfortunately, the fault matrix is not available during TSG. That is why it is necessary to find a way to apply the fault collapsing technique without searching the fault matrix. To see how this is done, let us revisit the test generation procedure for an isolated logic element.

Let us first consider a two-input **AND** gate whose input terminals are connected to nodes *A* and *B*, and output terminals to node *Z*. If we consider the fault *Z*/0, we find that the only test is 11/1, i.e., *AB*/*Z*. We find that this test simultaneously covers faults *A*/0 and *B*/0 and that it is the only test that covers these faults. The conclusion is that the group of faults {*A*/0, *B*/0, *Z*/0} constitutes a set of unrecognizable faults. If now, for the same **AND** gate, we consider the fault *A*/1, we will find that the only test is 01/0, that is $\overline{A}B/\overline{Z}$, and that test also covers fault *Z*/1. However, it is not the only test that covers the fault *Z*/1. It is also covered by tests 10/0 ($A\overline{B}/\overline{Z}$) and 00/0 ($\overline{AB}/\overline{Z}$). Based on these considerations, we conclude that the fault *A*/1 dominates the fault *Z*/1.

In order to reveal the properties of equivalence and dominance, it is necessary to analyze all gates in a similar way. The obtained results for the basic logic gates are

Table 7.5.3 Fault collapsing for basic logic gates

Gate type	Unrecognizable faults	Faults dominance
AND	$\{A/0, B/0, Z/0\}$	$A/1, B/1 \rightarrow Z/1$
OR	$\{A/1, B/1, Z/1\}$	$A/0, B/0 \rightarrow Z/0$
NAND	$\{A/0, B/0, Z/1\}$	$A/1, B/1 \rightarrow Z/0$
NOR	$\{A/1, B/1, Z/0\}$	$A/0, B/0 \rightarrow Z/1$
Inverter	$\{A/0, Z/1\}$; $\{A/1, Z/0\}$	None

shown in Table 7.5.3. By using such tables, the merging of faults can be done at the level of the description of logic circuits. Unfortunately, something else needs to be considered.

If we apply Table 7.5.3 for the circuit of Fig. 7.5.16 the following conclusions will be drawn:

Unrecognizable	Dominance
Gate 4: $\{E/1, F/1, Z/1\}$	$E/0, F/0 \rightarrow Z/0$
Gate 2: $\{A/0, D/0, E/0\}$	$A/1, D/1 \rightarrow E/1$
Gate 3: $\{B/0, C/0, F/0\}$	$B/1, C/1 \rightarrow F/1$
Gate 1: $\{B/0, D/1\}$; $\{B/1, D/0\}$	–

Now, if we compare the obtained result with the content shown in Table 7.5.1, we find that the conclusions we made for gates "4" and "2" are good, and those related to gates "3" and "1" are not. To investigate the cause of the problem we will run tests for $F/0$ and $B/0$, which we predicted were unrecognizable.

Following the usual procedure for developing a test for $F/0$ we get:

- $F = 1$ requires $B = C = 1$.
- $E = 0$ is required to allow fault propagation through the gate 4.
- Since $B = 1$, $D = 0$ so $E = 0$, independently of the value of A. Therefore, we can put $A = \text{X}$.

Based on this, we conclude that the test for $F/0$ is given by X11/1, which are tests t_3 and t_7. Similarly, the procedure for developing the $B/0$ test gives:

- $B = 1$ to establish a faulty condition.
- The fault spreads to the output in two ways: through gate "3" and through the gate "2." Therefore we have

  (i) In order to propagate through gate "3" it is necessary that $C = 1$. For further propagation toward the exit (through gate "4"), $E = 0$ is required. To maintain this value under faulty conditions (when $D = 0 \rightarrow 1$) $A = 0$ is required.

  (ii) To propagate through element "2" $A = 1$ is needed. To propagate through element "4" $F = 0$ is needed. Therefore, we need $C = 0$.

Therefore, for $B/0$ we have two possible tests: 011/1 and 110/1, which are tests t_3 and t_6.

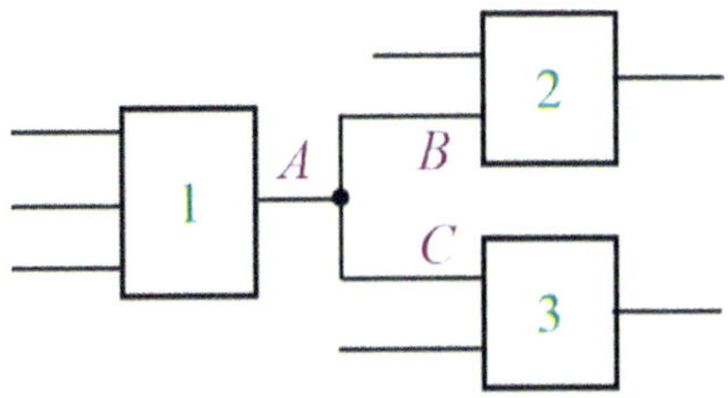

Fig. 7.5.17 Branching fault problem

Now we can conclude that the faults $B/0$ and $F/0$ are not unrecognizable. Two properties of the circuit that make these faults different can also be observed:

- t_7, which covers $F/0$, does not also cover $B/0$ due to negative reconvergence which results in the fault effect in node F being canceled by the fault effect arising in node E.
- t_6, which covers $B/0$, does not cover $F/0$ because it uses an alternative path through E instead of through F.

Both of these effects are caused by the existence of branching at node B. The agreement between the actual implications of the fault (as in Table 7.5.1) and those predicted by the analysis of individual gates can be re-established if we remove all implications related to the corresponding node in which branching occurs. In our example, the implications related to node B should be removed, the implications related to gate "1" will also disappear, and those related to gate "3" will become $\{C/0, F/0\}$; $C/1 \rightarrow F/1$.

Unfortunately, the story of branching does not end there. Figure 7.5.17 represents a part of a circuit in which the output of gate "1," marked with A, is simultaneously input B of the gate "2" and input C of the gate "3." Until now, A, B, and C were considered to be parts of the same node and to be tightly connected. An open-circuit fault in this node (see Fig. 7.3.12) can isolate one of these three points while leaving the other two connected. In doing so, two characteristic situations arise. If there is a break between the node and the output of gate "1" (break of the node's tree), the fault effect will propagate to gates "2" and "3." If, however, there is an open between the node and the input to gate "2" (or gate "3") (break of a node's branch) the fault effect will not propagate through its own node. This type of fault is called a *branching fault*. The main problem with the branching fault is related to reconvergence when there is a break in the node tree.

7.5.1.2 Pseudorandom Test Vector Generation

As already mentioned, based on the above considerations, deterministic algorithms for automatic test generation were developed. The best known are (with rising effectiveness) the D-algorithm, the 9-V algorithm, PODEM, and the FAN algorithm. Study of these algorithms goes beyond the scope of this book. All these algorithms, when covering a given set of faults, provide a test vector (a set of test words) of the smallest length. The disadvantage of this approach is the complexity of the procedure and the cost of generating the minimum test vector. On the other hand, a complete set of test words, as we have already seen, is practically inapplicable. Pseudorandom generation of test words is often used as an intermediate solution.

It is intuitively acceptable that fault coverage will increase as the number of test words increases. On the other hand, there is always the possibility that a randomly selected test word covers one or more faults in some complex circuit. Therefore, if a pseudorandom sequence of test words is used as the test vector (a random sequence in which there are no repetitions of samples as might be the case in a natural random process) a certain number of faults can be expected to be covered without the effort required for automatic test signal generation (ATSG).

Based on research reported in [22], the results of which are illustrated in Table 7.5.4, it is claimed that the following heuristic formula is valid

$$\mathrm{FC} = \left[1 - \mathrm{e}^{(-\lambda \cdot \log_{10} N)}\right] \cdot 100\%, \tag{7.5.7}$$

where FC is the fault coverage expressed in percentage, and λ is a constant that expresses the specificity of the circuit being tested. The qualitative flow of this curve is shown in Fig. 7.5.18. It can be seen that in the beginning the coverage grows rapidly and soon it reaches saturation, which means that the last faults, the most difficult to test, are very difficult to cover.

Table 7.5.4 Fault coverage (in percentage) with a pseudorandomly generated test vector for two circuits

Circuit	Number of input variables	Number of gates	Coverage in (%) obtained from N randomly generated test words		
			$N = 100$	$N = 1000$	$N = 10{,}000$
a	63	926	86.1	94.1	96.3
b	54	1103	75.2	92.3	95.9

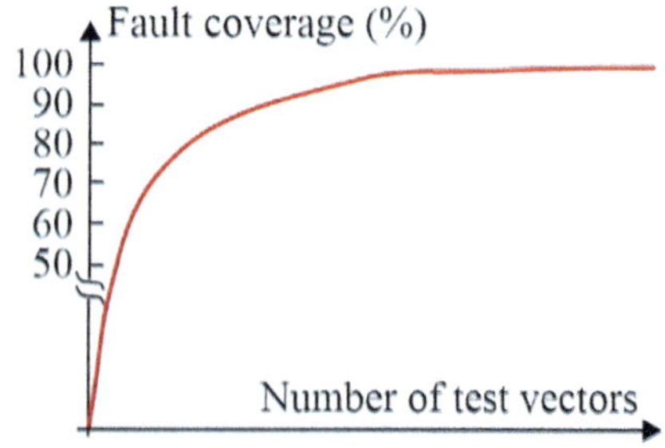

Fig. 7.5.18 Qualitative flow of the curve expressed by (7.5.7)

7.5.2 IDDQ Test for Digital Circuits

As with analog circuits, testing based on supply current can be an extremely good alternative to testing the logic function of a circuit. Usually, instead of the state at some primary output, the DC value of the supply current I_{DDQ} is observed.

The I_{DDQ} test is actually a comparison of the DC supply current of a fault-free and faulty circuit. With bipolar technologies, the DC component of the supply current can be relatively large and its variations due to faults are not so noticeable, although this does not always have to be the case. Also, the values of the direct current in the steady state depend on the excitation word considering that, depending on the logic state of the circuit, they "draw" a higher or lower current. In CMOS technology, in which by far the largest percentage of integrated circuits are produced today, the DC supply current is very small considering that, in the steady state, all transistors are non-conductive. It should not depend on the input word. A fault in such a circuit is very likely to be reflected in a change in the direct current of the power supply. Therefore, I_{DDQ} represents an alternative tool for testing digital circuits, especially CMOS circuits.

The question arises, if a test is generated for a given circuit, as described in the previous sections, what would the I_{DDQ} test be for? It is precisely the purpose of the presentation in this section to show that an I_{DDQ} test is always necessary even when the usual structural (logic) test shows that the circuit is correct. Namely, it turned out that depending on the technology and the way the circuit was tested, between 2 and 11% of the circuits that had already passed the usual structural test failed the I_{DDQ} test [23].

The reasons for this lie in the following:

- First of all, due to the very large number of possible faults, tests are usually not compiled for all potential faults in the circuit. The circuit is not tested for "all" faults. However, the I_{DDQ} test, since it is characterized by automatic observability, directly maps the fault effect onto the supply current and behaves as if all faults are tested simultaneously. Of course, one should not exaggerate here. Namely, not all faults are reflected in the supply current.
- Some faults of the "short-circuit" type, for which the resistance at the connection point is relatively high, do not change the logic function grossly (they change delay times, logic levels, etc.) and as such can remain unnoticed during a logic test. The supply current, however, will certainly change. An example of such a fault is the gate breakthrough, which we discussed in detail at the beginning of this book.
- It is similar with faults of the permanently open type. In such situations, due to interruption of the signal transmission path, logic testing is disabled. Often, however, such faults cause an increase in the supply current. The reason for this can be the accumulation of charges on the "hanging" gate, which forces the transistor whose gate is disconnected to conduct. Disconnected drain or source in CMOS memories is very often detected by the I_{DDQ} test.

- Faults of the "emergence of parasitic transistor" type over thin or even thick oxide in an integrated circuit are a frequent cause of increased supply current. At the same time, due to the small transconductance and inadequate threshold voltage, these transistors usually do not change the logic function of the circuit and cannot be detected by a structural test.
- Faults that represent "leakage" in places where signals are not transferred, such as leakage from power lines, also cannot be tested with a structural test. This category also includes faults related to breakdowns of some p-n junctions (drain-substrate, for example), which do not affect the logic function of the circuit.
- Finally, I_{DDQ} is an indispensable indicator of faults in redundant circuit parts that are latent and not normally used. This also applies to the various protection circuits that are built-in, which are inactive until the normal mode of use of the component is left.

I_{DDQ} testing of digital circuits can be done in three ways:

- Measurement of I_{DDQ} for each test vector. Here, during logic testing, for each test vector, the supply current is simultaneously sampled.
- Selective measurement of I_{DDQ}. Similar to the previous case with the fact that it is measured only for a selected set of test vectors.
- Additional I_{DDQ} testing. In this case, special (additional) test signals are developed that are exclusively intended for I_{DDQ} testing.

Measuring I_{DDQ} for each excitation vector is impractical. The number of logic test vectors is usually so large that it makes the measurement uneconomical, especially in mass production conditions.

Selective measurement of I_{DDQ} is more attractive due to the fact that the number of measurements is reduced. However, in this case, there is a probability that some faults remain untested. For example, this could happen if the vectors generated for logic testing were not developed for short-circuit type faults.

Therefore, there is a need for additional I_{DDQ} testing. The possible (or most of them) faults that cannot be tested easily (or at all) with a logic test, are selected and a separate I_{DDQ} test is developed for them.

In further presentations in this section, we will use the opportunity to illustrate the I_{DDQ} test, and at the same time, to show efficient ways of generating a test for short-circuit type faults that do not have to be mapped to permanent 0 or permanent 1 type static faults. To illustrate the development of an I_{DDQ} test, we will consider an example of generating a test for an internal short circuit (SC) inside a CMOS combination gate.

Short-circuit type faults are divided into two categories: internal and external. Internal are those faults that connect (short-circuit) two nodes within the logic operator, at least one of which is neither input nor output. An external fault of type SC is one that connects two nodes that are input or output nodes of a logic gate. For external faults, we also distinguish two categories: SC that does not form a feedback loop (SCNFL) and SC that forms a feedback loop (SCFL).

The essential difference in test generation procedures between internal and external faults is reflected in the difference in their number. Namely, for a complex circuit composed of a large number of logic gates, the number of internal faults is very large but, during the development of the test, the interior of only one logic gate is observed. In the case of external faults, we observe all circuit terminals at the same time, which makes an enormous number of combinations of short circuits. On the other hand, it is understandable that due to the fault effect, when generating tests using the I_{DDQ} test for these two categories of faults of the type of external SC, different procedures should be applied. There will be more to say about this later.

Let us first consider the I_{DDQ} test for internal SC faults in combination circuits. Illustrations will be made using examples of CMOS circuits. The structure of the combinational CMOS circuit in the general case can be shown in Fig. 7.5.19. Based on this representation, the I_{DDQ} test will detect an internal short circuit between two nodes in a combinational circuit if there is a conductive path from one of those two nodes to V_{DD} and from the other to V_{SS}. A test signal that detects a short-circuit type fault should cause these two nodes to be "pulled" toward opposite logic values (one toward 0 and the other toward 1 or vice versa). In this way, a low-ohm path is created between V_{DD} and V_{SS}, so a large value of the supply current is read as evidence of the presence of a fault.

The circuit of Fig. 7.5.20 requires a test for a short circuit between nodes α and β [24]. The test proposed here aims to bring node β to a higher potential than node α by opening the conduction path T_{P5}–T_{P1}–T_{P4}–T_{N4}–T_{N1}. The corresponding test signal is $ABCD = 0110$.

It should not be forgotten that when generating the I_{DDQ} test, observability was achieved automatically. This means that when generating the test, it is enough to ensure the controllability of the given nodes. When working with external faults of the SCNFL type, we ensure controllability with a test signal that meets the same requirements as for internal faults. In the circuit from Fig. 7.5.21a, for example, the test for *R*/1 which reads $ABCDEFG = 0101001$ is also a test for a short circuit between nodes *J* and *K*. In fact the test is $ABCDEFG = \text{X101XXX}$. Each of these tests forces $J = 1$ and $K = 0$.

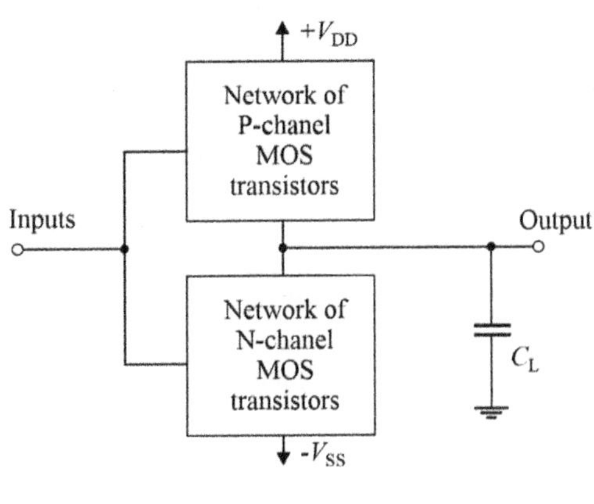

Fig. 7.5.19 Block diagram of a CMOS combinational circuit

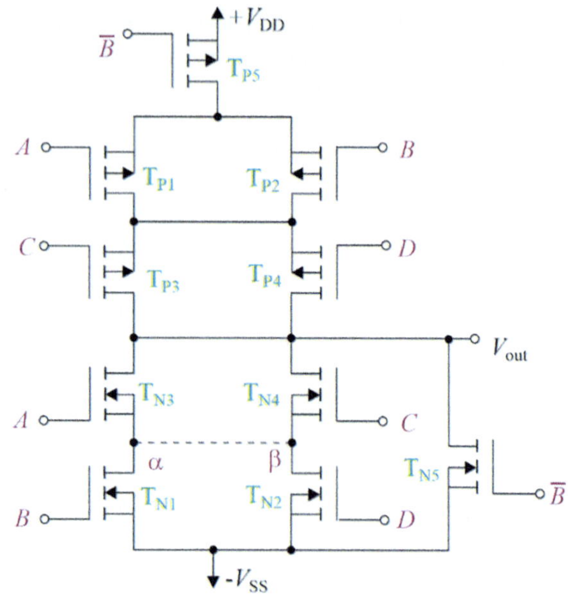

Fig. 7.5.20 CMOS logic circuit with internal short circuit

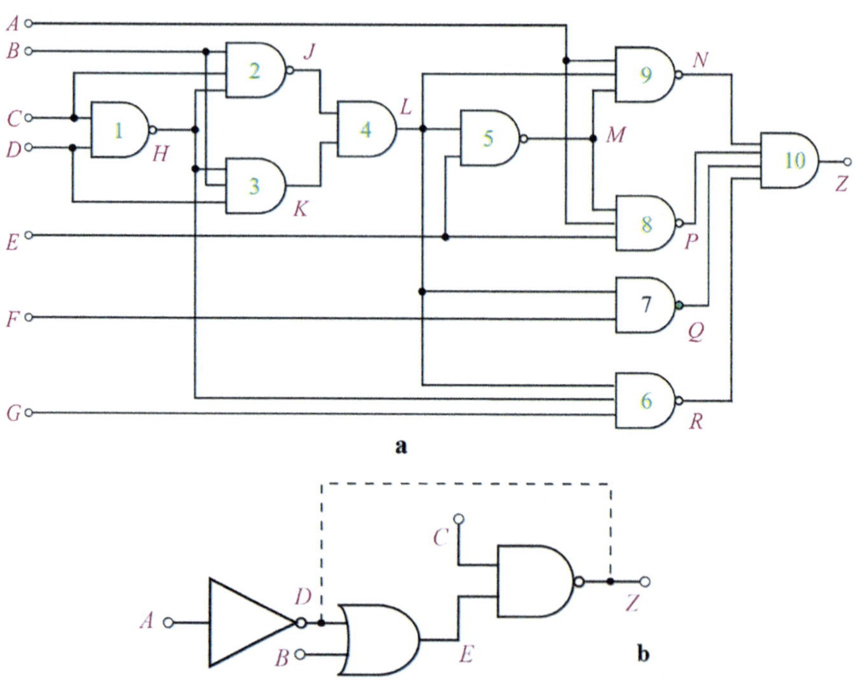

Fig. 7.5.21 **a** Circuit to illustrate test generation for SCNFL and **b** illustration for SCFL

Based on this, we conclude that the I_{DDQ} testability of a short circuit (between nodes X and Y) that does not form a feedback loop, in combinational circuits, can be realized if there is a test signal t that is such that, when applying that test, in the fault-free circuit forces $X \neq Y$. Finding that signal is achieved by the algorithms we have already mentioned.

The situation with SCFL is somewhat more complex. To explain this, we will consider the circuit from Fig. 7.5.21b.

Let the test signal for a short circuit between nodes Z and D be the subject of discussion in this circuit [25]. According to the mentioned criterion, the test for this fault would be $ABC = 101$. It creates $Z = 1$ and $D = 0$ in the fault-free circuit. Moreover, the values of B and C are independent of the value of D. If a fault is present in the circuit, the values of Z and D, which are equal, are indeterminate in the sense that when we look at the logic level, we do not know whether the result of this short circuit will be logic 0 or logic 1. Therefore it can be said that it is $Z = D = \mathrm{X}$. It follows that both $E = \mathrm{X}$ and $Z = \mathrm{X}$. Finally, as a consequence of everything, $D = \mathrm{X}$. The conclusion is that $ABC = 101$ is not a test for the given fault. This leads us to think about a stricter criterion for generating the test than the one that only requires that the difference between the logic states in the short-circuited nodes be realized. Such a stricter criterion is stated in [25] and can be briefly interpreted as follows.

Let it be necessary to test the SCFL between nodes A and B and let B' be the value at node B after applying the test t that puts A in state X. Then, t is the I_{DDQ} test for the SCFL between nodes A and B if $A \neq B'$ and $B' \neq \mathrm{X}$.

Applying this criterion for testing SCFL between A and B in Fig. 7.5.21b should use $ABC = 100$. It produces $D = 0$, $E = 0$, and $Z = 1$ in a fault-free circuit. In the faulty circuit, because $C = 0$, one gets $Z = 1$ independent on the state of E. This makes the difference between state D and Z.

Since we have considered the problem of test generation, it remains to consider the problem of the number of test signals. Considering the possible number of combinations of short circuits between connections, the number of possible faults is enormous. It is shown in [25], however, that when dealing with the I_{DDQ} test, this problem can be overcome. The idea is based on the equivalence of nodes from the point of view of the test. Namely, two nodes are equivalent from the point of view of the test if the response in them is the same for the entire test sequence. For a short circuit between those two nodes, the entire test sequence will not give any difference, so it is inapplicable for testing short circuits. All other (non-equivalent) short circuits can be tested with that sequence if they meet the above criteria. At the same time, the one previously developed for the logic (structural) test can be used as the original test sequence, and new tests will be developed for other faults of the SC type that remained untestable, if necessary. For more details on the implementation of this algorithm, the reader is referred to [25].

To illustrate, the example of testing of the circuit of Fig. 7.5.16 will be considered. For this circuit, which for convenience we show again in Fig. 7.5.22, in Sect. 7.5.1.1, a complete set of test signals is first developed, and then it is shown by optimization that all static faults can be tested with any element from a set of eight test sequences each having only four test words. One of those sequences is $t_0t_3t_4t_6$. The question arises whether all possible external faults of the SC type can be tested simultaneously with this sequence.

In order to answer the question, we first create a response table of the fault-free circuit from Fig. 7.5.22 for the given test run. It is shown in Table 7.5.5. In order to use this table, we will divide SC-type faults into three categories. To the first category we will assign faults of the SCNFL type, to the second those faults of the SCFL type that short-circuit one of the primary inputs with some other node, and to the third category we will assign the SCFL that form a feedback loop inside the circuit without any primary input. The division of SCFL into two categories is based on the fact that the excitation source generates a signal of unlimited strength so that in case of a short circuit it will determine the state on it and it will not be undetermined. Therefore, with these short circuits, a sufficient criterion for test verification will be the difference in the states of the nodes.

A list of short circuits and tests that force various states in short-circuited nodes is given in Table 7.5.6. Here *XY* means that nodes *X* and *Y* are short-circuited. Tests related to the first two categories can be accepted immediately. For the others, it should be checked whether they meet the criteria for SCFL that we mentioned earlier.

Accordingly, we conclude the following: t_0 is a test for SC between *D* and *E* because $A = 0$ and it forces 0 on *E*, independently of the state of *D*. Similarly, t_0 is a test for SC between *D* and *Z* because $F = 0$, which determines *Z*, as a consequence

Table 7.5.5 Responses within the circuit of Fig. 7.5.22 for the excitation $t_0t_3t_4t_6$

	t_0	t_3	t_4	t_6
A	0	0	1	1
B	0	1	0	1
C	0	1	0	0
D	1	0	1	0
E	0	0	1	0
F	0	1	0	0
Z	0	1	1	0

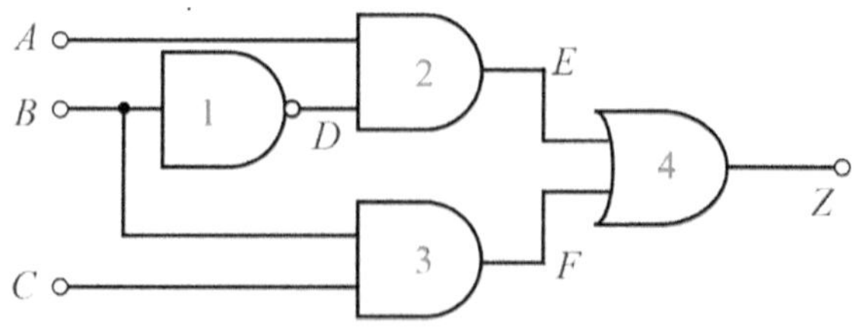

Fig. 7.5.22 Example of generating a test for an SC-type fault

Table 7.5.6 Tests for short circuits in the circuit of Fig. 7.5.22

SCNFL I		SCFL II		SCFL III	
Fault	Test	Fault	Test	Fault	Test
AB	t_3, t_4	AE	t_6	DE	t_0
AC	t_3, t_4, t_6	AZ	t_3, t_6	DZ	t_0, t_3
AD	t_0, t_6	BD	t_0, t_3, t_4, t_6	EZ	t_3
AF	t_3, t_4, t_6	BE	t_3, t_4, t_6	FZ	t_4
BC	t_6	BF	t_6		
CD	t_0, t_3	BZ	t_4, t_6		
CE	t_3, t_4	CF			
DF	t_3, t_4	CZ	t_4		
EF	t_0, t_3, t_4				

of the primary input C and is independent of B which determines D. t_3, however, is not a test for SC between D and Z because $Z = 1$ as a consequence of $F = 1$ while it is a consequence of $B = 1$ leading to a conflict that is described in Fig. 7.5.21. This test is marked with an asterisk.

By similar considerations we find that t_3 is a test for SC between E and Z, just as t_4 is a test for SC between F and Z.

In Table 7.5.6, there is one unfilled field. There is no test for SCFL-type fault between nodes C and F. The reason for this is their equivalence. This is checked by comparing the corresponding row in Table 7.5.5. For a given set of tests, the responses at nodes C and F are equal. In order to test this fault as well, it is necessary to introduce another test word. It should satisfy the criterion that C and F have different responses. Such is the $ABC =$ X01 test. This means that either t_1 or t_5 can be used.

A possible total test sequence will be $t_0t_1t_3t_4t_6$. It is easy to show that now all tests are individually essential. Summarizing the results of this example, we notice that by slightly expanding the set of tests that were intended for testing logic faults of the permanent state type, we can also obtain the I_{DDQ} test. We should not forget, however, that this is a completely different type of measurement. The same test signals are still used, but now the output voltage is not measured, but the current of the power supply. This would exhaust the description of this example.

In recent times, with the advancement of technology that tends to use smaller and smaller components and smaller distances between components on the chip, the problem of testing using I_{DDQ} becomes specific. There are two reasons for this, firstly, due to the enormous increase in the number of transistors on the chip, the DC supply current values even for CMOS technology become significant and leave the domain of currents of the order of 1 μA, and then, for transistors whose dimensions are deep in the submicron region, it is difficult to control the leakage currents. Nevertheless, it is still in use especially in mixed-signal circuits and, of course, the existing I_{DDQ}

test methods can be expected to adapt to new technology requirements. One such attempt is described in [26].

At the end of the discussion about the I_{DDQ} test, we should mention that testing based on the variable component of the supply voltage, the so-called v_{DDT} test [27]. It is expected that the importance of this test will increase with the increase in the total consumption of individual integrated circuits.

7.5.3 Testing Delay Faults

As part of the discussion on the testing of faults in digital circuits, we also mention the faults that are expressed in the change of the delay value of the logic gates, that is, about the *dynamic faults*. By *delay fault* we mean the deviation of the delay value from the set (nominal) value. Moreover, when the delay of the fault-free circuit, within the so-called *precise delay model*, is set as an interval, a delay fault arises when the value of the delay of the faulty circuit lies outside the limits of that interval, whether it is greater than the upper or less than the lower limit of the interval. The first discussion on the characterization and modeling of delay faults was given in [28].

The cause of a logic delay fault can be practically any physical fault. To see this, we will consider the circuit from Fig. 7.5.23 which illustrates the difference in the delay times of the rising and falling edges of the signal.

An error in the design or the tolerance of the mask manufacturing process or the tolerance in the exposure process (in alignment) which is manifested in the increase in the channel length of any N-channel transistor, increases the resistance in the

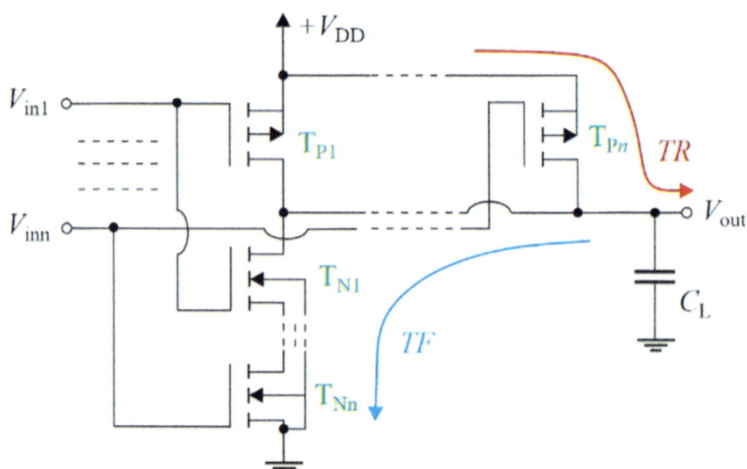

Fig. 7.5.23 CMOS **NAND** circuit and rising and falling edge delay

discharge branch of the load capacitor C_L and thus increases the duration of the falling edge of the pulse. Similarly, the reduction of the channel width of any P-channel transistor results in an increase in the charging time constant of the same capacitor, and thus in the duration of the rising edge of the pulse. Both of these phenomena result in a later reaching of the logic 0 threshold (or logic 1), which is interpreted as an increase in delay.

Short circuits would have similar effects. For example, shorting the source and drain of any N-channel transistor shortens the discharge path of the load capacitor and thus shortens the falling edge delay time. That short and open circuits can, among other things, have the effect of a fault and a change in signal delay can also be seen in [29].

The gate-to-channel breakdown discussed at the beginning of this chapter can also be mapped onto a fault effect expressed by a change in delay. If from the circuit of Fig. 7.1.3 the last inverter is dropped, the transfer characteristic would be given as V_{out} in Fig. 7.1.4 which, due to the decrease in slope, can be interpreted as an increase in delay.

Considering these examples, we come to the following conclusions. First, due to the different position of individual transistors on the path of charging and discharging the load capacitor, most likely the fault effect reflected in the delay change will also be different. This means that from the point of view of the delay fault, the input terminals of the combinational circuits become unequal. Second, the same physical fault is differently mapped into a delay fault of the falling and rising edge of the signal. Thus, the characterization of the delay fault at each (kth) terminal includes two values of the delay increment $(\mathrm{D}t_r)_k$ and $(\mathrm{D}t_f)_k$.

When considering testing from the standpoint of delay faults, however, those faults that cannot be otherwise tested will be considered here. If a physical fault is mapped simultaneously to both a static fault and a delay fault, it is likely that the static fault effect will be easier to test. On the other hand, when evaluating the dynamic properties of a system in which delay faults appear, it remains important to model and simulate the delay fault, despite the fact that there may be a static effect of the fault at the same time.

From the point of view of testing delay faults, as before, there is a difference in the approach to combinational and sequential circuits. We will handle here only the former.

In the case of combinational circuits, we distinguish the *delay fault of the logic gate* and *the delay fault of the propagation path* of the signal from the input to the output. Delay faults of logical gates are of a local character, and path delay faults can, due to the accumulation of delays on the path, lead to exceeding the limit imposed to the size of the delay. It is usual to test delay faults in combinational circuits on paths that exactly reflect the fault effects and can relate them to the correctness or malfunction of the circuit.

When it comes to generating a test for combination circuits the situation illustrated in Fig. 7.5.24 should be considered. At the moment t_0, the input word V_1 is applied to the input of the combinational circuit being tested. This initializes the circuit. When the signals in this circuit stabilize, at time instant t_1, a new test word V_2 is applied so

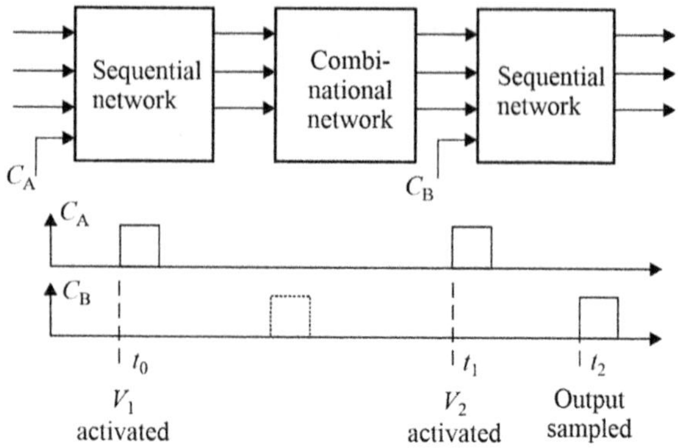

Fig. 7.5.24 Signal propagation through a combinational circuit built into a sequential system. C_A and C_B are the corresponding clock signals

that the rising or falling edge extends from the beginning of the path, along the path, to the end of the path under test. In other words, in order to detect a fault in a path, it is necessary to propagate a signal transition ($0 \rightarrow 1$ or $1 \rightarrow 0$) through that path. According to Fig. 7.5.24, a delay fault will occur if the fault effect is such that the transition does not arrive at the output terminal before the output signal is sampled (t_2).

For each signal path L that connects the primary input and the primary output of the combinational circuit, there are two corresponding transition paths. The rising path L_r is the signal propagation path that has a rising edge at the primary input, and the falling path L_f is the one that has a falling edge at the primary input. Of course, the direction of the transition (rising or falling) changes whenever the signal encounters an inverting logic gate. In the general case, when the direction of the transition is irrelevant, we will talk about the L_x path.

The time it takes for the transition to propagate along the path is called the *propagation delay* and is equal to the sum of all the delays the signal encounters on the path. In this context, we say that there is a path delay fault if the propagation delay is larger than the interval $Dt = t_2 - t_1$.

For a pair of excitation words $\{V_1, V_2\}$ we say that they make a robust test of the delay on the path L_x only if, when there is a fault on the path, at the time instant t_2 there is a difference between the expected and the measured state of the signal, independent of the delay size of the input terminals of the logic gates which (the terminals) are not on the path L_x. This considers the so-called initial delay model [29].

The importance of a robust test for delay faults can be seen on the basis of the following. A change (increase) in the delay value in circuits is often a result of component parameter variations caused by random fluctuations during the production process. Process variations can result in delay variations of multiple components in the circuit. It is therefore important to ensure that path delay faults are not masked by the delay characteristics of non-path elements.

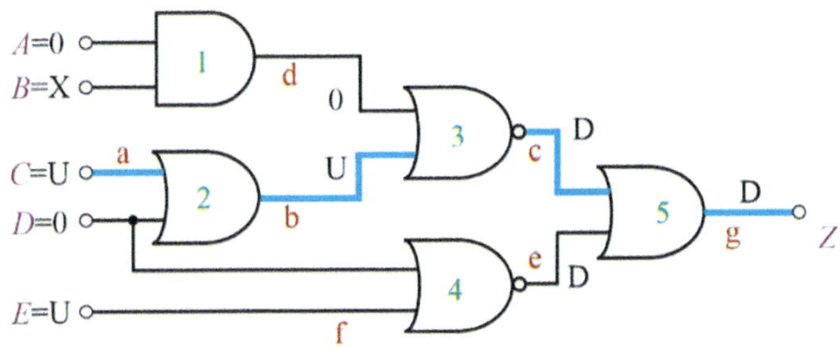

Fig. 7.5.25 Example of a path (a–b–c–g)

To that end, it is important to define the input terminals of the logic gate that is outside the sensed path. It is the terminal (a) of the logic gate (G) that is not on the sensed path if G is on that path. For the circuit of Fig. 7.5.25 the C–a–b–c–g–Z pathway was sensitized. The input terminal "d" of logic gate "3" is outside the sensed path. The same applies to input terminal "e" of logic gate "5," etc.

Before proceeding with test generation, it is necessary to consider the system of logic states that is needed to develop test generation procedures. The minimal set of states required for this purpose would be $A_5 = \{0, 1, X, U, D\}$, where states U (up) and D (down) have no duration. Such a system was used in the development of the test for the circuit of Fig. 7.5.25. Of course, it is possible to use all nine states used in the 9V-algorithm mentioned above.

Let us now consider test generation. To test the delay fault of the path C–a–b–c–g–Z in the circuit of Fig. 7.5.25 for rising excitation, it is first necessary to apply such excitation to input C. Then, since the **OR** circuit is excited, the input D should be 0 in order to open the sensed path. Furthermore, for the same reasons, input "d" of circuit "3" should be at zero, which gives two possible variants $AB = 0X$ or $AB = X0$. Finally, the sensitization of the path in branch "c" is realized by a logic 0 at the input terminal "e" of the gate "5," which, given that $D = 0$, requires $E = 1$ or at least $E = U$. The pair of test words is $\{V_1, V_2\} = \{0X\mathbf{0}01, 0X\mathbf{1}01\}$ or $\{V_1, V_2\} = \{X0\mathbf{0}01, X0\mathbf{1}01\}$. Therefore, similar to static fault testing, it is necessary that the logic states at the inputs of the logic gates located on the path being tested are such as to ensure signal propagation. For a robust test, it is necessary that the excitation at the primary input be U or D. The main difference compared to the algorithms we considered when generating the test for static faults refers to easier observability, given that the complete signal propagation path is known in advance.

For example, we will consider the problem of generating a test for a falling edge on the path A–a–b–c–t–Z of Fig. 7.5.26. The procedure goes by first setting $A = D$. In order to propagate this signal to the output, it is necessary to determine the signal states at the terminals of the logic gates that make up the path, namely "r," "g," and "q." It is necessary to set "r" to 1, "g" to 1, and "q" to 1. This would complete the first step of realizing the algorithm for generating the path fault test. It is fundamentally different from what we had with static faults. The rest comes down to *confirmation* (a term borrowed from the so-called PODEM algorithm for static test generation).

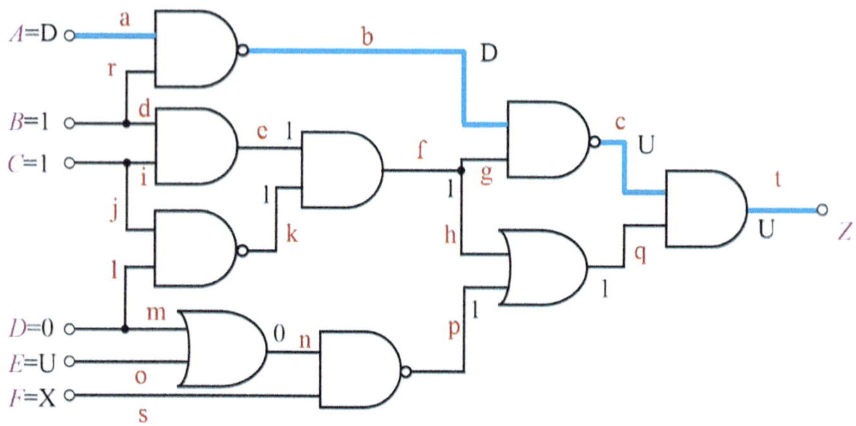

Fig. 7.5.26 Illustration of test generation for a delay fault

Therefore, in the second step, we choose the goal that has the highest priority, which is to confirm "q" is set to one. This is achieved by setting $D = 0$, $E = 0$, and $F = \mathrm{X}$. As a consequence of this choice we still have "m" set to 0, "k" to 1, "n" to 0, and "p" to 1.

The third step, second in priority, *validates* "g" which is set to 1. $B = 1$ is required. By simulation, we get that "e" is 1 and "f" is 1. Since the obtained result is without conflict, the test is generated. It reads $\{V_1, V_2\} = \{\mathbf{1}1100\mathrm{X}, \mathbf{0}1100\mathrm{X}\}$.

Analyzing the procedure just described on the example, several problems can be observed. Some of them will be discussed here. The first one refers to the testing of combinational circuits that also contain **EXOR** gates. As already mentioned, confirmation in this case is not unambiguous. Therefore, it is suggested that the **EXOR** circuit, just for this purpose, be modeled by the equivalent logic circuit of Fig. 7.5.27.

The other problem is related to testability. It is easy to see that with very complex circuits there will be conflicts that sometimes cannot be resolved, which means that not all paths are testable. In addition, here again, circuits containing reconvergence can cause problems. In the circuit of Fig. 7.5.28, for example, the actual delay on the path marked by the solid (blue) line, to the rising change of signal *A*, is 10 units. If the sampling moment is $t_2 = 12$, then $t_2 - t_1 = 2$, which means that all faults that increase the delay value by more than two units on the path are testable. However, as can be seen from the figure, due to the parasitic pulse, the delay faults whose delay

Fig. 7.5.27 $Z = \mathbf{EXOR}(A, B)$

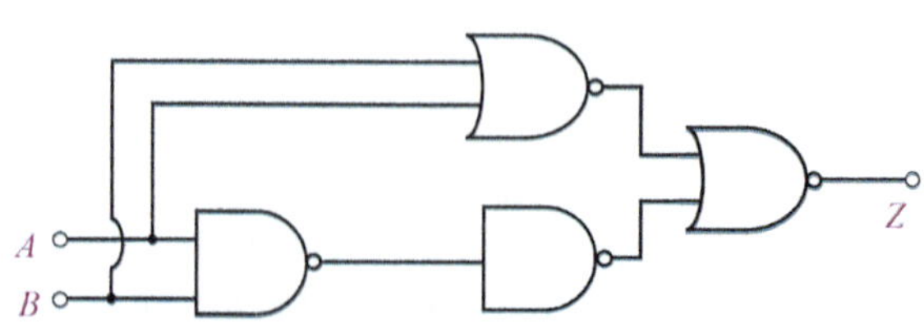

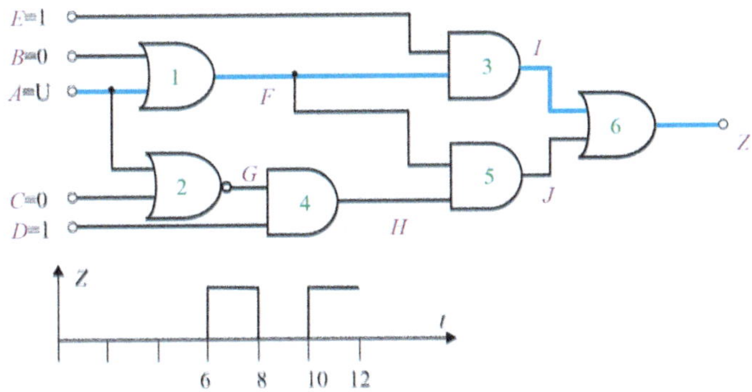

Fig. 7.5.28 Illustration of the impossibility of testing the delay faults. All gates have a delay of two units except gate "3" whose delay is 6

increment value is Dt = 4 or Dt = 5 will not be detected. Let us repeat, the reason for this is that two reconvergent paths are simultaneously initialized with a given signal.

Finally, one of the most complex problems we face when testing delay faults in combinational circuits is the number of paths. In the worst case, the number of paths in a combinational circuit is an exponential function of the number of nodes. Therefore, even in practical circuits, it becomes impossible to number all the paths and all the path faults to be tested. The similarity with static faults and their number is obvious. To overcome this problem some methods consider only subsets of paths such as the set of longest paths. Others form a subset by selecting those paths that contain at least one of the "highest" delays in the circuit. In modern circuits, however, by applying optimization from the point of view of delay, it is achieved that most paths tend to have similar nominal delay values. Therefore, the number of paths that need to be numbered is still large.

The path counting procedure should precede any path delay test generation procedure. Here we will describe one such procedure as given in [30]. It starts from the outputs to the inputs and each *branch of the node* (g) is assigned a number (label) that is equal to the number of paths from "g" to the primary outputs. An example of node branch numbering and assigned labels is given in Fig. 7.5.29a. The numbers in curly brackets are the ordinal numbers of the branches of the nodes that are numbered (arbitrarily) so that one goes from the input to the output and that the ith branch is excited by the jth, and that $j < i$. Behind them, outside the brackets, are assigned labels indicating the numbers of paths. As we will see shortly, the number of paths is calculated as the sum of the labels assigned to the primary inputs. In the example from Fig. 7.5.29 that is NP = 1 + 2 + 4 + 3 + 1 = 11.

The procedure for assigning labels to node branches is described by the algorithm in Fig. 7.5.30.

Since we have determined the total number of paths, it is of interest to establish the number of path delay faults detected by a single test signal (of course, composed of two test words). This would create conditions for evaluating the coverage of delay

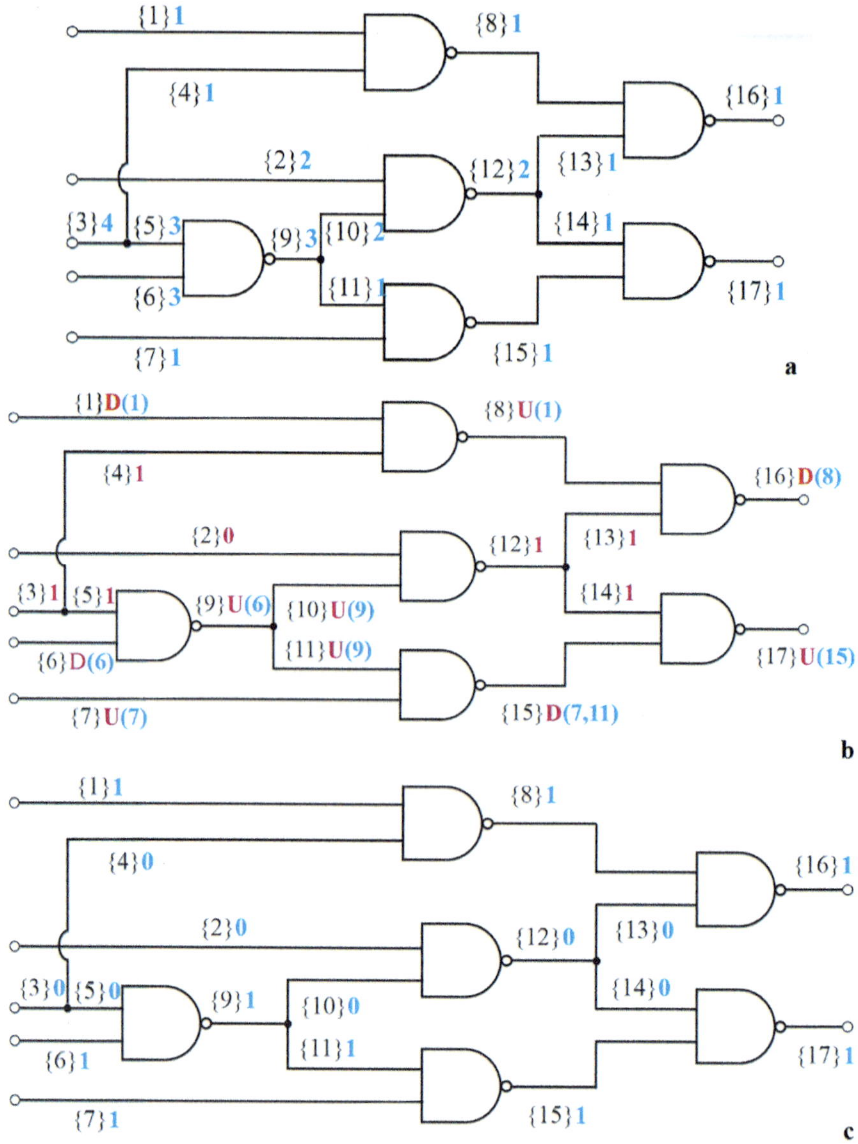

Fig. 7.5.29 Path delay testing. **a** Calculation of the number of paths, **b** response to the test signal (bold letters in red), and **c** paths sensitized by the test signal under (**b**)

faults with the existing test. For this purpose, it is necessary to carry out a simulation and to establish (by analyzing the results) by which paths the transition (excitation signal) is transmitted robustly from the input to the primary outputs. An example of the simulation results for the test $t_{\{1\}\{2\}\{3\}\{6\}\{7\}} = \text{D01DU}$ is given in Fig. 7.5.29b. The states in the nodes are marked in bold (red).

1. Assign the label "1" to each primary output. Set i=L, where L is the total number of branch nodes in the circuit. For the circuit of Fig. 7.5.29 we have L=17.

2. If branch i is the input terminal of element G whose output terminal is g, and the label of branch g is $N(g)$, label branch i with $N(g)$. For example, in Fig. 7.5.29a logic 1 is transferred to all input terminals of logic gates whose outputs are primary.

3. If branch i is at the same time a tree of the node whose branches are i_1, i_2, ..., i_k, which are already marked with $N(i_1)$, $N(i_2)$, ..., $N(i_k)$, the branch i should be marked with $N(i)=\Sigma^k N(i_j)$.

4. Put i-1→i and, if i>0, continue with step 2.

5. The number of paths is $\sum N(i_p)$, equal to the sum of labels of primary paths. For the circuit of Fig. 7.5.29 for primary inputs we have:

$$N(1)+N(2)+N(3)+N(4)+N(5) = 1+2+4+3+1=11$$

Fig. 7.5.30 Procedure for calculating the total number of signal propagation paths through the combinational circuit

By simulation, it was established that the signal D in the branch of the node {6} robustly extends to the branch {9} if the branch {5} is kept at logic 1. That is why the label (6) is added to branch {9}. Furthermore, the state change in branch {9} robustly propagates (by itself) to branches {10} and {11}. That is why the label (9) is assigned to these branches. Similarly, since the transition U extends robustly to the branch {15}, it bears the labels (7, 11).

For each branch, we write down a set of branches from which transitions are transferred robustly to it. Thus, for each branch, a set $R(i)$ is formed. To count the delay faults covered by a given pair of test words, we need to count the number of paths passing through i so that $R(i)$ is not an empty set. In Fig. 7.5.29c this number of paths is given for each branch when the test from Fig. 7.5.29b is applied. We notice that this test sensitized three path faults, one from the primary input {1}, the second from the primary input {6}, and the third from the primary input {7}.

Based on this, conditions are created for the application of the improved version of the procedure for generating the overall test of the given combinational circuit as shown in Fig. 7.5.17. The speeding up of this procedure, which is proposed in [31], is to observe not individual delay faults, but individual paths, of which, of course, there are significantly fewer.

This would exhaust the considerations related to testing delay faults in combinational circuits. The reader can find methods that are not based on path counting in [32–35].

References

1. Zwoliński M (2003) Digital system design with VHDL (2nd edn). Pearsons
2. Roth PJ (1980) Computer logic, testing and verification. Pitman, London
3. Abramovici M, Breuer M, Friedman AD (1990) Digital system testing and testable design. Computer Science Press, New York
4. Rice R (1982) Computer-aided design, testing, and packaging. IEEE Computer Society Press, Piscataway
5. Bleeker H, van den Eijnden P (1993) Boundary-scan test, a practical approach. Kluwer Academic Publishers, Dordrecht
6. Cortner JM (1987) Digital test engineering. Wiley, New York
7. Wilkins BR (1986) Testing digital circuits. Van Nostrand Reinhold (UK), Wokingham
8. Liu R-W (ed) (1991) Testing and diagnosis of analog circuits and systems. Van Nostrand Reinhold, New York
9. Williams TW (1986) VLSI testing. North-Holland, Amsterdam
10. Williams TW (1993) IEEE standard test access port and boundary-scan architecture. IEEE, New York
11. Williams TW (1995) P1149.4 mixed-signal test bus standard. Working group meetings, minutes, and associate materials. IEEE, New York
12. Williams TW (1995) Supplement to IEEE Std. 1149.1-1990, IEEE standard test access port and boundary-scan architecture. IEEE, New York
13. Williams TW (1995) P1149.4 mixed-signal test bus. IEEE Standard Project. IEEE, New York
14. Williams TW (2003) International technology roadmap for semiconductors. 2003 Edition modeling and simulation. https://www.semiconductors.org/resources/2003-international-technology-roadmap-for-semiconductors-itrs/
15. Williams TW (2023) Testing of electronics. https://www.jtag.com/testing-of-electronics-tutorial-jtag/
16. Roth PJ (1966) Diagnosis of automata failures: a calculus and a method. IBM J 278–291
17. Cha CW, Donath WE, Ozguner F (1978) 9-V Algorithm for test pattern generation of combinational digital circuits. IEEE Trans Comput C 27(3):193–200
18. Goel P (1981) An implicit enumeration algorithm to generate tests for combinational logic circuits. IEEE Trans Comput C 30(3):215–221
19. Fujiwara H, Shimono T (1983) On the acceleration of test generation algorithms. IEEE Trans Comput C 32(12):1137–1144
20. Iyer MA, Abramovici M (1996) FIRE: a fault-independent combinational redundancy identification algorithm. IEEE Trans Very Large Scale Integr (VLSI) Syst 4(2):295–301
21. Savir J (1998) Redundancy revisited. IEEE Trans Very Large Scale Integr (VLSI) Syst (6)4:620–624
22. Eichelberger EB, Bardell PH (1983) Random-pattern coverage enhancement and diagnostics for LSSD self-test. IBM J Res Dev 27:265–272
23. Soden JM (1996) I_{DDQ} testing: issues present and future. IEEE Des Test Comput 13(4):61–65
24. Isern E, Figueras J (1995) I_{DDQ} test and diagnosis of CMOS circuits. IEEE Des Test Comput 12(4):60–67
25. Cheng K-T (1996) Gate-level test generation for sequential circuits. ACM Trans Design Autom Electron Syst 1(4):405–442
26. Sachdev M (1997) Deep sub-micron I_{DDQ} testing: issues and solutions. In: Proceedings of the European design and test conference (ED&TC 97), Paris, France, pp 271–278
27. Cole EI et al (1997) Transient power supply voltage (v_{DDT}) analysis and detecting IC defects. In: Proceedings of the international test conference, Washington D.C., pp 23–31
28. Breuer MA (1974) The effects of races, delays and delay faults on test generation. IEEE Trans Comput 23(10):1078–1092
29. Litovski VB (2000) Electronic circuit design, simulation, optimization, testing, and physical design. Nova Jugoslavija, Vranje, Serbia (in Serbian)

30. Pomeranz I, Reddy SM (1992) An efficient non-enumerative method to estimate path delay fault coverage. In: Proceedings of the international conference on computer-aided design, ICCAD'92, pp 560–567
31. Pomeranz I, Reddy SM (1995) NEST: a nonenumerative test generation method for path delay faults in combinational circuits. IEEE Trans Comput-Aided Design Integr Circuits Syst 14(12):1505–1515
32. Litovski V, Maksimović D (1999) Tuning the logic simulator for timing analysis. Electron Lett 35(10):800–802
33. Maksimović D, Litovski V (2002) Logic simulation methods for longest path delay estimation. IEE Proc Comput Digit Tech 149(2):53–59
34. Padmanaban S, Tragoudas S (2003) An implicit path-delay fault diagnosis methodology. IEEE Trans CAD Integr Circuits Syst 22(10):1399–1408
35. Soković M, Litovski V, Zwoliński M (2008) New concepts of worst case delay and yield estimation in asynchronous VLSI circuits. Microelectron Reliab 49(2):186–198

7.6 Mixed Signal Testing

Abstract At the beginnings of the modern era of integrated electronics, due to the potential interference of the signals of the digital and analog part, separate designs were in use. In many cases that was a consequence of the differences in the frequency bands used for analog and digital circuits. For example, audio signal produced by a microphone, when converted into a digital one, will need much higher sampling frequency. The sampling-related noise, if both parts are on the same silicon die and not protective measures are undertaken, may distort the analog signal due to intermodulation. On the other side, the amplitudes in analog circuits usually span over several decades (sometimes even more than six) while the amplitude of the digital signal is fixed (and may be set to an arbitrarily low value exceeding the noise). In modern CMOS integrated VLSI circuits, one uses guard rings (as described in LNAE_Book 1) to prevent interference while increasing some distances also helps. So, mixed-signal integrated circuits are nowadays a reality allowing realization of complex signal processing systems on chip. As it happens, however, in this new technology, new difficulties and challenges are imposed to testing and diagnosis. This chapter is intended to slightly open the curtain and to allow the reader to become aware of the complexity of the subject of testing and to learn the main solutions.

7.6.1 Introduction

There are electronic circuits that, by their very nature, contain both digital and analog parts that are connected to each other via A/D or D/A converters. Often, however, the coupling is such that it is impossible to characterize the digital and analog parts independently of each other. As an example of such a circuit, consider Fig. 7.6.1. It is the so-called sigma-delta modulator or analog-to-digital converter. It is also an example of the fact that the converters themselves can be treated as mixed-signal circuits. It contains a number of elements whose nature is different, that is, whose models are expressed in different ways. We recognize electrical elements that can be described using one or more Kirchhoff equations and for which electrical signals are valid, then digital elements that can be described by logic functions and are described

V. B. Litovski, *Lecture Notes in Analog Electronics*, Lecture Notes in Electrical Engineering 1265, https://doi.org/10.1007/978-981-97-8257-4_6

discretely by logic signals, and finally, "pulse elements" whose function is given by a special description containing Kirchhoff's equations and/or logical functions that are applied only if some special conditions are met. An example of such a pulse element is a switch that is an analog element (short circuit or open circuit) but is controlled by a digital signal. A similar example is a comparator that compares analog voltage levels and generates an output signal expressed in the form of a logic state. Modern electronic circuits, however, in addition to these contain other types of signals related to integrated and discrete subsystems that are often encountered in electronics. Thus, we have microelectromechanical elements that are described by quantities such as force, pressure, force coupling, etc., and also other elements such as optoelectronic, magnetic, and others. All these elements are described by analog quantities, which, however, by their nature, and thus by their absolute values, are often found in completely different value ranges than the electric ones.

The architecture of modern application-specific integrated circuits (ASIC) is considered to be similar to that shown in Fig. 7.6.2 [1]. As many as 20% of modern ASIC circuits have mixed-signal functions and this percentage is constantly increasing [2] since integration encompasses more and more electronic applications [3] including space [4] and RF systems [5].

Testing the analog part of the ASIC becomes one of the most expensive activities in the ASIC circuit manufacturing process. This is illustrated in Fig. 7.6.3 where a vision is given of the ratio of the price of the production itself, the price of testing the digital part, and the price of testing the analog part of the ASIC circuit. The main reasons for such estimates lie in the following facts:

- lack of tools for automating test generation for the analog part of the circuit and
- difficulties in defining the test for non-electric stimuli such as temperature, illumination, and pressure.

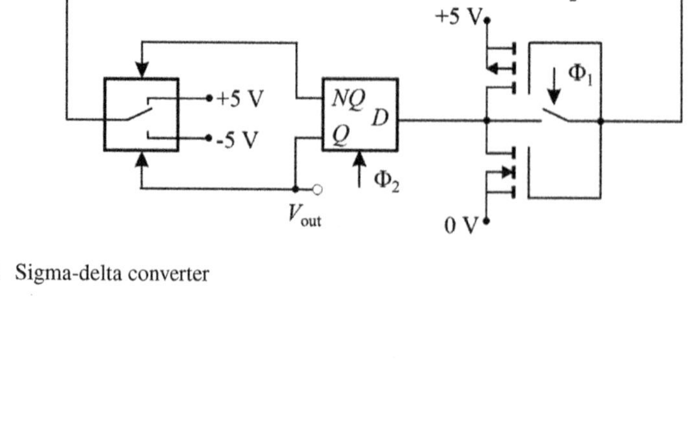

Fig. 7.6.1 Sigma-delta converter

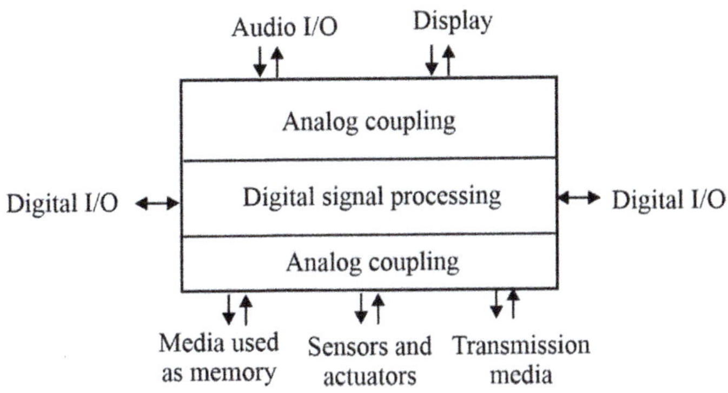

Fig. 7.6.2 Architecture of a modern ASIC circuit

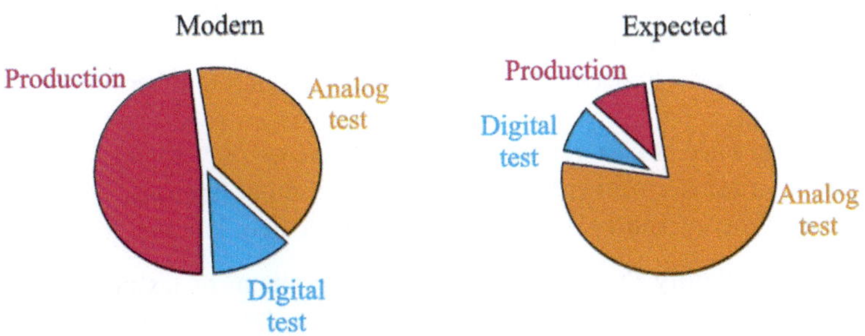

Fig. 7.6.3 State and prediction of the relative cost of production and testing of ASIC circuits

The differences in testing the analog and digital parts of the ASIC circuit can be seen even more clearly based on Table 7.6.1.

Table 7.6.1 Comparison of some aspects of testing the analog and digital parts of mixed-signal ASICs

	Digital part	Analog part
Test signals	Unique type	Several different types
Response analysis	Direct	It is usually necessary to interpret the results with the help of post-processing
Measurements	Low precision	High precision
Fault effects	Usually, catastrophic	Both catastrophic and parametric
Response of the fault-free circuit	Binary word	Range of tolerances in a multidimensional space
Reconfiguration and installation of test outputs	Easy	Difficult

In addition to the above, the most important characteristics of testing the digital part of the ASIC circuit would be the following:

- The signal usually has a unique propagation direction.
- The function of the circuit is expressed easily, with a state table.
- A structural test is usually applied.
- There are already standardized configurations for design for testability.
- Synthesis for testability is also possible.

With the analog part, the situation would be as follows:

- The signal does not propagate in one direction.
- There are complex couplings of subsystems that are particularly pronounced under the conditions of different excitation signals.
- Testing relies mostly on functionality, that is, on given specifications.
- Sensitivity to parameter changes makes it difficult to reach a conclusion about the correctness of the circuit even with a successful test.
- It is also necessary to design for testability with a small number of standardized solutions, but
- There is also a need for designing for measurement, i.e., developing structures that are easier to measure either by introducing new output terminals or by enabling isolation of the measuring system and the ASIC to avoid mutual loading.

In order to simplify the understanding of the problem of testing such circuits and systems, we will consider Fig. 7.6.4. The system depicted here starts from the configuration given in Fig. 7.6.2.

From a testing point of view, however, some specific problems arise that cannot be seen at first glance. Namely, one gets the impression that both the digital and analog parts are easily accessible for excitation with test signals and for response measurement, given that each subsystem has its own input and output set of terminals. We do not come to that conclusion if we look at the circuit from Fig. 7.6.1, however. It is easy to see that access to some subassemblies is obstructed. For example, the

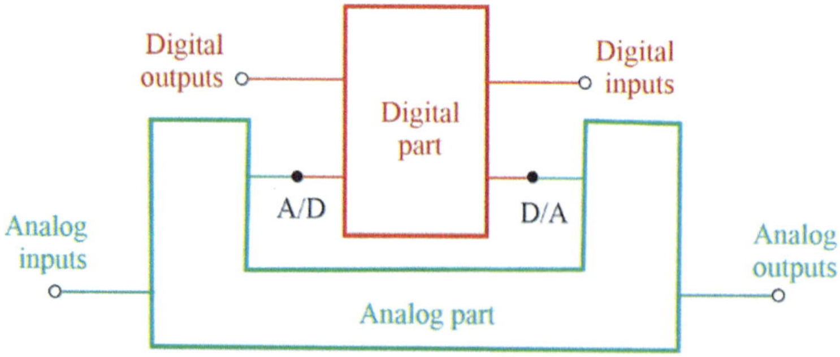

Fig. 7.6.4 Functional representation of a mixed-signal circuit

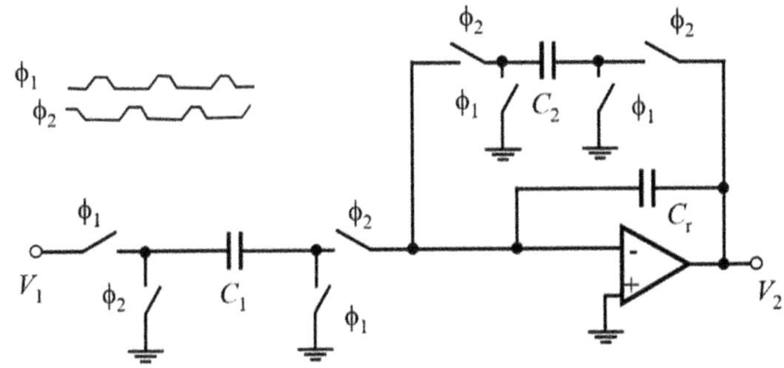

Fig. 7.6.5 A simple switched capacitor filter with the waveforms controlling the switches indicated

operational amplifier is practically inaccessible for testing and the question is whether there are test signals that would be used to test structural faults in the operational amplifier.

Therefore, designing for testability is crucial here as well. This topic will be discussed later, but here we will give some basic insights that point to possible solutions.

Figure 7.6.5 shows a switched capacitor filter. During normal operation of the filter, switches with an even index and switches with an odd index open and close alternately according to the control signals shown in the figure. This creates the function of a filter that transmits samples of the input signal V_1. If we were to test the operational amplifier, the switches should be placed as shown in Fig. 7.6.6. There is a direct connection from the input to the operational amplifier and from the operational amplifier to the output. Therefore, controllability and observability are realized. It can be seen, however, that this position of the switch cannot be achieved with the existing signal waveforms, given that the switches φ_1 and φ_2 at the very input of the circuit, have the opposite position from the rest of the switches controlled by the same signals. This indicates the need to design for testability, where, during testing, in a proper way, the circuit from Fig. 7.6.6 is created. In such a case, $A_r = -C_1/(C_2+C_4)$ is obtained for the fault-free gain of the circuit.

7.6.2 Testing of D/A and A/D Converters

Signal conversion at the point of D/A and A/D coupling is most often achieved by special circuits called digital–analog and analog–digital converters. Testing such circuits is a special task due to their specific mode of operation. In this section, we will specifically address both the principles of operation of these circuits and general testing methods. Namely, it should be kept in mind that there are countless ways of realizing such circuits (a simple A/D converter circuit is shown in Fig. 7.6.1), and an exhaustive analysis of testing is practically impossible. The interested reader is advised to consult some or all of the following references [6–10].

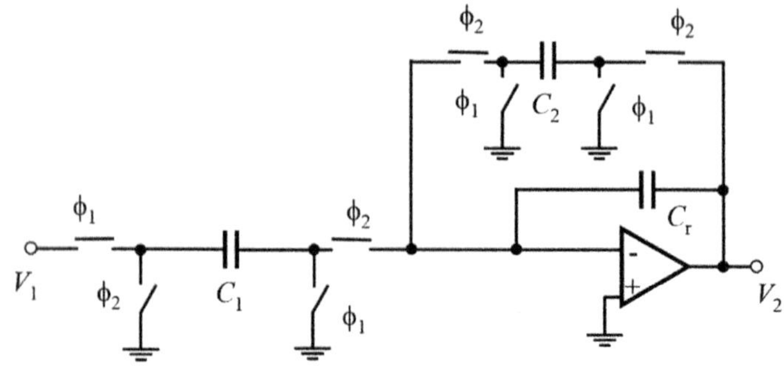

Fig. 7.6.6 Required switch positions in the filter of Fig. 7.6.5 to test the operational amplifier

7.6.2.1 Testing a D/A Converter

A parallel binary signal generated from a digital subsystem is fed to the input of the D/A converter. The task of the circuit is to convert it into an equivalent analog signal with the appropriate normalization performed on the basis of the value of the preselected reference voltage V_R. A voltage is generated at the output of most D/A converters, and rarely a current.

The input digital word will be labeled by the sequence $\mathbf{b} = \{b_0, b_1, b_2, \ldots, b_{N-2}, b_{N-1}\}$. b_0 is called the digit of the highest weight and is marked with most significant bit (MSB), and b_{N-1} is called the digit of the least weight and is marked with least significant bit (LSB).

The output voltage can be expressed as

$$v_{out} = K \cdot V_R \cdot D, \tag{7.6.1}$$

where K is the scale coefficient, and

$$D = \frac{b_0}{2^1} + \frac{b_1}{2^2} + \frac{b_2}{2^3} + \cdots + \frac{b_{N-1}}{2^N}. \tag{7.6.2}$$

N is the total number of bits in **b**.

The characteristics of D/A converters can be grouped into static and dynamic. Here we will first consider the static ones. The resolution of the D/A converter is expressed as N bits. Figure 7.6.7 shows, with a solid (blue) line, the transfer characteristic of a 3-bit ($N = 3$) D/A converter [11] It represents the dependence of the normalized value of the output voltage on the digital word. Analog levels are separated by one LSB. In the analog domain, the numerical value of the LSB is

$$\text{LSB} = V_R/2^N. \tag{7.6.3}$$

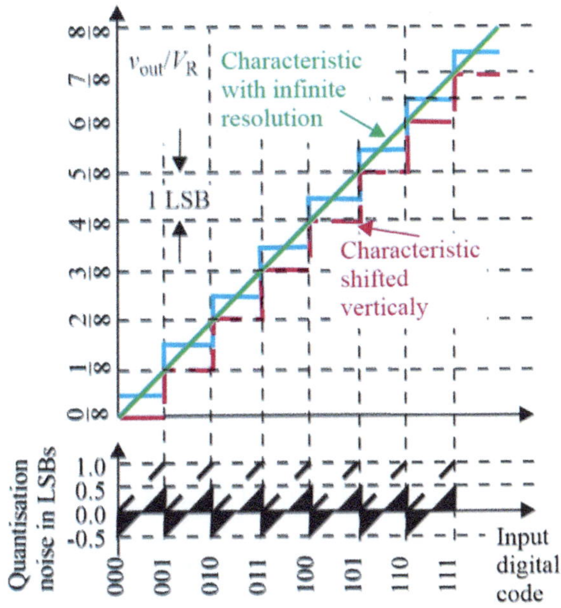

Fig. 7.6.7 Static characteristic of the D/A converter

As the digital word increases by a bit, the analog value increases by LSB. It is a matter of agreement which values will be used to start. Thus, the solid (blue) line shows the static characteristic starting with an analog value of $0.0625 \cdot V_R$ (V). The dashed line, however, shows the transfer characteristic starting with a zero-analog value.

Since the resolution of the D/A converter is finite, the highest analog value does not reach V_R. This parameter is known as the full value of the D/A converter and is denoted by FS (full scale), and is calculated from

$$\mathrm{FS} = V_R - \mathrm{LSB} = V_R \cdot \left(1 - 1/2^N\right). \tag{7.6.4}$$

Quantization noise is the uncertainty that arises during the digitization of an analog signal. In order to understand this quantity, Fig. 7.6.7 also shows the characteristic with infinite resolution (strait line in green). It is the line that occurs when the number of bits N tends to infinity. Quantization noise can now be defined as the difference between an analog response with infinite resolution and an analog response with a finite number of bits. When this definition is applied to a three-bit D/A converter, the bottom part of Fig. 7.6.7 arises. It can be seen that the quantization noise has a saw tooth shape and that its value is between $+0.5 \cdot \mathrm{LSB}$ and $-0.5 \cdot \mathrm{LSB}$, i.e., $1 \cdot \mathrm{LSB}$ from peak to peak. When this is expressed in volts, we get $V_R/2^{N+1}$. This is an important feature of the D/A converter because it determines the effort we need to make during the conversion. Namely, any uncertainty in the value that is less than $0.5 \cdot \mathrm{LSB}$ is masked by this noise and can only be reduced by increasing the resolution.

The dynamic range of the D/A converter is denoted by dynamic range (DR). It represents the quotient of the reference voltage and the LSB

$$\mathrm{DR} = \frac{V_\mathrm{R}}{\mathrm{LSB}} = \frac{V_\mathrm{R}}{V_\mathrm{R}/2^N} = 2^N \tag{7.6.5a}$$

or expressed in decibels

$$\mathrm{DR(dB)} = 6.02 \cdot N \text{ dB}. \tag{7.6.5b}$$

The signal-to-noise ratio is denoted by SNR and is calculated as the quotient of the reference voltage and the effective (RMS) value of the quantization noise. For the effective value of the quantization noise, by definition, we have

$$v_n = \sqrt{\frac{1}{T}\int_0^T \mathrm{LSB}^2 \cdot \left(\frac{t}{T} - 0.5\right)^2 \cdot dt} = \frac{\mathrm{LSB}}{\sqrt{12}} = \frac{V_\mathrm{R}}{2^N\sqrt{12}}. \tag{7.6.6}$$

Now we have

$$\mathrm{SNR} = \frac{v_{\mathrm{out\ rms}}}{V_\mathrm{R}/\left(2^N\sqrt{12}\right)}. \tag{7.6.7}$$

In the case of a sinusoidal voltage, the effective value is $V_\mathrm{R}/\left(2\sqrt{2}\right)$, so for the highest value of the signal-to-noise ratio one gets

$$\mathrm{SNR_{max}} = 2^{N-1}\sqrt{6} \tag{7.6.8a}$$

or in decibels

$$\mathrm{SNR_{max}}\ \mathrm{(dB)} = 6.02 \cdot N \text{ dB} + 1.76 \text{ dB}. \tag{7.6.8b}$$

The effective number of bits is denoted by effective number of bits (ENOB) and is calculated as

$$\mathrm{ENOB} = (\mathrm{SNR} - 1.76)/6.02. \tag{7.6.9}$$

Thus, for a 10-bit D/A converter, the signal should be about 62 dB above the noise, regardless of its origin.

Any deviation of the characteristics of the D/A converter from that shown in Fig. 7.6.7 is characterized as a static conversion error. There are the following static conversion errors: offset error, gain error, integral nonlinearity, differential nonlinearity, and monotonicity. The offset error represents a constant distance between the real characteristic with finite resolution and the ideal characteristic with finite resolution, which is illustrated in Fig. 7.6.8a. The gain error, as shown in Fig. 7.6.8b, represents the difference between the actual characteristic with finite resolution and the ideal characteristic with infinite resolution measured at the rightmost vertical jump. The gain error is proportional to the magnitude of the output voltage of the D/A converter.

Integral nonlinearity INL is the largest difference between the real characteristic with finite resolution and the ideal characteristic with finite resolution measured vertically. It is expressed through the number of LSB or as a percentage of V_R. For a 3-bit D/A converter it is shown in Fig. 7.6.9. INL can be positive and negative as shown in the figure.

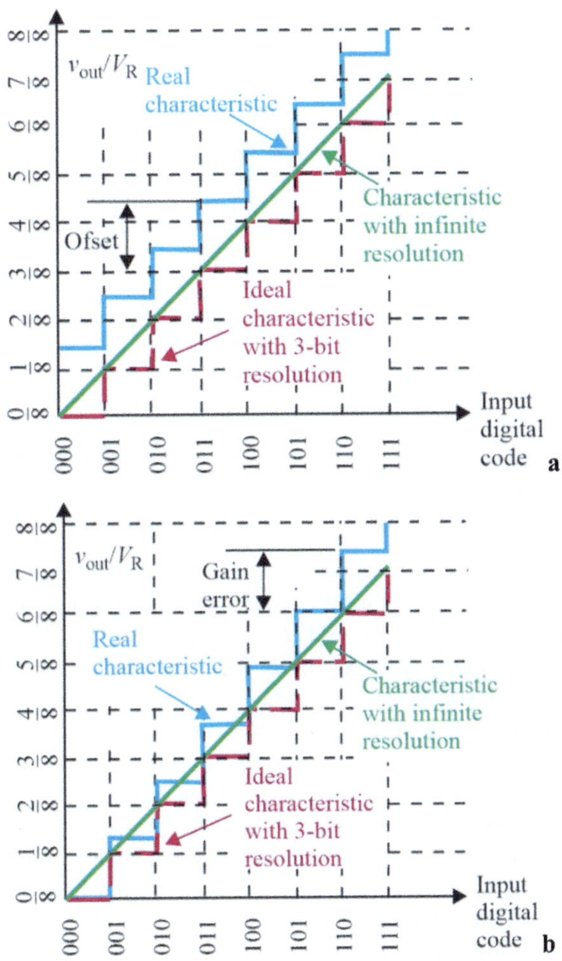

Fig. 7.6.8 **a** Offset error and **b** gain error

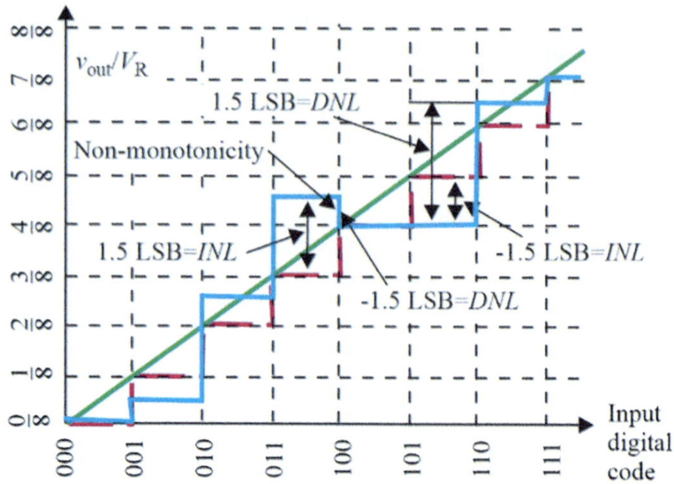

Fig. 7.6.9 Illustration of INL, DNL and non-monotonicity

Differential nonlinearity, denoted by DNL, is defined as the deviation of the difference between adjacent analog levels for each bit separately from the ideal jump value. If we denote the actual increase by V_{cx}, and the ideal by V_s, then the differential nonlinearity can be expressed as

$$\mathrm{DNL} = \frac{V_{cx} - V_s}{V_s} \cdot 100\% \tag{7.6.10a}$$

or

$$\mathrm{DNL} = \left(\frac{V_{cx}}{V_s} - 1\right)[\mathrm{LSB}]. \tag{7.6.10b}$$

In that, it is $V_s = V_R/2^N$.

The differential nonlinearity is illustrated in Fig. 7.6.9. From that figure we can see that the highest value of DNL is obtained at the transition from 101 to 110. It amounts to $\mathrm{DNL}_{max} = 1.5 \cdot \mathrm{LSB}$. At the same time, $V_{cx}/V_s = 2.5$. On the other hand, the smallest DNL value occurs at the transition from 011 to 100. $\mathrm{DNL}_{min} = -0.5$ LSB is obtained. At the same time, $V_{cx}/V_s = -0.5$. It is interesting that the mentioned definition is a little deficient in the sense that it cannot characterize situations when the increment of the analog value is equal to zero. Thus, at the transition from 100 to 101, the value of the actual response does not change, so $V_{cx} = 0$, but it is clear that DNL $= -1$ LSB.

Monotonicity is the property that the analog value never decreases if the digital value is constantly increasing. In Fig. 7.6.9 the property of non-monotonicity is observed at the transition from 011 to 100. Of course, if the D/A converter is non-monotonic then its DNL is definitely noticeable.

For the characterization of the D/A converter, its dynamic characteristics are important. The most important dynamic feature is its conversion rate. The conversion rate is the time required to establish a DC level at the output from the moment of creation of the excitation digital word. Depending on the type of D/A converter, the rate ranges from milliseconds to nanoseconds. The cause of the delay is found in the capacitances of the transistors and connections in the circuit, as well as in the time-domain properties of operational amplifiers that are often used in D/A converters. That is why the analog response to the step function (from a digital word to another digital word) is usually monotonic (has no overshoot).

The goal of D/A converter testing is to verify the static and dynamic characteristics. One method for checking static characteristics is the so-called input–output test. For this purpose, an accurate A/D converter is first required. The test configuration is shown in Fig. 7.6.10. In this system, a digital input word is incremented from 00…0 to 11…1 and drives the D/A converter under test and the subtraction circuit simultaneously. The resulting voltage V_{out}, of the converter under test, is now fed into the A/D converter which generates a new digital word which also drives the subtraction circuit. As a result, a digital error is obtained at the output. The key is that the A/D converter has higher resolution (at least by one bit) than the D/A converter. In that way a potential error of the A/D converter will not affect the testing result. A/D converters with two redundant bits are usually used. A correct A/D converter generates 00…0 at the output. INL will be expressed as 1 in the appropriate place. If a 1 appears at the output in the Nth bit, then the INL is greater than ±0.5 LSB. DNL will be expressed as the change in each pair of successive responses at the output of the subtraction circuit. Additional bits of the A/D converter can be used to detect errors smaller than ±0.5 LSB.

Alternatively, spectrum analysis can be used to test A/D converters. This concept is depicted in Fig. 7.6.11. The system is excited by a digital signal generator that generates a digitally coded waveform of the test signal, for example, a sinusoid. It is important that the signal has a significant fundamental frequency value that is at least $6 \cdot N$ dB above the harmonics, where N is the resolution of the D/A converter. It is also important that the excitation sequence is of sufficient length to establish an analog signal at the output of the D/A converter under test. The output of the D/A converter is fed to a spectrum analyzer which actually measures the harmonics of the output voltage of the D/A converter. If the THD (Total Harmonic Distortion) is $6 \cdot N$ dB less than the amplitude of the fundamental harmonic, then it can be claimed that both INL and DNL of the D/A converter are less than ±0.5 LSB.

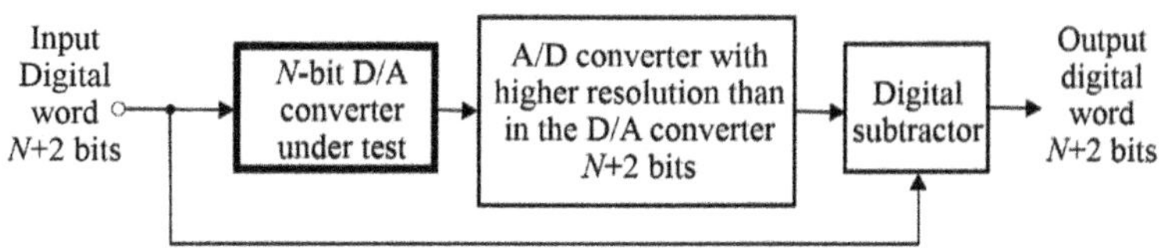

Fig. 7.6.10 Test setup for a D/A converter

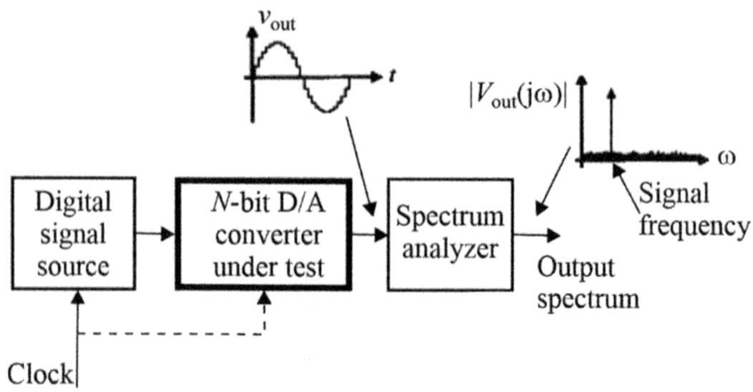

Fig. 7.6.11 Testing the D/A converter using a spectrum analyzer

In Fig. 7.6.11 a very important property of the spectral test is given. Namely, if the noise of the voltage source V_R is insufficiently small, since it also appears at the output, it can limit the dynamic range of the converter.

The test system of Fig. 7.6.11 can also be used for the dynamic test of the D/A converter. This can be achieved if the period of the excitation signal is gradually reduced, that is, if its frequency is increased. Due to the frequency dependence of the response, i.e., due to the limited operating speed of the converter, the noise level will gradually increase in relation to the main harmonic. The SNR that will be determined by this measurement can be used to calculate the effective number of bits ENOB.

7.6.2.2 Testing A/D Converters

Figure 7.6.12 represents a block diagram of a general A/D converter. A prefilter called an anti-aliasing filter (AA) is required. This prevents the overlapping of the high-frequency spectrum of the output signal with the basic frequency band of the A/D converter. Often the AA filter is realized by the very frequency characteristic of the A/D converter. The AA filter is followed by an S/H (from sample and hold) sampling circuit that keeps the input analog signal of the A/D converter constant while it is being converted into an output digital code. This time interval is called the *conversion time* of the A/D converter. The conversion is realized based on the *quantization step*. By its very nature, the quantizer separates the reference voltage into segments or sub-bands. Usually there are 2^N sub-bands where N is the number of bits in the output digital code. A quantization step that amounts to

$$QS = FS/2^N \tag{7.6.11}$$

finds the sub-band corresponding to the sampled analog signal. Knowing this subrange enables the digital processor to encode the corresponding digital bits, which

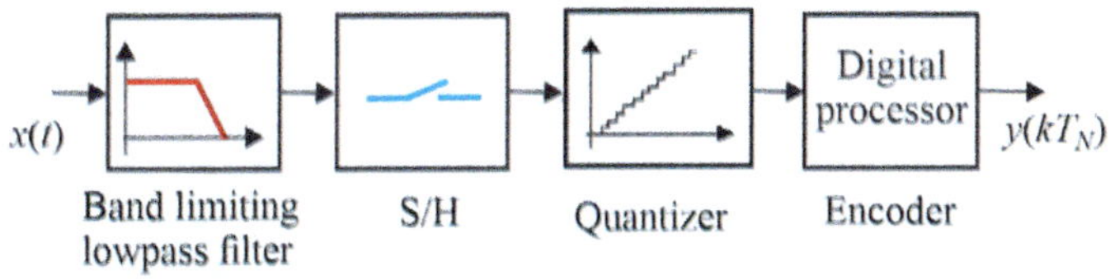

Fig. 7.6.12 General schematic of a D/A converter

is the function of the encoder. Overall, during the conversion time, the sampled analog signal is converted into an equivalent digital output code.

It is important to understand the frequency range of the A/D converter of Fig. 7.6.12. Suppose that the analog signal being processed has a frequency spectrum as in Fig. 7.6.13a. Let us also assume that f_B is the highest frequency in the spectrum of the signal of interest. When the analog signal is sampled, with frequency f_s, a new signal is obtained whose spectrum is shown in Fig. 7.6.13b. The spectrum of the input signal is repeated at the sampling frequency and its multiples. If the spectrum of the input signal (f_B) were wider than $f_s/2$, the spectra would begin to overlap as in Fig. 7.6.13c. In such a case, it is not possible to recognize the input signal. This aspect of the conversion is expressed by the concept of the Nyquist frequency or rate, whereby the sampling frequency must be at least twice the spectral band of the signal in order for the signal to be recognized in the samples. Therefore, if the sampling frequency is selected in advance, it is necessary to apply an AA filter before starting the processing, which ensures that all the frequencies of the converted signal are below $f_s/2$. It is illustrated in Fig. 7.6.13d.

The static characteristic of the A/D converter, for a three-bit converter, is shown in Fig. 7.6.14. In this particular case, the input is shifted so that ideally a step change occurs at the output every $0.5 \cdot (2 \cdot i{-}1) \cdot$ LSB, where for an N-bit converter, $i = 1, 2, \ldots, N$.

Fig. 7.6.13 Illustration of band overlap and AA filter effect. **a** Spectrum of the input signal, **b** spectrum of the output signal for $f_B < f_s/2$, **c** spectrum of the output signal for $f_B > f_s/2$, and **d** effect of the AA filter

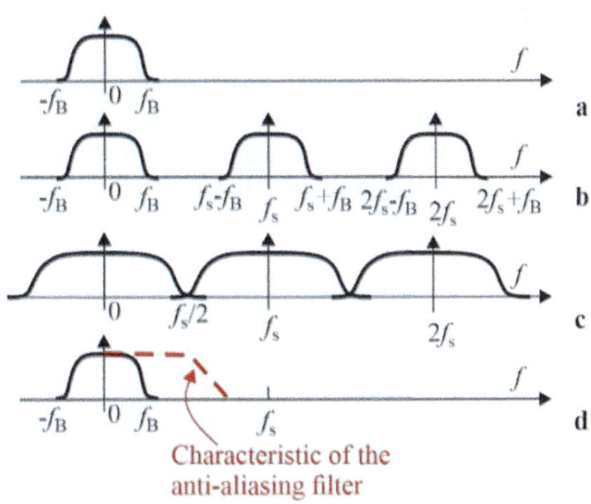

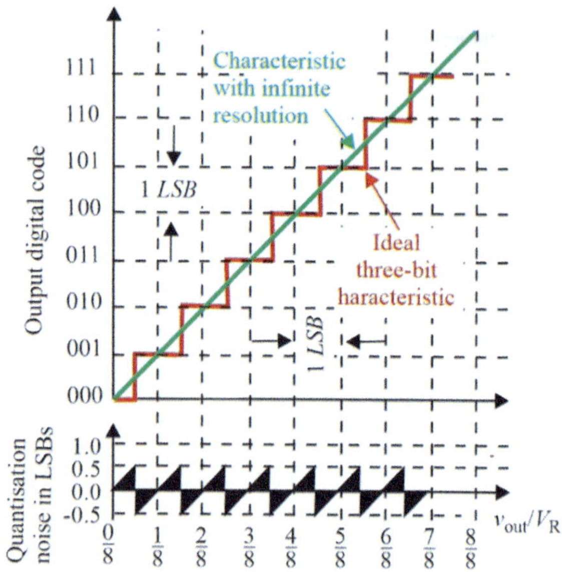

Fig. 7.6.14 Ideal output characteristic of a three-bit A/D converter

Below the transfer characteristics of the three-bit A/D converter, in Fig. 7.6.14, the quantization noise as a function of the input analog signal is shown. Quantization noise now represents the difference between a transfer characteristic with infinite resolution and an ideal transfer characteristic. An ideal transfer characteristic has a quantization noise that lies within $\pm 0.5 \cdot \text{LSB}$.

The definitions of dynamic range, signal-to-noise ratio, and effective bit rate are the same as those given for the D/A converter. There they were related to an analog variable, and now they are related to a digital word at the output.

The resolution of an A/D converter is the smallest analog change that the A/D converter can recognize. Resolution can be expressed in percentage of full scale (V_R), but it is more common to express it in the number of bits, N, where the converter has 2^N possible states.

The basic characteristics that define the static properties of A/D converters are offset error, gain error, integral nonlinearity, and differential nonlinearity. To determine the offset error (in absence of other errors) we move the characteristic with infinite resolution horizontally until we obtain that the quantization noise is symmetric when defined along that line. The horizontal distance between the new line with infinite resolution and the one passing through the origin is the offset error. An illustration is given in Fig. 7.6.15a.

The gain error is the difference between the actual characteristic and the characteristic with infinite resolution and is proportional to the input voltage. It can be considered as the change in the slope of the infinite gain characteristic whose value is greater or less than 1. The gain error is illustrated in Fig. 7.6.15b. Similar to the D/A converter, it can be measured as the horizontal difference between the actual and ideal transfer characteristics at the level of the highest digital value, i.e., between 110 and 111.

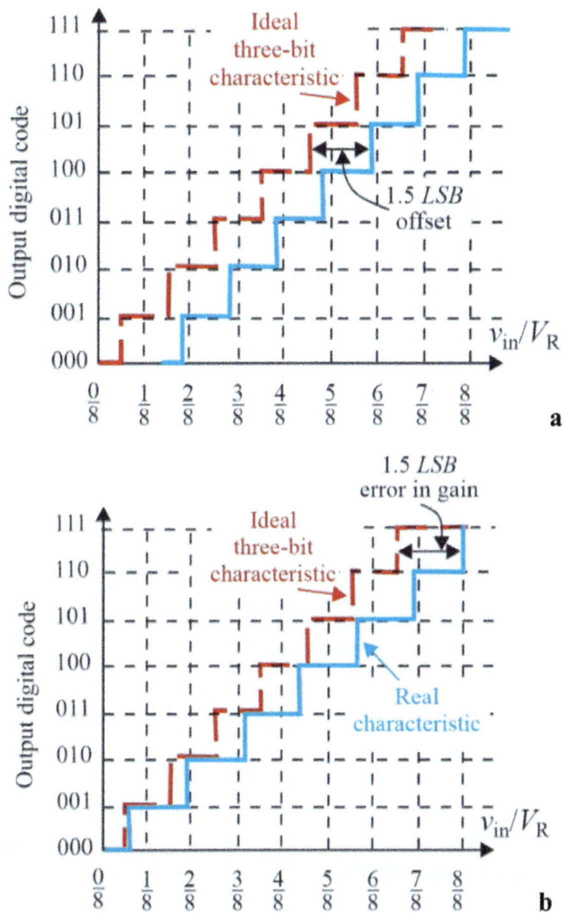

Fig. 7.6.15 **a** Example of offset error in a three-bit A/D converter and **b** example of gain error in a three-bit A/D converter

The integral nonlinearity (INL) of an A/D converter is defined as the difference between the ideal transfer characteristic and the actual transfer characteristic measured vertically as a percentage of LSB. With this definition, we get only integer values (since the output digital code is represented by discrete values). This will not be a problem if the resolution increases and the LSB becomes a small value.

The differential nonlinearity (DNL) of an A/D converter is defined as a measure of the distance between adjacent codes measured at each vertical step in percentage of LSB. The differential nonlinearity can be expressed as

$$\mathrm{DNL} = (D_{cx} - 1) \cdot \mathrm{LSB} \tag{7.6.12}$$

where D_{cx} is the actual vertical step size. Illustration of DNL for a three-bit A/D converter is given in Fig. 7.6.16. It can be seen that the maximum and minimum

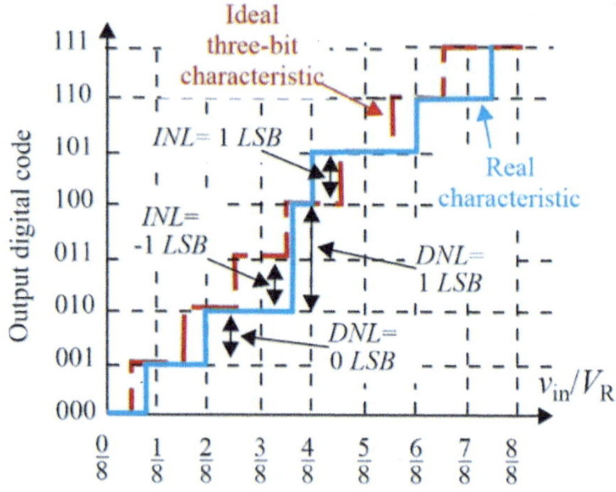

Fig. 7.6.16 Examples for INL and DNL for a three-bit A/D converter

values of INL are +1 LSB and −1 LSB, respectively. The highest and lowest values of DNL are +1 LSB and 0 LSB, respectively.

Non-monotonicity in the A/D converter occurs when the vertical jump is negative. Non-monotonicity can only be detected by DNL. If the vertical jump is less than 0 LSB for the A/D converter we say it is non-monotonic. The characteristic shown in Fig. 7.6.17 is an example of a non-monotonic A/D converter. Non-monotonicity is most often manifested at the bit with the highest weight, i.e., at the transition from 01111... to 10000...

The dynamic characteristics of the A/D converter are determined by the same causes as those of the D/A converter. In addition to the capacitances that we find everywhere, in all A/D converters we find at least one comparator whose static and dynamic characteristics affect the properties of the A/D converter.

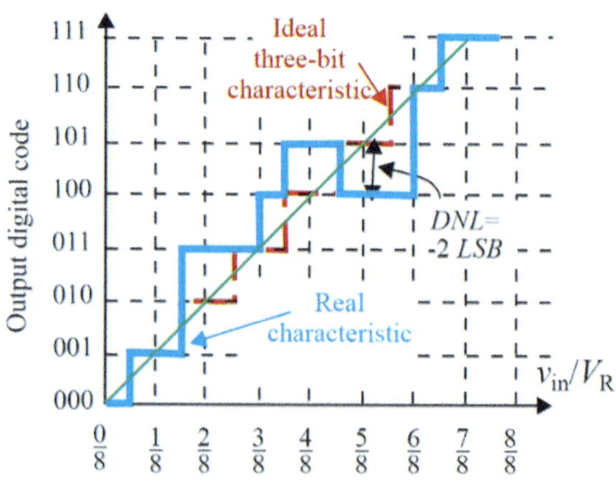

Fig. 7.6.17 An example of a non-monotonic three-bit A/D converter

In some cases, the A/D converter can use an operational amplifier, whereby again its characteristics are mapped into the dynamic properties of the A/D converter.

A/D converter testing has the same goals as D/A converter testing. These are the verification of the static and dynamic characteristics. The simplest test that is applied is the input–output test. The configuration of the testing system in this case is shown in Fig. 7.6.18. The input voltage varies linearly from zero to V_R and the resulting digital code is fed to a D/A converter that is more accurate than the A/D converter under test. Usually, it is enough that the accuracy of the D/A converter is two bits higher than the accuracy of the A/D converter. The obtained analog signal is further subtracted from the input signal and thus an error signal is formed which is displayed graphically at the output.

The dependence of the subtraction result on the magnitude of the input voltage should be equal to the quantization noise. When converting, the ideal difference of the amplitudes should be in the interval of ±0.5 LSB. Otherwise, this test can also be used to measure offset, gain error, INL, and DNL. Figure 7.6.19 shows one possible result of testing a four-bit A/D converter using an input–output test. INL and DNL errors are shown. If there was an error in the gain, it would appear as a constant increase or constant decrease in the saw tooth diagram when increasing v_{in}. The offset would be expressed as a constant shift above or below the 0 LSB line.

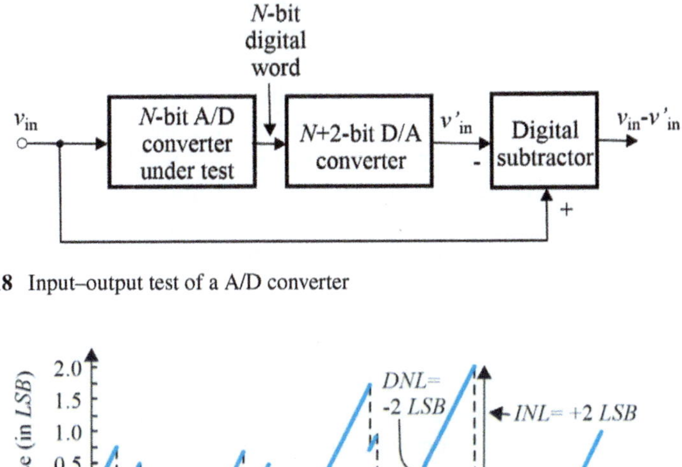

Fig. 7.6.18 Input–output test of a A/D converter

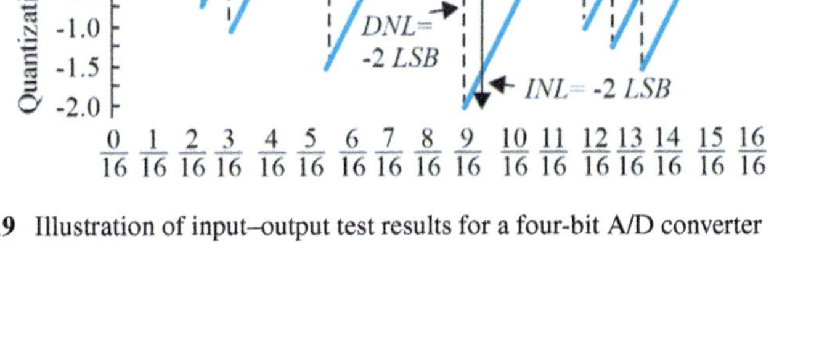

Fig. 7.6.19 Illustration of input–output test results for a four-bit A/D converter

Functional testing of A/D converters is usually done with a sinusoidal waveform. Before considering sine-based A/D converter testing methods, however, we will consider the important relationship that exists between the A/D converter's input signal frequency and the converter's internal (sampling) frequency, which controls when individual discrete words are generated. If the input AC signal is of very low frequency compared to the internal frequency, then it is intuitively clear that for all steps in the output digital circuit to be performed, so theoretically, only one period of the input signal is sufficient. As the signal frequency increases, however, one period of the input signal will be less and less sufficient to fully verify the conversion.

Take as an example an eight-bit A/D converter with 256 output codewords (from 0 to 255) driven by a 5 kHz test sinusoidal waveform and assume that the internal sampling rate of the converter is 25,000 samples per second, which is above the limit required to avoid band overlap. Let the input period be repeated 200 times. In that case, we can say that the following applies:

- the input frequency is $f_{\text{in}} = 5$ kHz,
- number of input periods $M = 200$,
- duration of testing $\text{TT} = (M \text{ periods at frequency } f_{\text{in}}) = M/f_{\text{in}} = 40$ ms,
- sampling frequency $f_s = 25$ kHz, and
- number of samples for the duration of testing $n = f_s \cdot \text{TT} = 1000$.

Therefore, if we also assume that the amplitude of the input signal is large enough, it seems that the number of samples is sufficient to test the selected converter. If we consider the data a little more carefully, however, given the values of $f_{\text{in}} = 5$ kHz and $f_s = 25$ kHz, we conclude that due to the integer frequency ratio, the input signal will be constantly sampled at the same place and that it will constantly (in all periods of the input signal) get the same sample values, that is, there will be only five different output codes that are repeated two hundred times. Therefore, in order for this test procedure to be satisfactory, the ratio of the frequencies f_{in} and f_s must be changed so that the sampling points within the period of the input signal are moved from period to period.

Two possibilities arise, known as *coherent* and *incoherent sampling*. In coherent sampling, the period ratio is not an integer, but sampling synchronization is maintained during the test period. This is achieved in the following way.

Starting from the desired number of samples within the testing period n, and from the sampling frequency f_s, it can be written

$$n = f_s \cdot \text{TT} \tag{7.6.13a}$$

or

$$f_s = n \cdot (1/\text{TT}) = n \cdot \delta, \tag{7.6.13b}$$

where δ is called the *primitive frequency* of the test. The frequency of the input signal should be

Table 7.6.2 Choice of the test signal frequency

M	$f_{in} = M \cdot \delta$ (Hz)
197	4925
199	4975
200	Not allowed
201	5025
203	5075
...	...

$$f_{in} = M \cdot \delta, \tag{7.6.13c}$$

where M is the number of periods of the input signal with the fact that now M and n have no common divisor.

For the above example it is

$$\delta = \frac{f_s}{n} = \frac{25{,}000}{1000} = 25 \text{ Hz} \tag{7.6.13d}$$

which allows one to select an input signal frequency that is close to the one before ($f_{in} = 5$ kHz) according to Table 7.6.2.

If we choose, for example, $f_{in} = 5025$ Hz, the complete coherent conversion would be performed with the following parameters: $f_{in} = 5025$ Hz, $M = 201$, $\mathrm{TT} = 40$ ms, $f_s = 25$ kHz, and $n = 1000$.

One thousand samples are now distributed over the entire range of digital codewords with no specific value on the sinusoid being sampled more than once in all 201 periods. On the other hand, however, this does not mean that the output codewords are not repeated. The reason for this is the finite resolution of the converter, whereby different analog values belong to the same quantization step.

For sampling, we say that it is incoherent when f_{in} is shifted from a value for which f_s is an exact integer multiple, and there is no mutual relationship between those frequencies. Since there is no longer a recognizable form of synchronization between the input signal and the sampling frequency, the sampling process becomes practically stochastic and, in order to achieve a complete and satisfactory test, a very large number of input signal periods may be required.

Incoherent testing underlies a procedure known as histogram testing. In this procedure, after a large number of periods of the input sinusoid, a histogram is formed that has individual (all) codewords on the abscissa, and the number of their occurrences during sampling (at the output of the A/D converter) on the ordinate. The sampling process is illustrated in Fig. 7.6.20a. The sinusoid in the figure is chosen so that it overshoots (or undershoots) the highest (lowest) threshold by a significant margin. The reason for this is the need to ensure that all code thresholds will be covered by the excitation signal.

For the case when the converter has no faults, a histogram is created as in Fig. 7.6.20b which is actually a graphical representation of the probability function

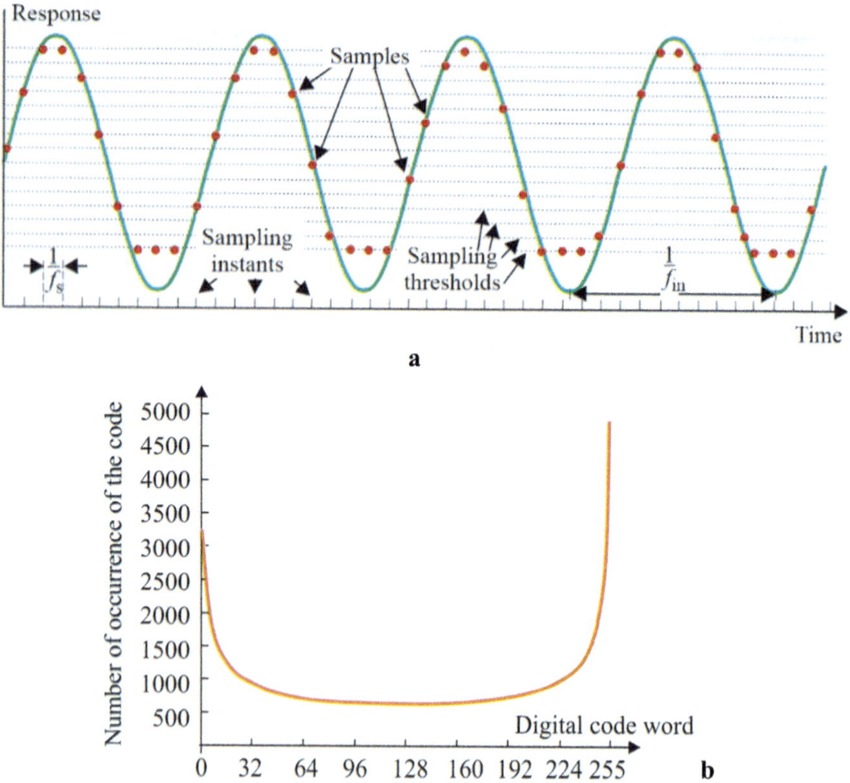

Fig. 7.6.20 Testing with histograms. **a** Sample generation example for the case of a four-bit converter and **b** histogram of a correct A/D converter under sinusoidal excitation and incoherent testing

$$p_{\mathrm{i}}(V) = 1/\left(\pi\sqrt{A^2 + V^2}\right), \tag{7.6.14}$$

where A is the maximum instantaneous value (amplitude) of the sinusoidal signal usually numerically given by FS/2, and $p_{\mathrm{i}}(V)$ is the (ideal) probability that the codeword V will occur in case of an ideal conversion. It should be noted the equivalence of V and the ordinal number of the code m.

When this way of representing probability is used in the presence of faults, histogram distortions occur. Nonlinearity leads to the appearance of spikes on this line, and missing codes are expressed by segments whose value drops to zero. The offset leads to vertical asymmetry of the histogram, and the gain error leads to a change in the width of the histogram. An example of a histogram obtained by measuring a real circuit is given in Fig. 7.6.21. The number of samples in this example was 262,164.

When the measurement is made, a relatively long field of codewords is created. By searching that field and counting the samples that have the same values, a histogram

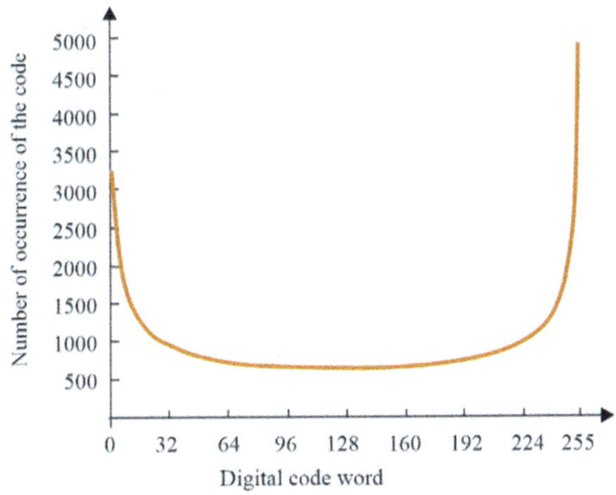

Fig. 7.6.21 A histogram for a real circuit

for the measured circuit is formed. To determine the probability of occurrence of individual codes, by integrating the probability density function for sinusoidal excitation by the ordinal number of the code, *m*, we obtain the following relation

$$p(m) = \frac{1}{\pi}\left[\arcsin\frac{\mathrm{FS}\cdot\left(m-2^{N-1}\right)}{A\cdot 2^{N}} - \arcsin\frac{\mathrm{FS}\cdot\left(m-1-2^{N-1}\right)}{A\cdot 2^{N}}\right]. \quad (7.6.15)$$

The deviation between the measured and the ideal histogram at each point (for each *m*) can be used to determine DNL as

$$\mathrm{DNL(LSB)} = [p(m)/p_{\mathrm{i}}(m)] - 1. \quad (7.6.16)$$

For the sake of illustration, on Fig. 7.6.22, the dependencies of DNL and INL are given, which were generated based on the diagram from Figs. 7.6.20 and 7.6.21.

Unfortunately, testing with histograms requires generating a large amount of information. The number of required samples depends on the resolution of the converter, the required level of certainty in the measurement results, and the size of the DNL error. For example, a ten-bit A/D converter with an error DNL (β) of 0.1 LSB and a confidence level ($z = 1.96$) of 95% requires about half a million samples, *n*. Increasing the confidence level ($z = 2.58$) to 99% would lead to the need for more than a million samples which is calculated according to the formula

$$n = \pi \cdot 2^{N-1} \cdot z^2/\beta^2. \quad (7.6.17)$$

When numerical values are substituted, $n_{|z=1.96} = 6176.920$ and $n_{|z=2.58} = 1070.678$ are obtained.

As the resolution increases, the number of samples required to maintain the confidence level doubles for each new bit. For a twelve-bit converter, the number of

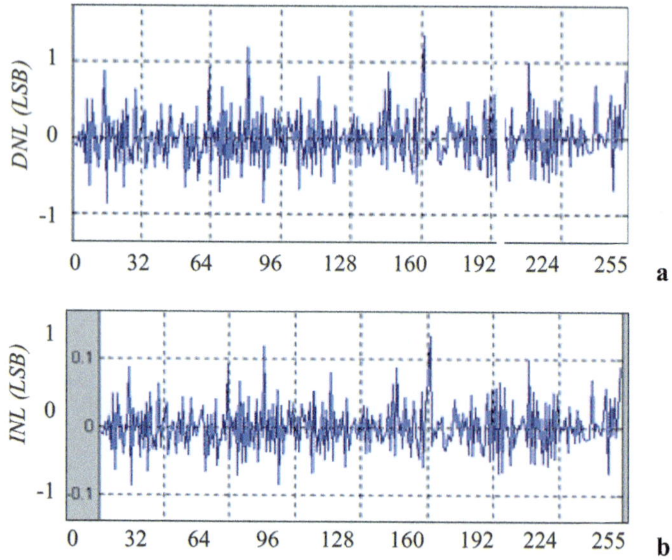

Fig. 7.6.22 **a** DNL and **b** INL of real circuit as a function of the digital code at the output obtained by the method of histogram

samples reaches over four million and gradually becomes a limiting factor for the application of this method. As an alternative, the so-called cumulative histogram method [12] which we will not consider in detail here may be used.

Finally, let us mention that a fast Fourier transform (FFT) can be applied to the converted output signal. For this purpose, a sinusoid with a DC component is used as excitation: $\mathrm{FS}/2 + \mathrm{FS} \cdot \sin(\omega t)$. At the output, 2^N samples are taken, using the maximum sampling frequency. After applying the FFT, the resulting spectrum is examined to determine the characteristics of the converter. By checking the amplitudes of the harmonics found in the spectrum, which are the image of nonlinearities, it is possible to judge the correctness of the converter. Of course, the frequency of the excitation sinusoid should be chosen so that the harmonic frequencies do not overlap with the sampling frequency and its harmonics. It was found that if the ratio of the fundamental to the highest harmonic is greater than $6 \cdot N$ dB, the nonlinearity of the converter is negligible compared to the converter's own quantization noise.

References

1. Maloberti F, Davies AC (2016) A short history of circuits and systems. River Publishers, Aalborg
2. Frostholm B (2015) Demystifying analog & mixed-signal ASICs. https://www.eeweb.com/demystifying-analog-mixed-signal-asics/

3. Casier H (1994) Mixed signal ASIC design for automotive and industrial applications. In: Sansen W (ed) Analog circuit design. Kluwer Academic Publishers
4. Bacci M, Bigongiari F, Boatella Polo C, Dittrich R, Inversi M, Jansen R (2014) Assessment of mixed signal technology. In: 5th Analogue and mixed-signal integrated circuits for space applications workshop, Geneve (CH)
5. Rashidzadeh R, Ahmadi M, Miller WC (2007) Test and measurement of analog and RF cores in mixed-signal SoC environment. IEEE Trans Comput Aided Des Integr Circuits Syst 26(10):1855–1865
6. Huang J-L, Ong C-K, Cheng K-T (2000) A BIST scheme for on-chip ADC and DAC testing. In: Proceedings design, automation and test in Europe conference and exhibition 2000 (Cat. No. PR00537), Paris, France
7. Hongzhi L (2004) A BIST (built-in self-test) strategy for mixed-signal integrated circuit. PhD thesis, University of Erlangen-Nurnberg, Germany
8. Kester W, Sheingold D (2004) Testing DACs. In: Kester W (ed) Analog-digital conversion. Analog devices
9. Kester W (2004) Testing ADCs. In: Kester W (ed) Analog–digital conversion. Analog devices
10. Stratigopoulos H-G, Kaminska B (2016) Analog and mixed-signal test. HAL Id: hal-02175745. https://hal.science/hal-02175745
11. Allen PE, Holberg DR (2002) CMOS analog circuit design. Oxford University Press, Oxford
12. Bankman I (2009) Handbook of medical image processing and analysis, 2nd edn. Elsevier, Amsterdam

7.7 Design for Testability

Abstract The concept of design for testability (DFT) for digital and analog circuits is described hereafter. After explanations of the necessity for this approach, some ad hoc ideas are presented. These are used to justify the implementation of the scan-test concept which eventually becomes IEEE 11.41 standard. The implementation of this standard to digital circuit (IEEE 1141.1) is briefly reviewed first. That should allow understanding of the very concept. Then the version intended to be used for mixed-signal systems (IEEE 1141.4) is introduced. Examples are given for the implementation of IEEE 1141.4 standard. For both standards, the concept of DFT implemented at circuit schematic level is explained and exemplified.

7.7.1 Introduction

When developing a test signal for different circuits it may be observed that two similar circuits may present problems of different orders of magnitude. The importance of the concept of circuit testability emerges from such considerations. Namely, we are looking for a quantitative measure that will talk about the ease of generating a test signal. Of course, such a request is justified and easy to understand, but defining testability and developing rules for its evaluation is not such a simple problem.

When a test signal is developed, it is usual to count the faults covered by that test and to express that number as a part (percentage) of the total number of faults. This number is often called a measure of testability. There are, however, several reasons why this quantity cannot be considered satisfactory information. Among them we will mention the following.

Each fault can be associated with a binary value that indicates whether the fault is testable or not. Based on it, the total measure of testability is generated. That binary number, however, does not say anything about the effort needed to find that test or the effort (and equipment) needed to run the test.

On the other hand, the test is usually not generated for the complete set of faults. Therefore, the measure of testability would refer only to those faults that are on the list of faults and will not say anything about faults that are not on the list. There is a possibility that some of them are not testable at all.

V. B. Litovski, *Lecture Notes in Analog Electronics*, Lecture Notes in Electrical Engineering 1265, https://doi.org/10.1007/978-981-97-8257-4_7

Very often, one test signal covers several faults. In such cases, if it turns out that the faults are in a part of the circuit that does not change the static function but only the delay, for example, the measure of testability gives a more optimistic picture than the testability problem really is.

Because of all this, here we will not deal with the derivation of special formulas for testability measures.

What are the causes that prevent testability? They have already been indicated in previous chapters, and only a brief overview will be given here. These are redundancy in combinational circuits, problems related to testing sequential circuits related to initialization, the impossibility of achieving observability in circuits with mixed signals, etc.

Solutions for increasing testability that are used in modern electronics should be studied within the synthesis procedures of analog and digital systems. Many synthesis procedures, however, do not take testability into account, and the knowledge in the field of testing that is presented in this book is usually not available within the study of synthesis methods. That is why we will briefly state some of the basic ideas here, which we hope will enable the reader to continue studying this area independently.

7.7.2 Basic Concepts for Design for Testability of Digital Circuits

There are two approaches to increasing testability: ad hoc and a systematic approach.

When we say ad hoc, we mean countless design ideas and tricks that can be applied to increase testability, and which arose during the development of certain solutions. Their application is, of course, irreplaceable even with a systematic approach. Without intending to give an exhaustive overview of the set of such ideas, we will list here only some illustrative examples.

Increasing controllability and observability is achieved by increasing the number of external terminals. An example of such a solution is given in Fig. 7.5.14. In this circuit, the controllability is increased by the introduction of the input terminal T, and the observability is increased by the introduction of the output terminal Y.

Applying this method to complex circuits can lead to the requirement for too many external terminals of the printed circuit board or the chip. This would limit its application. One method that makes it possible to overcome the limitation related to the number of external terminals is the use of multiplexers, which would enable the existing primary inputs and outputs to be assigned alternative tasks that would be related to testing.

The principle is illustrated in Fig. 7.7.1. One control pin (M_1) can select an arbitrary number of internal nodes (inputs to the L_2 logic), disconnect them from their normal excitation, and connect them directly to the primary inputs. The second control signal (M_2) allows an arbitrary number of primary inputs to be connected to internal nodes (outputs of the L_1 logic), which enables observability at those nodes. There are three operating modes with this scheme:

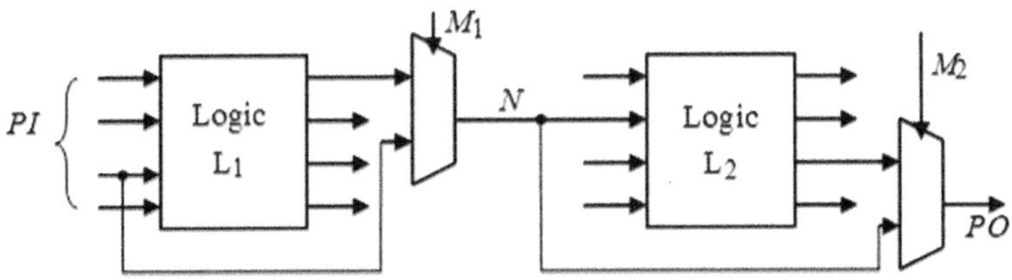

Fig. 7.7.1 Use of a multiplexer to enable the dual role of input/output terminals (*PI* primary inputs, *PO* primary outputs)

1. When $M_1M_2 = 00$, we are in normal working mode.
2. When $M_1M_2 = 01$, we test L_1. Node N is now connected to the output.
3. When $M_1M_2 = 10$, we test L_2. Node N is directly controlled by the primary input.

A similar solution is applied in situations where the number of primary inputs is so large that the number of combinations (test words at the input) is unacceptably large. Namely, it is estimated that the cost of generating the test is proportional to the square of the number of input terminals n. Thus, if it is possible to decompose a circuit with n inputs into m subcircuits so that each has n/m inputs, then the total cost will be $m \cdot \left(\frac{n}{m}\right)^2/n^2 = \frac{1}{m}$-th fraction of the cost of generating a test for an un-decomposed circuit.

Decomposition is achieved by using a multiplexer that will allow to break connections within the circuit and to ensure the necessary controllability and observability. The general principle is shown in Fig. 7.7.2 where a circuit is divided into two subcircuits A and B. Half of the primary inputs are directly connected to each subcircuit. In addition, through the multiplexer, the other half of the primary inputs and part of the outputs generated from the other half of the circuit are fed to each subcircuit. By incorporating multiplexers and "links" shown in dashed lines, circuit A can be tested in isolation from circuit B by borrowing as many primary inputs and outputs of circuit B as it needs to achieve observability and controllability. Similar modifications can be introduced to enable testing of subcircuit B with the condition that it is isolated from subcircuit A.

When working with sequential circuits, which were not considered in this book, asynchronous circuits should be avoided in order to avoid the problems of hazard or race in the circuit. Also, it is recommended to avoid situations when signals generated in the circuit are used as trigger signals of flip-flops. Here, too, the concept of decomposition is applicable in its own way. This refers to dividers that divide with a large number, and the time required for testing is long. A suitable reorganization (which will not be discussed here) enables observability in some parts inside a counter.

The problem of initialization of sequential circuits is of crucial importance but we do not intend to discuss this problem in detail here since it goes beyond the scope of this book.

Summarizing the considerations on ad hoc methods for increasing testability, we conclude that the difficulty with most solutions that are applied is that they actually

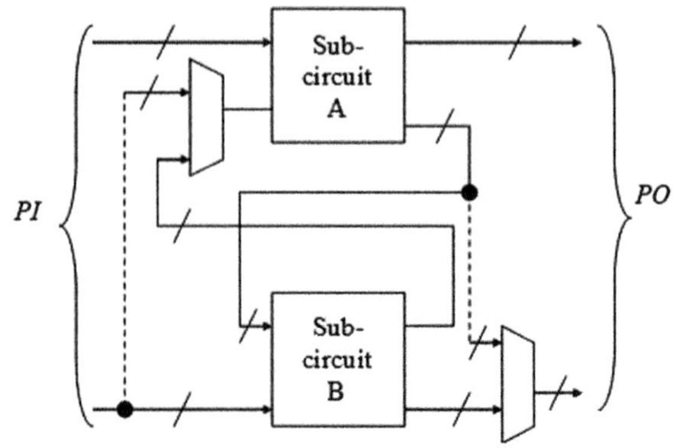

Fig. 7.7.2 Circuit separation using multiplexers

represent modifications of solutions that are bad from the point of view of testability. That is why a systematic approach to ensuring testability is desirable, which will enable testability to be controlled in the stage of circuit synthesis, and not in the stage of improving testability.

The basic problems of testability of sequential circuits can be illustrated by the circuit of Fig. 7.7.3. It is not difficult to establish that

- in order to test the combination part of the circuit, a special test sequence should be applied. However, this is not enough considering that the combinational part of the circuit is also excited by state variables which, with this kind of test vector, we cannot fully control.
- Even when we apply an excitation signal to combinational logic, its outputs are not observable since some of them constitute internal states.
- Finally, state variables are neither directly observable nor directly controllable.

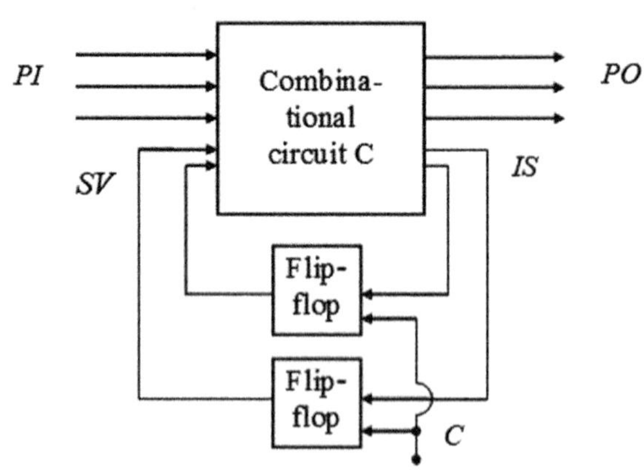

Fig. 7.7.3 Canonical structure of a synchronous finite state machine (*PI* primary inputs, *PO* primary outputs, *SV* variable states, *IS* internal states, and *C* clock)

In order to overcome these problems of controllability and observability, we should apply some of the mentioned ideas about ad hoc improvements. For realistically complex circuits, however, there are small possibilities to introduce additional input/output terminals even if the price of the whole job is acceptable. Our goal should be to establish complete controllability and observability, while requiring the minimum number of new input and output terminals. This can be done if a new design concept is introduced, which in some cases will also require new specific logic gates.

All formal methods of *design for testability* (DFT) of digital circuits that have been proposed so far rely on two basic principles:

(a) The structure of the system becomes such that parts that store states and combinational parts, in the testing phase, can be isolated from each other.
(b) Observation and control of state variables are achieved by serial access in order to minimize the number of input–output terminals intended for testing.

Due to the property of serial access, these concepts are known as scan-in, scan-out (SISO) methods, and the *serial path* is known as *scan-path*. Under SISO, we should understand the search to (or toward) the fault, and the search from the fault to the output.

If we assume that all sequential gates are D-type flip-flops, the modified circuit of Fig. 7.7.3 for applying the SISO concept would look like Fig. 7.7.4b (clock signal not shown). As can be seen, a multiplexer is introduced in front of the D input of each flip-flop (Fig. 7.7.4b). It is controlled by unique control signal M that tells about the mode of operation. When $M = 0$, the upper input to the multiplexer is selected and taken to the flip-flop, and when $M = 1$, the lower input is selected. Now the output of each flip-flop, in addition to being directly connected to the combinational logic, is also led to the next flip-flop through the multiplexer. This is how a chain of flip-flops is formed, with its ends (SD_{in} and SD_{out}) becoming input/output terminals of the system. Their labels refer to the words: scan data (SD).

The system has two operating modes that are selected by the M signal.

When $M = 0$ the system is in normal operating mode where the combinational logic drives the flip-flops.

When $M = 1$, the system is in scan mode, where the combinational logic is separated from the flip-flops. The flip-flops are now configured into a unique shift register whose input is SD_{in} and output is SD_{out}. The shift register input and output are directly accessible. Therefore, the state of the circuit is directly controllable by moving it to any desired state expressed by a series of SD_{in} values. Also, the state is directly observable in that the content can be moved and expressed as a series of SD_{out} values.

With this circuit architecture, the problem of generating a test sequence is greatly simplified. A set of flip-flops can be thought of as a shift register with direct input and output access. Combinational logic can be considered isolated with the fact that all inputs can be controlled and all outputs can be observed.

The testing procedure for the SISO structure would be as follows:

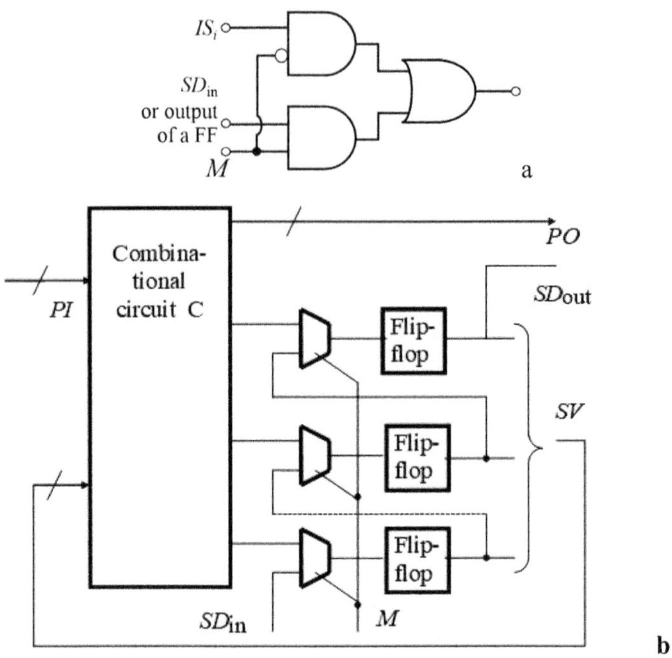

Fig. 7.7.4 SISO concept. **a** Multiplexer and **b** reorganized synchronous finite state machine

1. By setting $M = 1$ and setting SD_{in} and SD_{out} as system ports, it is possible to test the set of flip-flops as a shift register. One possible test would be to communicate the sequence (on SD_{in}) 00110011—in accordance with the clock. This requires each flip-flop to perform all four transitions.
2. *Testing combinational logic*. For this case, it is necessary that the developed tests consider the controllability of all inputs (both primary inputs and state variables) and the observability of all outputs (inputs of flip-flops as well as primary outputs). There are three phases to each test:

   2.a Set $M = 1$ and write the desired set of variables into the shift register. These variables are, of course, part of the set of input variables in the combinational circuit.

   2.b Set $M = 0$. Apply appropriate signals to the primary inputs that make up the rest of the input variables in the combinational circuit. Observe the resulting primary outputs that form the first part of the output set of the combinational circuit. Trigger the flip-flops once in order to remember the states of the other outputs of the combinatorial circuit in the flip-flops.

   2.c Set $M = 1$. Move the contents of the flip-flops and observe the result on the SD_{out}.

This fundamental concept has experienced mass application in recent decades and many different variants. Among the most famous concepts based on the SISO principle, we will mention two here, and only one will be described in more detail.

The concept of level-sensitive scan design (LSSD) was, and still is, in use mostly by IBM for dozens of years [1]. It will not be discussed here.

Instead, the boundary scan concept, which has been developed to the level of a standard [2–6] will be briefly described.

At the end of this section, the problem of designing for testability at the level of the electrical schematic and the synthesis of logic gates that are compatible with the scan-path concept that was described will be discussed.

7.7.3 Boundary Scan

In this section, a concept of DFT will be briefly described, which has gained the status of a standard and is increasingly being applied to the design of electronic systems, systems on printed boards, and thus also to the design of complex integrated circuits. The presentations will be mainly based on [7] and [2].

In order to explain the design concept based on the boundary scan standard, we will consider the problems of testing a system consisting of three integrated circuits that are "strongly" connected as in Fig. 7.7.5.

We will impose the following requests. In the *testing phase*, the new architecture should enable:

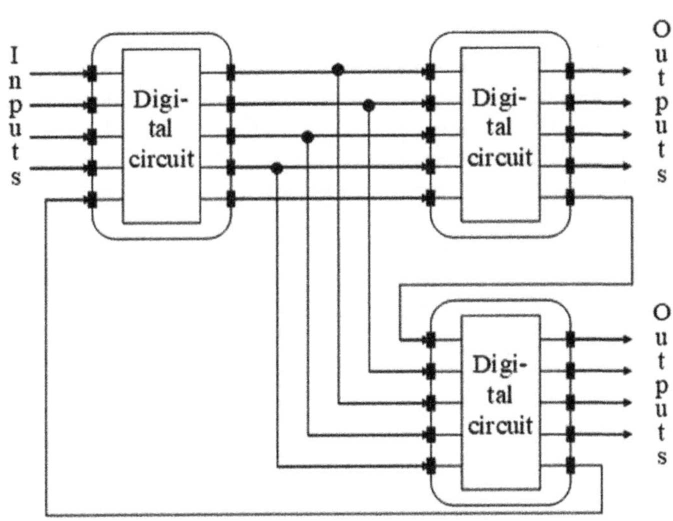

Fig. 7.7.5 Three chips on a printed circuit board with a large number of interconnections

- Testing connections between system components, which will be called *extest*.
- Testing of the system components themselves, which will be called *intest*.
- Identification of system components (optional) to check proper installation.
- Bypassing (ignoring) a system component to speed up testing.

Of course, during the *phase of normal system operation*, the system extension that performs the above functions should be "invisible" or transparent to useful signals.

The basis of the idea that enables the realization of these requirements consists in adding shift register cells to each input/output terminal of the component. By component here we mean mostly a complete integrated circuit, but also a complex cell within the integrated circuit that was realized with this concept.

During test mode, these cells are used to check the state of (each) input port on the one hand and to read the state of (each) output port on the other. The physical realization of this idea in a rudimentary form is shown in Fig. 7.7.6. As one can see, between each input/output terminal and the digital circuit that performs the basic function, a cell (small square) is installed. In the phase of normal operation, when the control signal test mode select (TMS) is at logical zero, it is transparent and directly connects the terminal and the digital circuit (from left to right and vice versa).

In the test phase, when the TMS is at logic 1, the cells isolate the input/output pins (small black rectangles) and the digital circuit and, under the control of a separate and completely independent clock signal TCK (Test clock); the signals are transferred

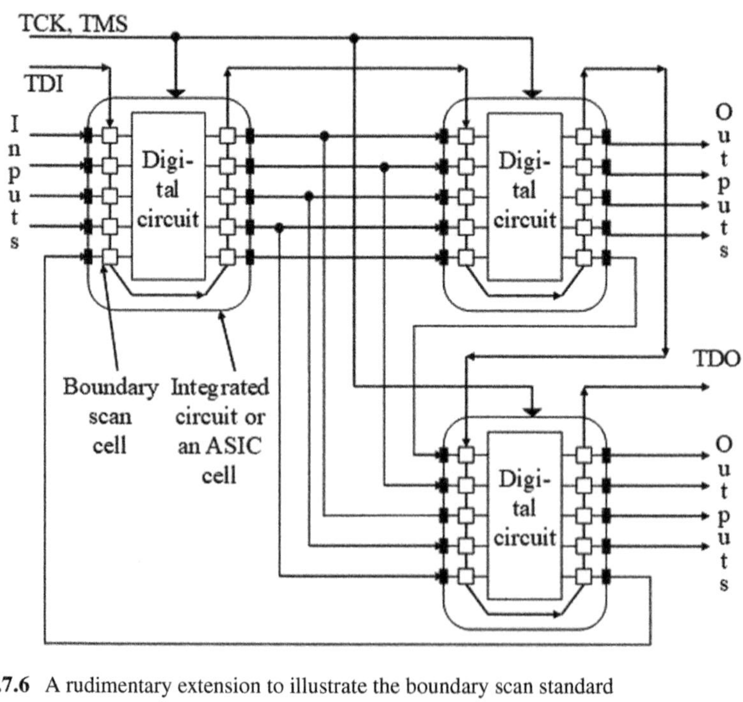

Fig. 7.7.6 A rudimentary extension to illustrate the boundary scan standard

on the way from the TDI (Test Data In) to TDO (Test Data Out). In addition, they perform a number of other functions, which will be discussed here later.

Since the cells of the shift register are located around the perimeter (periphery, boundary) they were named boundary scan cell (BSC), and the complex shift register they make up is called boundary scan register (BSR). We will use the abbreviation BS for the entire concept. All the above-mentioned abbreviations are created according to the IEEE 1149.1 standard [2] and as such are immutable. At all times it should be kept in mind that the test and control signals are generated externally by the test device and assembled by the test engineer according to each sample being tested.

The way this concept works in the testing phase will be illustrated by a simple example of testing one connection on a printed circuit board. Figure 7.7.7 will serve as the first illustration. A very simplified version of the entire system is shown here. One can see only two "cells" that are connected by two connections. The task is to test one of the connections as indicated in the picture. TMS and TCK signals are not shown because their operation is assumed. Namely, the TMS is on the logic 1 (test), and each action (moving data along the BSR) takes place only along the rising edge of the TCK.

According to the boundary scan concept, the following operations should be performed:

1. *Shift-DR* (DR stands for Data Register which in this case is used as a substitute for BSR): Shift the excitation signal from the input TDI port to the cell that is connected to the output port of the "cell" that excites the connection under test. The result of this operation is shown in Fig. 7.7.7. The solid (blue) line shows the displacement path of the test signal. The content of the BSC that will trigger the connection is logic 0.
2. *Update-DR*: Bring the excitation signal to the output of the BSC connected to the connection to be tested.
3. *Capture-DR*: Remember the state of the signal at the end of the connection in the receiving BSC. The result of the last two activities is shown in Fig. 7.7.8. Logic 0, if the connection is correct, will be transmitted to the receiving BSC. The active part of the circuit is marked with a solid (blue) line.

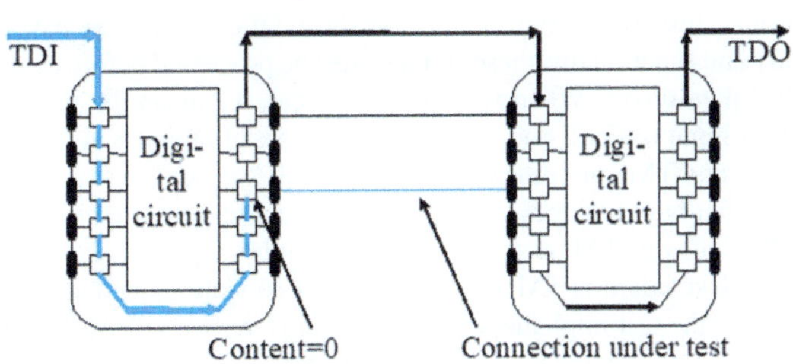

Fig. 7.7.7 First step. Move the test data to the output cell

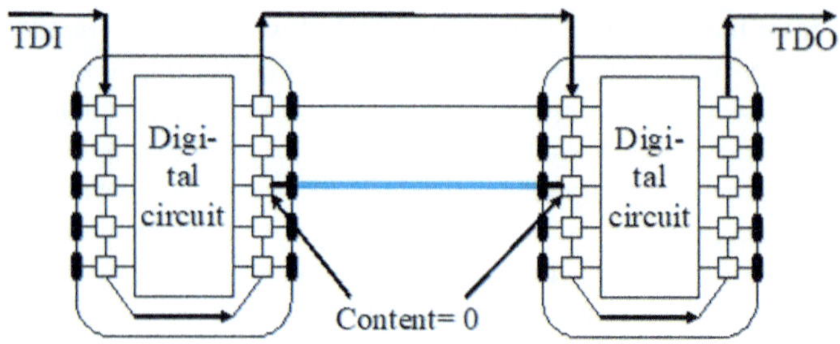

Fig. 7.7.8 Second and third steps. Apply the test signal and remember the result

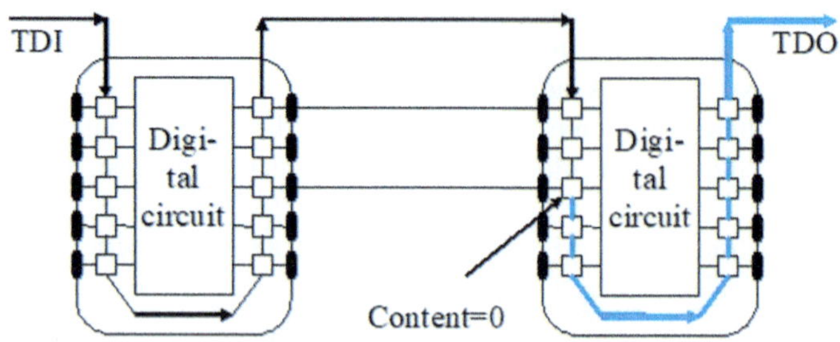

Fig. 7.7.9 Fourth step. Export the memorized data

4. *Shift-DR*: Move the contents of the receiving BSC to the TDO output itself. The result of this testing phase is shown in Fig. 7.7.9. The solid (blue) line shows the path of the measured signal to the TDO, that is, to the test device.

This would describe the procedure of connection testing in a simplified form, which we said is called extest within this standard.

Based on this simple example, we can draw deeper conclusions about the structure of the system that would be designed with the BS concept.

It should contain at least four additional terminals per printed circuit board (or per chip) related to TDI, TDO, TMS, and TCK signals. This set of ports is mandatory and is known as Test Access Port or TAP for short. Of course, as we have seen both topologically and functionally, these signals must be processed before being directed to the "cells" themselves. When we say topological, we mean the fact that TDI is taken as serial input to BSR, and TDO is taken as serial output. On the other hand, however, TCK and TMS are run parallel to each BSC. When we say functional, we mean the fact that based on these signals, others are also generated (update-DR, for example). That is why TAP is followed by a complex digital circuit whose name according to the standard is TAP controller. It is a 16-state sequential machine that actually realizes the test implementation protocol. Here, we will not go into details about the physical realization of the TAP controller.

Another important element that makes up the structure that realizes the BS concept is the Instructional Register (IR). It enables instructions related to testing to be moved to each "cell" or chip. During the test, these instructions determine which register (TDR) the test data is in, and which tests should be performed.

In order to speed up the testing, it is often necessary to exclude from the BSR all the cells that are connected to the parts of the system that are not tested. The so-called bypass register is used for this purpose. It is a single stage shift register. It ensures the minimum length of the serial data path that moves from TDI to TDO.

Let us sum up. The most important elements of the standard BS architecture are BSR, TAP controller, IR, and bypass register. The logic schematics and physical layout of these cells realized in 1 μm technology, together with the results of their simulation, can be found in [8]. In addition to these elements, other subsystems related to testing can be incorporated into the system being tested. One of them is the component identification register (Device ID register), which enables the test designer to recognize the products that have been installed. The second would be a subsystem that performs built-in self-testing (BIST).

A simplified picture of the complete architecture of the BS concept is shown in Fig. 7.7.10. Signals S_{in} and S_{out} refer to some additional test structures that, according to their design practice and needs, the designer incorporates into the circuit being tested. Among such structures may be those intended for self-testing (BIST).

It remains to describe the BS cell itself in a little more detail. One solution of a BS cell is shown in Fig. 7.7.11. As one can see, it consists of two multiplexers and two D-latches. Abbreviations of the names of the cell terminals are listed in the figure itself, and the full names are in the caption to the figure. It is important to keep in mind that the configuration of all BS cells is the same, both those connected to the chip inputs and those connected to the outputs. The direction of the data flow is determined by the control signals indicated.

Fig. 7.7.10 A simplified representation of the architecture of a digital system designed according to BS standard IEEE 1149.1

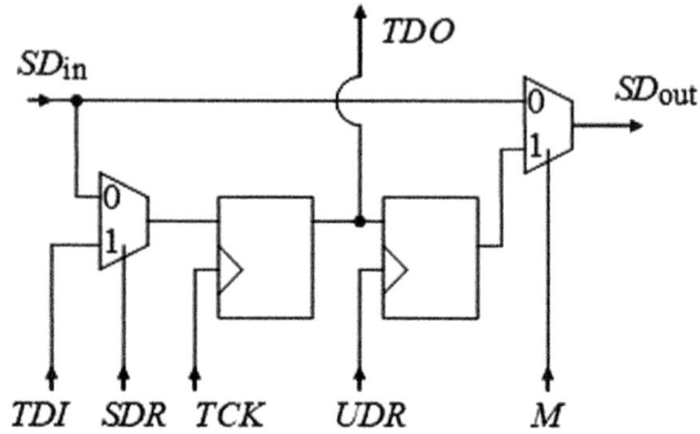

Fig. 7.7.11 Digital boundary scan cell. *SD* system data, *TDI* test data serial input, *TDO* test data serial output, *TCK* clock, *UDR* refresh, *M* mode (test or normal operation), *SDR* mode (shift/store)

The BS cell has two data inputs, four control inputs, and two outputs. The input TDI and output TDO are serial and when the first and last BS cells are excluded, they transport information through the BSR. The rest of the inputs and outputs are parallel. This means, for example, that the control signals TCK, TMS, SDR, and UDR are common to all BS cells.

SD_{in} is the input data signal. It is parallel in the sense that each cell has its own input signal at its input. If it is a cell connected between the external pad and the logic being tested, SD_{in} will be the external input signal that is used during normal operation of the system, and if the input of the BS cell is connected to the output terminal of a circuit inside the system, then SD_{in} is the response of the tested part of the system to be output on the TDO. Finally, in the mode of operation known as SAMPLE, when the BSR is used only to monitor the operation of the system under test, the input signal of the BS cell is one of the responses in the system under test selected for observation or one of the input signals being selected to be observed.

SD_{out} is the data output signal. For the BS cell that is connected between the system's input pad and the digital circuit being tested in normal operation of the system and in SAMPLE mode, it can be the excitation signal itself. In test mode it can be the test signal that excites the digital circuit. For the cell that is connected between the digital circuit within the system and the connection being tested, SD_{out} will be the test signal that tests the connection. Finally, for the cell that is connected between the digital circuit being inside the system and the output pad of the system in normal operation, SD_{out} will be the response of the system in normal operation.

The TDI and TDO signals do not need to be re-explained much.

The first multiplexer, which is controlled by the signal SDR, decides whether TDI or SD_{in} will be forwarded through the BS cell to the first D-latch. If TDI is passed, it means that the cell is working in shift mode. According to the clock, the TDI signals are transferred to TDO, without being transferred to the other multiplexer which is in test mode and which would transfer the signal from the lower terminal to the output. Hence, the need for the second latch which will transfer signals only if the UDR signal allows it. In the event that the SD_{out} excites a connection or some logic gate under test, the UDR signal enables transfer of TDI to SD_{out}.

If the data from the SD_{in} is forwarded through the first multiplexer, the BS cell is in the "capture" mode, which means that the input data is temporarily stored in the first latch. In the SAMPLE operation, it is passed to the TDO by itself, without bringing it to the SD_{out}. Most of these situations are covered by the illustrations given in Figs. 7.7.12, 7.7.13, 7.7.14 and 7.7.15.

The first of them (Fig. 7.7.12) shows the operation of part of the BSR in SAMPLE mode. In this mode, as already mentioned, the system works normally, but at the same

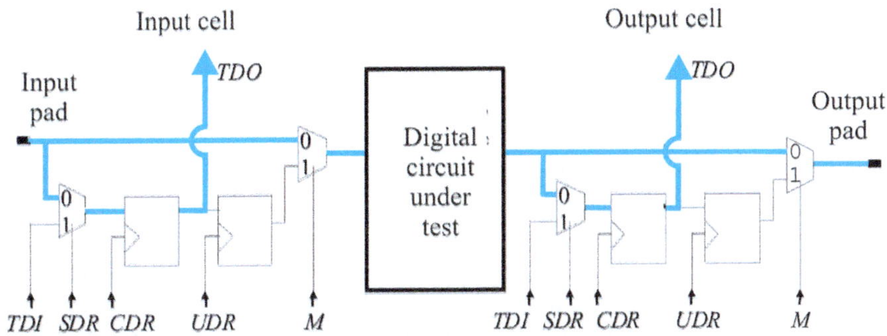

Fig. 7.7.12 Data flow during the SAMPLE instruction

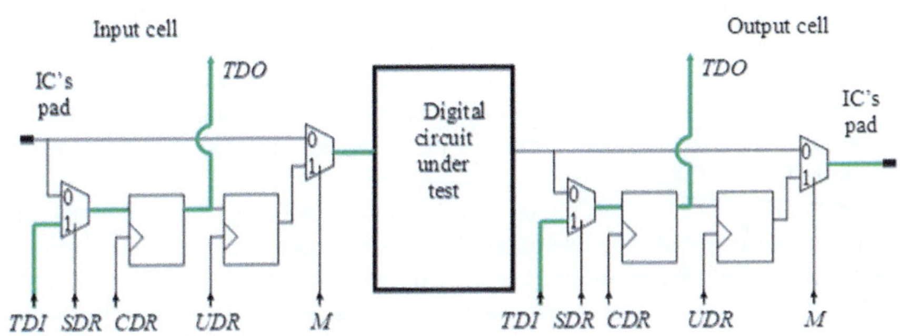

Fig. 7.7.13 Data flow during the PRELOAD instruction

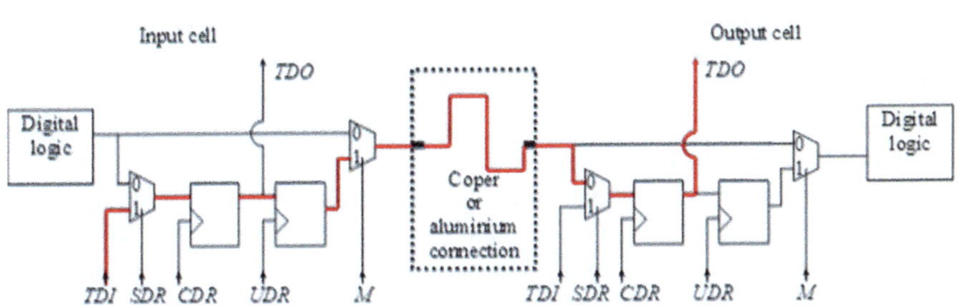

Fig. 7.7.14 Data flow during the EXTEST instruction

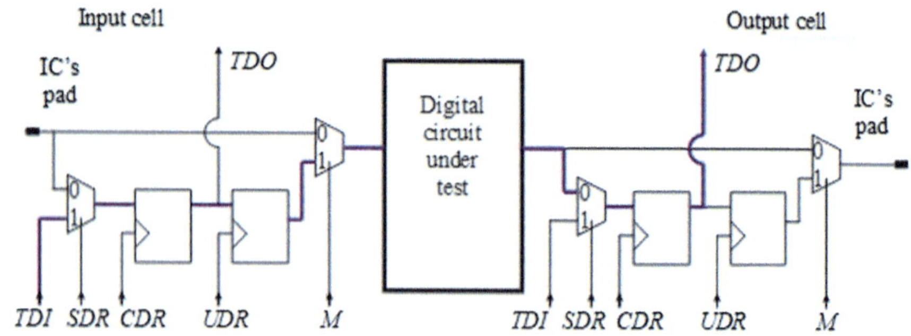

Fig. 7.7.15 Data flow during the INTEST instruction

time its operation is observed through TAP, which can be interesting for diagnostic purposes or for documentation. The data flow is shown in solid (blue) lines. This is not a test mode but the control signal SDR and clock CDR are still active.

The following figure (Fig. 7.7.13) shows the flow of data in the PRELOAD operating mode. In this phase, the test signal is fed through the TDI and is "locked" in the first latch. It should not be forgotten that the cells make up a register and in order to assign the right content to each one, a large number of content shifting steps are required.

The flow of data and the EXTEST phase, which refers to the testing of internal connections on printed boards and in integrated circuits, is shown in Fig. 7.7.14. Now, under the control of the UDR signal, in the exciting cells, the test signal prepared in the PRELOAD phase is transferred to the output and excites the connections, and in the excited cells, under the control of the SDR, the response is "captured" and transferred to the TDO.

Finally, Fig. 7.7.15 shows the flow of data in the INTEST phase. Now only the digital circuit is tested. The operation of the excitation cell is the same here as in connection testing: the memorized signal from the PRELOAD phase is transferred to the output of the BS cell, which excites the logic circuit. A set of such BS cells forms a parallel input to the circuit under test and the output cell behaves like in EXTEST. The response(s) of the digital circuit under test are taken in series under SDR control.

This would end this short story about the boundary scan standard, noting that it is becoming more and more popular and its use is becoming more and more widespread. One proof of that trend is the application of this standard in the design of microprocessors [9] immediately after its publication. The reader is referred to the previously mentioned literature for much more details.

It should be noted that this standard is further developed in two main directions. One refers to his improvement in the field of testing digital circuits [10], and the other to the application of the same concepts to testing of mixed, analog, and digital circuits [3, 5, 11], about which there will be more words.

At the end of the presentation on the IEEE 1149.1 standard and its extensions, it should be noted that the testing of electronic systems, printed boards, or integrated circuits designed according to the requirements of this standard is a complex engineering task. That complexity arises simultaneously from the complexity of the function being tested and from the complexity of the testing procedure prescribed by the standard itself. In order to overcome these problems, the standard was immediately accompanied by a language for describing testing and synthesized test structures. The language received a name BSDL (Boundary Scan Design Language) and a description in [4]. This language is based on VHDL and is therefore acceptable for designers. Examples of the application of this language can be found in [7] and [11]. There are two most important applications of this language. One refers to the documentation of the project for testability, which can be provided together with the documentation of the digital system that is the subject of testing. This enables exchange of solutions and the protection of copyrights from that point of view. Another, perhaps even more important application, relates to the programming of testers. This standardizes the process of describing the test signal and the system designed for testability and at the same time facilitates portability and documentation.

7.7.4 DFT of Digital Circuits at the Circuit Level

So far, the consideration of procedures for DFT has aimed to adapt the existing solutions of digital circuits and systems to DFT by adding structures that enable detectability and observability. There are situations, however, where some faults can be made detectable or observable only if one intervenes in the very electrical schematic that realizes the given logic function. Therefore, designing for testability in such situations means adjusting the electrical schematic in order to achieve the following goals:

- observability and detectability of the selected set of faults and
- the simultaneous ability to be incorporated into a more complex system where testing is based on a scan-path concept such as the boundary scan just described.

The application of this concept will be described on the example of D-latch from Fig. 7.7.16. Faults of the type of break or open circuit will be considered for it. The presentations will be mainly based on [12].

CMOS logic circuits consist of two groups of transistors: a group of P-channel MOS transistors (PMOS or PT) that play the role of load and a group of N-channel MOS transistors (NMOS or NT) that play the role of an amplifier. The general structure is shown in Fig. 7.7.19. The logic function is realized by establishing logic 1 (or 0) at the output by bringing the PT (the NT) into a conducting state. In order to detect a break fault in the source or drain of a PT (NT), the test signal must have two phases. First, an initialization signal t_1 is needed that will set 0 (or 1) at the output, and then, a test signal t_2 should be applied that will establish one or more conduction

Fig. 7.7.16 D-latch

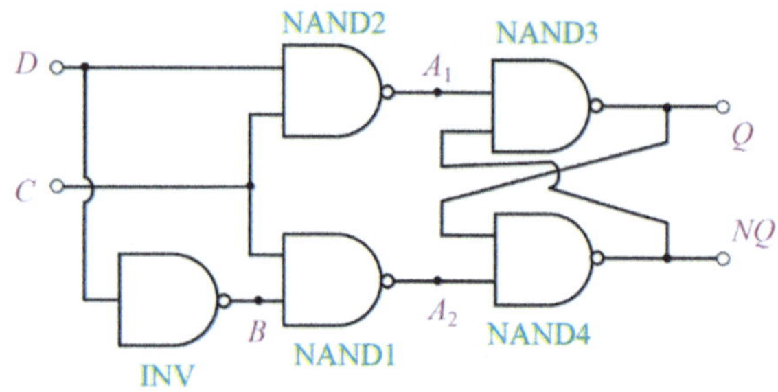

paths through the P-channel (N-channel) transistors. At the same time, we should be sure that no conduction path has appeared in the N-channel (P-channel) transistors.

When t_2 is applied and it is found that the transistor assumed to be cut-off is not, the output will be at logic 1 (or 0). Otherwise, the output will be at logic 0 (or 1) and a high impedance state will be established so that the logic state will be determined by the charge in the load capacitance that is accumulated during the application of t_1.

The test sequence $\{t_1, t_2\}$ is called a two-step test. Applying this concept to the scan-path technique would require a modification to the scan-path flip-flops to remember t_1 while applying t_2.

Figure 7.7.17 shows the electrical schematic of the D-latch from Fig. 7.7.16. Suppose that we want to detect a break in the P-channel transistor P7 of the NAND3 circuit whose output is Q. In order to detect such a fault, we need an initialization test signal t_1 that will set Q to logic 0, that is, its inputs (A_1 and NQ) to the value 11. This should be followed by the test signal t_2 which will set $A_1 \cdot \text{NQ} = 10$. Now comes the problem. NQ can be set to logic 0 only if Q is set to logic 1. This would erase the initial value of Q. Therefore, this fault (P-channel transistor P7 cut-off) cannot be detected. For similar reasons, the break in P9 cannot be detected either.

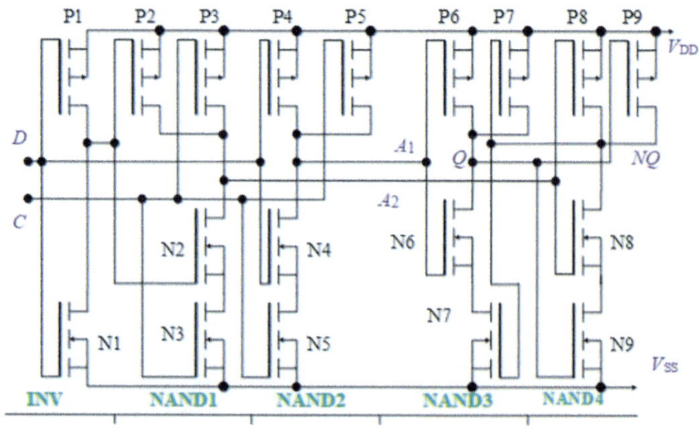

Fig. 7.7.17 Electrical schematic of D-latch realized according to Fig. 7.7.16

By similar considerations, it can be shown that even the breaks in P4 and P5 are not detectable. In all cases, the reason for this is that there is a contradiction. The application of t_2 cancels the effect of t_1.

One additional problem can be observed. Assume that P4 is broken. A two-step test would be $t_1 = CD = 11$ and $t_2 = CD = 10$. When this sequence is applied, however, in a fault-free circuit, after applying t_1, $Q \cdot \text{NQ} = 10$ is obtained, and after applying t_2, $Q \cdot \text{NQ} = 01$. In the faulty circuit, after applying t_1, $Q \cdot \text{NQ} = 10$ is obtained, and after applying t_2, $Q \cdot \text{NQ} = 11$. Therefore, the fault effect extends only to Q. A similar result is obtained if a break occurs in P2. The fault effect only appears on NQ. Based on this, we conclude that, except in cases when the fault effect is transferred from both outputs of the latch to the primary (or test) outputs of the system, some of the faults in the memory elements may be undetectable. In an application of the scan-path technique, elimination of this problem would require that both outputs Q and NQ be tied to the data inputs of the scan register. Moreover, in order to detect some of the undetectable faults, both outputs of the last scan register should be observed.

Keeping all this in mind, we can set some design criteria that should be applied when developing the circuit diagram of a new D-latch. These would be:

- to synthesize the circuit in such a way that the faults are detectable,
- to enable the fault effect to be observable at the selected output, and
- to apply the usual procedure provided by the scan-path technique that has been selected.

Now we will try to describe the design of the new D-latch, which is based on the mentioned requirements. The D-latch obtained in this way will be able to be used for the construction of pulse-triggered MS or scan-flip-flops in which open-circuit type faults will be detectable. For systems using a single-phase clock, the proposed flip-flops will require an additional input terminal for test mode. For two-phase clocked systems, the additional test mode input terminal can be avoided if we force the clock signals to overlap (in the test phase). If static *set* and *reset* inputs are also introduced, CMOS flip-flops can be developed in which all open-circuit faults are detectable.

Difficulties in detecting this type of faults are mainly due to the impossibility of initializing the node in the memory element with the given initializing signal and then applying the test signal. The first reaction to this difficulty would be to add new controllable inputs to set the selected nodes in the desired states. Related to that, of course, two problems arise. First, by designing in this way we cannot guarantee that all faults will be detectable. Second, the number of such input terminals would be too large.

Therefore, a better solution is to develop a completely new electrical circuit of the flip-flop, which will be testable. As a basis for development, the circuit diagram of the D-latch given in Fig. 7.7.18 is used. For normal operation of this circuit, the data input (D) needs to be stable when the clock signal changes state. It is also required that all fault effects be made observable at the Q output. Moreover, the fault effect is required to be observable when the clock is inactive.

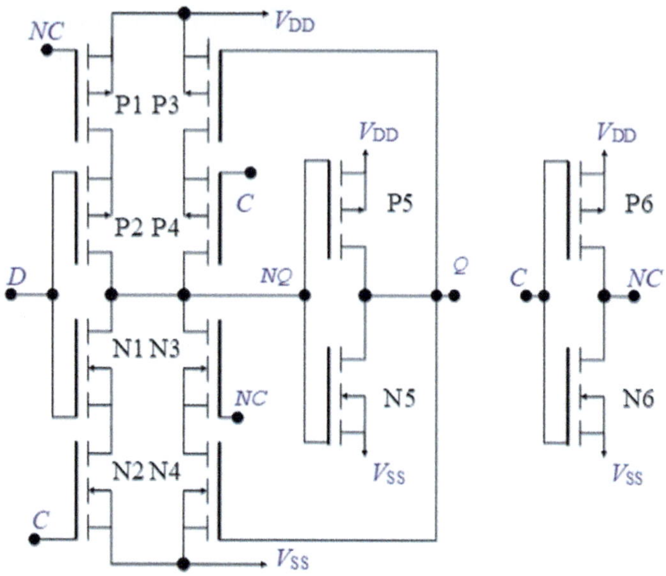

Fig. 7.7.18 A new version of the CMOS D-latch

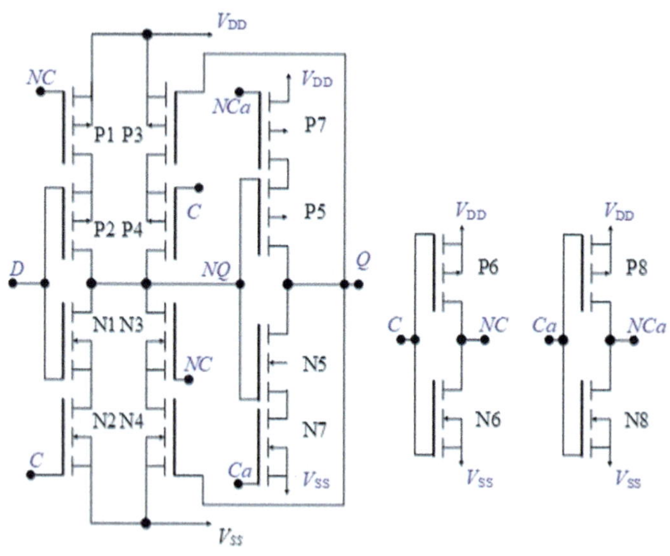

Fig. 7.7.19 Testable D-latch

Table 7.7.1 shows tests for all open-circuited transistors in the circuit shown in Fig. 7.7.18, except for transistors P3 and N3. Faults related to these transistors are undetectable. Z0 and Z1 indicate high impedance states that occurred after the state of logic 0 or logic 1, respectively. When using these results, it is assumed that the clock is fast enough so that the load capacitor is not discharged until a new condition occurs. Verification of these tests is left to the reader.

In order to make the open-circuit faults in transistors P3 and N3 detectable, the circuit shown in Fig. 7.7.19 was developed. An additional controllable input C_a was introduced. During normal operation of the circuit, C_a is kept at logic 1, and during testing it can be changed independently of the clock state. When the tests of Table 7.7.1 are applied to the new circuit, with $C_a = 1$, they can still detect faults located in the unchanged part of the circuit.

Table 7.7.1 Testing the D-latch from Fig. 7.7.19 Part 1

Open circuit in	Test sequence			Fault-free latch			Faulty latch		
	t	C	D	NC	NQ	Q	NC	NQ	Q
P1 or P2	t_1	1	1	0	0	1	0	0	1
	t_2	0	0	1	0	1	1	0	1
	t_3	1	0	0	1	0	0	Z0	1
	t_4	0	X	1	1	0	1	0	1
N1 or N2	t_1	1	0	0	1	0	0	1	0
	t_2	0	1	1	1	0	1	1	0
	t_3	1	1	0	0	1	0	Z1	0
	t_4	0	X	1	0	1	1	1	0
P5	t_1	1	0	0	1	0	0	1	0
	t_2	0	1	1	1	0	1	1	0
	t_3	1	1	0	0	1	0	0	Z0
	t_4	0	X	1	0	1	1	1	0
N5	t_1	1	1	0	0	1	0	0	1
	t_2	0	0	1	0	1	1	0	1
	t_3	1	0	0	1	0	0	1	Z1
	t_4	0	X	1	1	0	1	0	1
P6	t_1	1	1	0	0	1	0	0	1
	t_2	0	0	1	0	1	Z0	1	0
N6	t_1	0	1	1	X	X	1	X	X
	t_2	1	1	0	0	1	Z1	0	1
	t_3	0	0	1	0	1	1	0	1
	t_4	1	0	0	1	0	Z1	0	1
	t_5	0	X	1	1	0	1	0	1

By using such a testable D-latch, testable flip-flops can be built. Figure 7.7.20 shows the schematic diagram of a scan-path flip-flop that uses the previous results. *M* denotes the mode signal, and NM is its complement. SD_{in} is the test signal input.

Table 7.7.2 contains only tests for the rest of faults in the modified D-latch.

In order to ensure the testability of faults of the "open-circuit" type in flip-flops that will be constructed from such a latch, when they are integral parts of synchronous sequential circuits, we set certain limitations to the test signals. If the tests meet the set of requirements when testing open circuit in the D-latch, then they will be applicable when testing MS and scan-flip-flops.

In this circuit, only the *D* signal is used, not its complement. On the other hand, both the clock signal *C* and its complement $NC = \overline{C}$ are required.

Similar considerations lead to solutions for testable circuits in other technologies. Examples of the development of testable BiCMOS circuits are described in [13, 14].

7.7.5 DFT in Mixed-Signal Circuits

First, the design procedure for testability based on the boundary scan concept will be described. It starts from the usual chip layout where the analog and digital parts are usually physically separated, and sometimes with special measures to prevent crosstalk. Accordingly, the IEEE 1141.4 standard was developed as an extension of the older 1141.1 standard [15]. The architecture of the chip designed according to this standard is given in Fig. 7.7.21.

The following notations, used in Fig. 7.7.21, have the same meaning as in the IEEE 1141.1 standard:

TAP Test access port
TDI Serial input of test data
TDO Serial output of test data
TMS test mode select
TCK Clock
TRST*n* test reset (optional).

As can be seen from Fig. 7.7.21, the system was expanded with analog boundary scan cells as well as additional terminals and functions. Thus, it can be said that an ATAP (from Analog Test Access Port) was formed, which is represented by two terminals in the figure:

AT1 which serves to supply the analog test excitation and
AT2 which accepts the analog output signal.

Optionally, inverse AT1 (AT1*n*) and inverse AT2 (AT2*n*) can be added, which are used in formation of differential signals.

As one can see, AT1 and AT2 are connected to the analog boundary scan cells via the subsystem labeled TBIC (from Test Bus Interface Circuit). Namely, there are two buses in the circuit, one for the analog subsystem and the other that includes the

Fig. 7.7.20 Testable scan-path flip-flop with a single clock

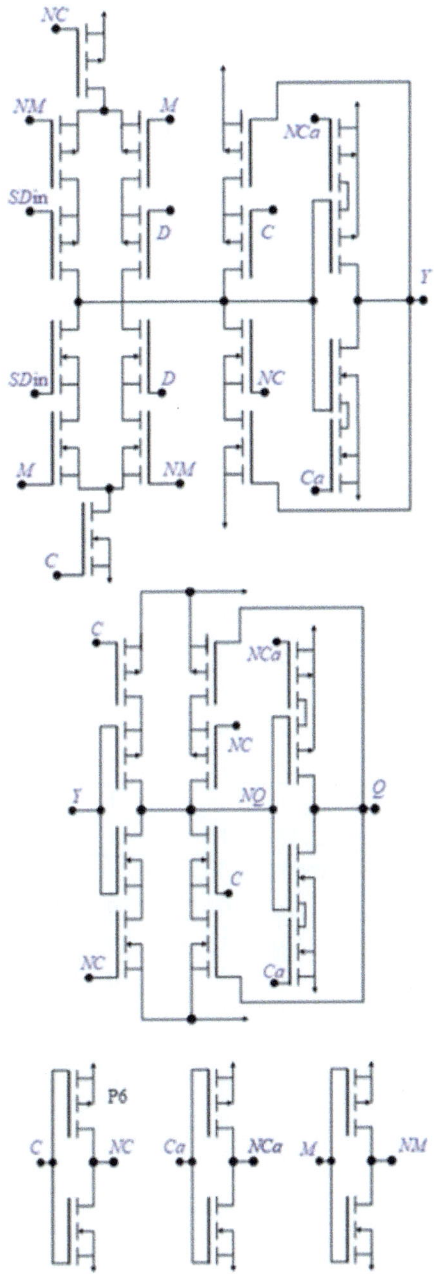

Table 7.7.2 Testing the D-latch from Fig. 7.7.19 Part 2

Open-circuit in	Test sequence				Fault-free latch				Faulty latch			
	t	C	D	C_a	NC_a	NC	NQ	Q	NC_a	NC	NQ	Q
P3 or P4	t_1	1	0	1	0	0	1	0	0	0	1	0
	t_2	0	1	1	0	1	1	0	0	1	1	0
	t_3	0	1	0	1	1	1	Z0	1	1	1	Z0
	t_4	1	1	0	1	0	0	Z0	1	0	0	Z0
	t_5	0	X	0	1	1	1	Z0	1	1	Z0	Z0
	t_6	0	X	1	0	1	1	0	0	1	0	1
N3 or N4	t_1	1	1	1	0	0	0	1	0	0	0	1
	t_2	0	0	1	0	1	0	1	0	1	0	1
	t_3	0	0	0	1	1	0	Z1	1	1	0	Z1
	t_4	1	0	0	1	0	1	Z1	1	0	1	Z1
	t_5	0	X	1	1	1	0	Z1	1	1	Z1	Z1
	t_6	0	X	1	0	1	0	1	0	1	1	0
P7	t_1	1	0	1	0	0	1	0	0	0	1	0
	t_2	0	1	1	0	1	1	0	0	1	1	0
	t_3	1	1	1	0	0	0	1	0	0	0	Z0
	t_4	0	X	1	0	1	0	1	0	1	1	0
N7	t_1	1	1	1	0	0	0	1	0	0	0	1
	t_2	0	0	1	0	1	0	1	0	1	0	1
	t_3	1	0	1	0	0	1	0	0	0	1	Z1
	t_4	0	X	1	0	1	1	0	0	1	0	1
P8	t_1	1	0	1	0	0	1	0	0	0	1	0
	t_2	0	1	1	0	1	1	0	0	1	1	0
	t_3	0	1	0	1	1	1	Z0	Z0	1	1	Z0
	t_4	1	1	0	1	0	0	Z0	Z0	0	0	1
	t_5	0	X	0	1	1	1	Z0	Z0	1	0	1
N8	t_1	1	0	1	0	0	1	0	X	0	1	0
	t_2	0	1	1	0	1	1	0	X	1	1	0
	t_3	0	1	0	1	1	1	Z0	1	1	1	Z0
	t_4	0	1	0	1	1	1	Z0	1	1	1	Z0
	t_5	1	1	0	1	0	0	Z0	1	0	0	Z0
	t_6	0	X	1	0	1	0	Z1	X1	1	1	0

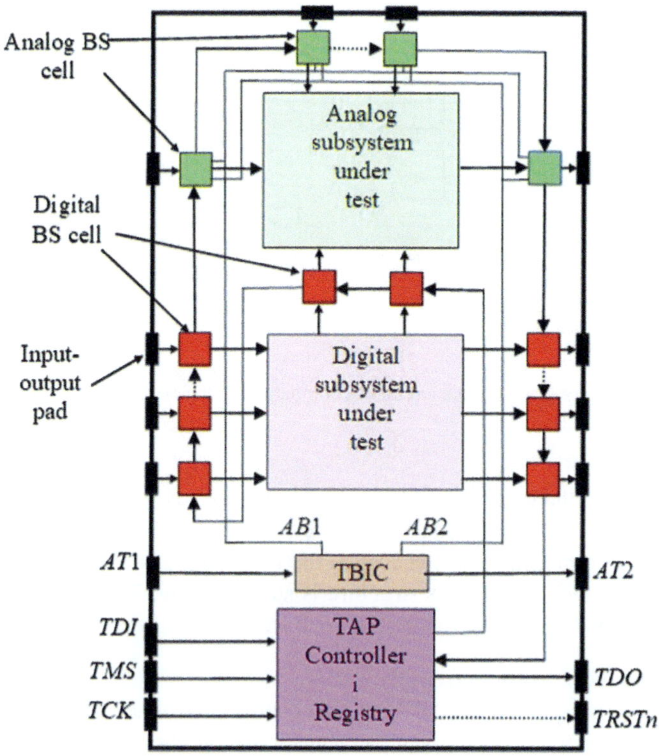

Fig. 7.7.21 A simplified representation of the architecture of a system designed according to BS standard IEEE 1149.4

entire system, with the fact that it first visits the digital boundary scan cells, then the analog ones, and finally, the digital ones again. The analog bus is implemented as two wires AB1 and AB2. By means of AB1, the signal is led to the analog subsystem or to the output port, and by means of AB2, the response is led from the analog subsystem to the output port.

In particular, it should be noted that there are digital boundary scan cells that represent the coupling between the digital and analog subsystems and are unilateral.

The TBIC controls the connection between the ATAP and the internal analog bus. The switching architecture of this cell is shown in Fig. 7.7.22. As one can see, this architecture makes it possible:

- that AT1 or AT2 to be connected to V_H or V_L (constant DC thresholds where V_H is a high level and V_L is a low level)
- that AT1 or AT2 to be connected to AB1 or AB2
- to connect AT1 or AT2 to the internal reference voltage V_C
- to form a one-bit representation of AT1 or AT2 by comparing with V_{TH}.

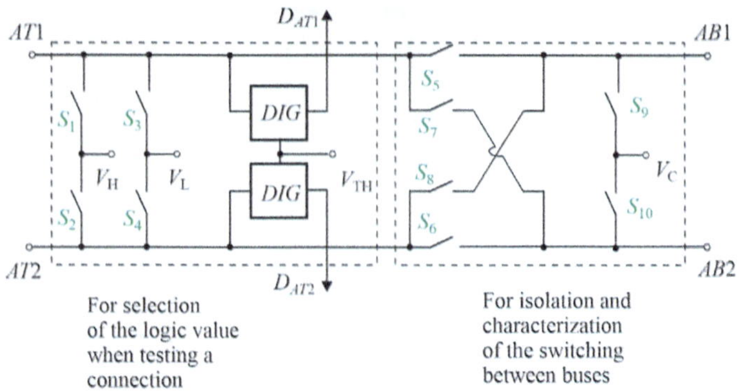

Fig. 7.7.22 TBIC (test bus interface circuit). A comparator that generates one-bit digital information is labeled DIG

The analog boundary scan module (ABM) (or cell) is the heart of the P1149.4 standard. Supplying the test signal and extracting the response through the ABM cell is achieved by combining serial access to the boundary scan register and access to the ATAP analog excitations. Its structure is shown in Fig. 7.7.23. The state of the switches is defined by the control procedure. It should enable the following functions:

- to separate the subsystem from the pad
- that the pad is excited with AB1 (which ensures controllability)
- that AB2 is excited from a given pad (thus ensuring observability)

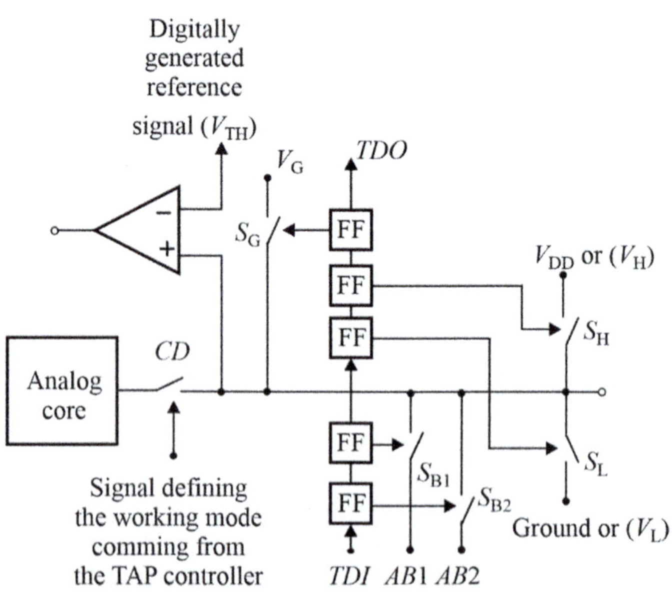

Fig. 7.7.23 Analog boundary scan cell

- to display an one-bit representation of the voltage on the pad compared to V_{TH}
- to bring V_H or V_L to the pad (when testing connections)
- to bring a quality reference voltage V_G to the pad (for parametric measurements).

The reader is advised to visit [16] for some advances in ABM synthesis.

When it comes to the set of instructions introduced by the IEEE 1149.4 standard as compared with 1149.1, the difference is small. Here we will list only the MEASURE, PROBE, and CLAMP instructions as the most important new features. Thereby:

- the MEASURE instruction is used to test a common wire (connection), connection with passive components, differential or extended connections,
- the PROBE instruction allows the analog pins of the integrated circuit to be connected to AB1 and AB2. This mode of operation prevents feedback and other unwanted effects,
- the CLAMP instruction disconnects all pins from the subsystem under test and freezes the analog pins in the found state. In addition, it freezes the TBIC in the found state and keeps the circuit in a steady state as long as voltage and current measurements are made.

The application and effectiveness of the P1149.4 standard can best be seen through examples. First, the procedure for measuring the impedance connected between a pad and the ground will be described. Only one analog BS cell is used. The way the measurement is set up is shown in Fig. 7.7.24. The measurement flow would be as follows:

- Current I_T is supplied to the impedance Z_L via AT1 and switches S_5 and S_{B1}, while
- the impedance voltage, V_T, is measured at AT2 by connecting Z_L through S_6 and S_{B2}.
- Of course, we get the impedance value as $Z_D = V_T/I_T$.

As the second example, the complete measurement of the analog part will be shown. Figure 7.7.25 shows the analog core being tested and two analog boundary

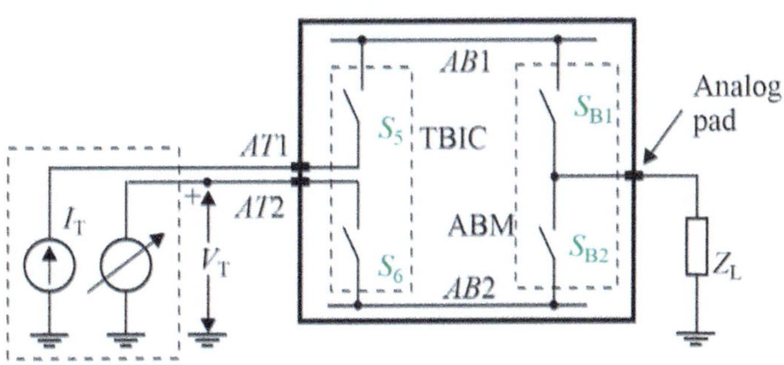

Fig. 7.7.24 Application of analog boundary scan for measurement of grounded impedance

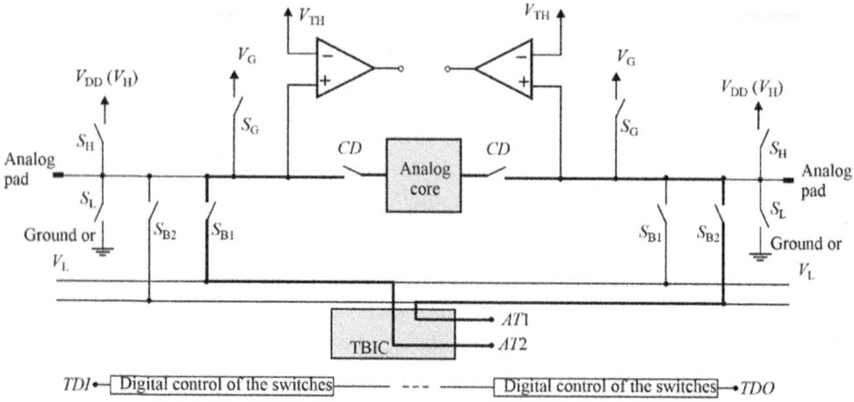

Fig. 7.7.25 By closing the switches S_{B1} and CD in the left cell and CD and S_{B2} in the right cell, the conditions are created for the signal AT1 to be supplied as an excitation, and the signal AT2 to be extracted as a response

scan cells. An excitation signal is fed through one, and the resulting response is fed to the chip output through the other. The signal establishment path is marked with a thick line.

In conclusion, we can say that the P1149.4 standard allows DFT of circuits with mixed signals mostly because it enables controllability and observability. It should not be forgotten that in the absence of such a concept, the designer would be forced to separate a large area of the chip in order to provide additional test points for analog signals.

The price paid for achieving testability refers to the specifics of the analog parts of the circuit. Namely, due to finite resistances (series resistance of a switch, for example) and capacitance (of the connections, for example) in the circuit, it can be said that the average error in the measurement is about 5%. Moreover, if the signal propagation path is too long, it may be necessary to regenerate it (digitally).

The following usually applies to a switch in CMOS technology: its resistance in the on-state is between 100 and 1000 Ω, its resistance in the off-state is about 10^{12} Ω. The time required for state change (commutation) is less than 1 ms, and the required silicon surface is about 20 mm^2. The last quantities, both, get reduced with the advance of the technology, i.e., by reducing the sizes and parasitics of the transistors.

When measuring integrated circuits that are designed in this way, some other effects should be considered, such as

- the switches themselves exhibit a nonlinear characteristic when changing state, which introduces a specific transients,
- appearance of unwanted impedances during testing,
- occurrence of leakage currents towards the ground, and
- the bandwidth of the 1149.4 bus at the time of the introduction of the standard was 1 MHz only.

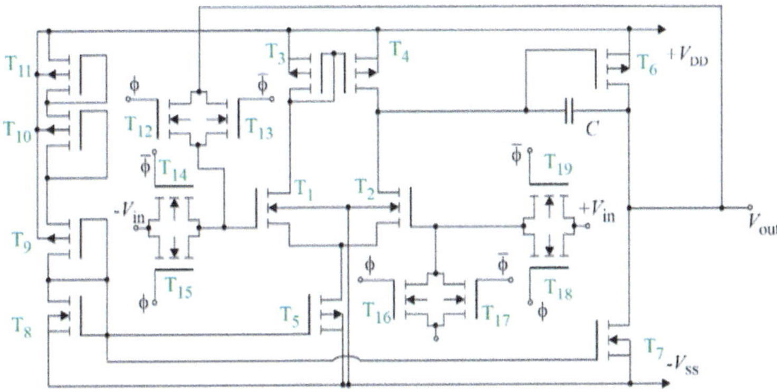

Fig. 7.7.26 Switched operational amplifier (circuit diagram)

7.7.6 Designing for Testability of Analog Circuits at the Electrical Circuit Level

The application of the DFT concept of analog circuits at the electrical circuit level will be illustrated by the example of an operational amplifier, considering that it is the most commonly used analog block.

In Sect. 7.7.2, we discussed the selection of test signals for testing an operational amplifier. The most important version of the op-amp circuit that allows testing is the unity gain or the isolation amplifier because it contains no other elements than the op-amp and thus the fault effects of the amplifier are not masked. Details related to the generation of test signals for this configuration were also provided.

Having this in mind, when designing for testability, the aim is to enable "configuration" of the operational amplifier when in the test mode of operation, so that it is transformed into a schematic that represents a unity gain amplifier. An example of such a solution is the circuit shown in Fig. 7.7.26. It can be seen that the circuit is expanded with four analog transmission gates, i.e., with four CMOS transistors connected in parallel that act as switches. The state of the switch is determined by the auxiliary signal, which is denoted (in the usual way) by ϕ. The test signal V_T is fed to a new terminal introduced for this purpose.

Figure 7.7.27 illustrates how a switched operational amplifier behaves. When $\phi = 0$, the feedback circuit and the test signal are turned off so that a normal amplifier remains (Fig. 7.7.27b). When switching for testing, $\phi = 1$ is set. Then the feedback loop and the test signal are activated, and the normal excitation signals are turned off (Fig. 7.7.27c).

As mentioned in the description of the IEEE 1149.4 standard, switches are imperfect and, in fact, at $\phi = 1$, we get a finite resistance between the input and the output. Considering, however, that in this mode the inverting terminal is isolated, the influence of this resistance will be expressed only at the highest frequencies of circuit application, which means that its influence is negligible for practical applications.

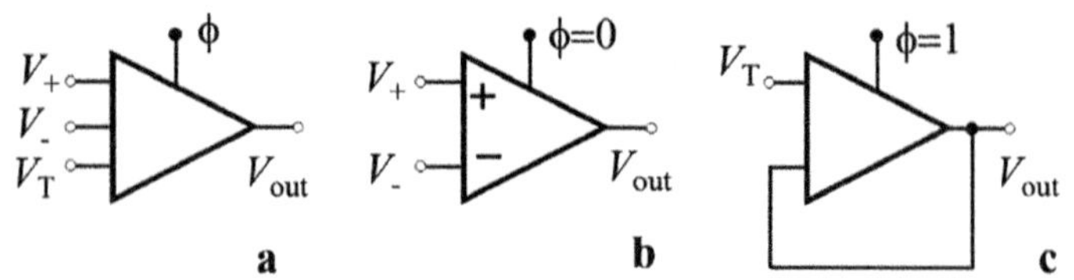

Fig. 7.7.27 Switched operational amplifier. **a** Symbol, **b** normal operation mode, and **c** testing operation mode

References

1. Williams TW (1986) Design for testability. In: Williams TW (1986) VLSI testing. North-Holland, Amsterdam
2. Williams TW (1993) IEEE standard test access port and boundary-scan architecture. IEEE, New York
3. Williams TW (1995) P1149.4 Mixed-signal test bus standard; working group meetings, minutes, and associate materials. IEEE, New York
4. Williams TW (1995) Supplement to IEEE Std. 1149.1-1990; IEEE standard test access port and boundary-scan architecture. IEEE, New York
5. Williams TW (1995) P1149.4 mixed-signal test bus. IEEE Standard Project. IEEE, New York
6. Williams TW (2023) Testing of electronics. https://www.jtag.com/testing-of-electronics-tutorial-jtag/
7. Bleeker H, van den Eijnden P (1993) Boundary-scan test a practical approach. Kluwer Academic Publishers, Dordrecht
8. Panić V, Janković S, Milovanović D, Litovski VB (2000) Cell design for boundary-scan implementation. In: 22nd International conference on microelectronics, MIEL'99, Niš, Serbia, pp 719–722
9. Bruce WC et al (1991) Implementing 1149.1 on CMOS microprocessors. In: Proceedings of the international test conference, Nashville, USA, pp 879–886
10. Tegethoff MMV, Parker KP (1995) IEEE Std. 1149.1: where are we? Where from here? IEEE Des Test Comput 12(2):53–59
11. Parker KP, McDermid JE, Oresjo S (1993) Structure and metrology of analog testability bus. In: Proceedings of the IEEE international test conference, Baltimore, MD, USA, pp 309–322
12. Reddy MK, Reddy SM (1986) Detecting FET stuck-open faults in CMOS latches and flip-flops. IEEE Des Test Comput 3(5):17–26
13. Menon SM, Jayasumana AP, Malaiya YK (1995) Manifestations of faults in single—and double-BJT BICMOS logic gates. IEE Proc Comput Digital Tech 142(2):135–144
14. Menon SM, Malaiya YK, Jayasumana AP (1995) Testable design of BICMOS circuits for stuck-open fault detection using single patterns. IEEE J Solid State Circuits 30(8):855–863
15. Scheffer L, Lavagno L, Martin G (2006) EDA for IC system design, verification, and testing. Taylor and Francis, Boca Raton
16. Kač U, Novak F, Azaïs F, Nouet P, Renovell M (2003) Extending IEEE Std. 1149.4 analog boundary modules to enhance mixed-signal test. IEEE Design Test Comput 20(2):32–39

7.8 Built in Test

Abstract It is well known that the designer bares the responsibility of the whole life cycle of the product he/she designs no matter what the product is. In that, the main concern is the functionality of the product which here is an electronic circuit or system. To ensure life-long successful performance one is to use as good technology as it is available. On the other side to ensure warning that a malfunction is lurking one is to continuously test the functionality be it in the phase of production or in phase of exploitation. Here we discuss that issue in the context of very complex circuits in which the access to the potentially faulty sites from outside becomes practically not possible. Implementation of scan-path techniques, due to the complexity, becomes also not feasible. The solution is found in the idea to redesign the circuit so that, in addition to normal function, to be capable to perform testing based on incorporated additional circuitry and to export signatures reporting the status. If these structures are activated from outside we call this concept built-in test (BIT) and if it gets activated periodically (or non-periodically, based on internal signals) it will be named built-in self-test (BIST). In this way some crucial parts of the system can be tested at low price as compared to use complex test equipment including specific test signal synthesis. To BIT of analog circuits is this chapter devoted while the issue of BIT of digital circuit is out of the scope of this book since it needs complete digital system synthesis to be involved.

7.8.1 Introduction

The growth in the number of transistors per integrated circuit is significantly faster than the growth in the number of pins, so that the possibility of accessing individual transistors and groups of transistors from the outside is decreasing more and more. If the number of transistors per pin is called *Index of testing complexity*, then in the year 2002 its value reached about 75,000 [1].

Another particularly important aspect that complicates testing of large-scale integrated circuits is related to the "number of machines" that are built into such a chip. Namely, if the circuit were combinational, regardless of the number of transistors, it

V. B. Litovski, *Lecture Notes in Analog Electronics*, Lecture Notes in Electrical Engineering 1265, https://doi.org/10.1007/978-981-97-8257-4_8

would be enough to supply signals to the input and to measure them at the output. In real chips, however, due to the complexity of the function, it may turn out that multiple sequential subsystems are embedded, some of which even have their own clock signal different from the main clock. Therefore, with the increase in the number of such subsystems, testing is more complex, and the test results are increasingly important.

The limited number of pins, which can be expressed as a limitation on input/output approaches, is more than a difficulty in accessing the transistors. Namely, the growing disproportion between the frequency of the internal (silicon) clock signal and the upper cut-off frequency of the signal that passes through the package, makes testing with external signals difficult and sometimes impossible.

If the input/output bandwidth (IOB) is defined as the product of the number of pins and the maximum frequency of the signal, then based on the available data, it can be argued that its value increases much more slowly than the internal bandwidth (IB), which is defined as the product of the number of transistors and the maximum operating frequency for silicon. If the ratio of these two bandwidths in 1997 was equal to (IB/IOB) one, in 1999 it increased to (3.165:1.97), and in 2002 reached (19.52:3.59).

At the same time, the price of the package (pins) is falling much more slowly than the price of the transistor. After all, the fact that the dimensions of the input/output terminals remain at the macroscopic level while the dimensions of the transistors slowly move into the nanometer region is a consequence of the problems related to the assembly of the chip and the package, and the package and the printed circuit board.

In today's circuits, we still encounter the concept of testing that was illustrated in the first chapter and in the previous section. Because it is considered that the difference in the value of the internal and external bandwidth is small, it is possible that the test data is supplied externally from the excitation source, and that the response obtained from the input/output pins is led to the external measuring equipment. The test equipment, as stated, would contain a signal generator, a response meter, a control computer, and a coupling probe.

In the actual technological generation of integrated circuit production, the number of transistors per pin has long exceeded 10,000. In such conditions, relying only on an external signal would be out of the question. The idea of changing the architecture of the chip in order to adapt it to testing became dominant. One concept that is most popular today is the scan-path technique and boundary scan as its version. In this way, the transfer of test data from the pins to the chip is facilitated and access to selected transistors and groups of transistors is enabled as discussed in the previous chapter.

However, it should be borne in mind that the size of the test vector (number of test words) increases with the number of faults we test, and the number of potential faults with the number of transistors on the chip. If one works with chips whose complexity exceeds a million transistors, the size of individual subsystems that make up the boundary scan becomes a bottleneck.

The first alternative would be the implementation of the built-off concept [2]. It uses built-off test interface (BOTI) which is a specialized hardware intended to

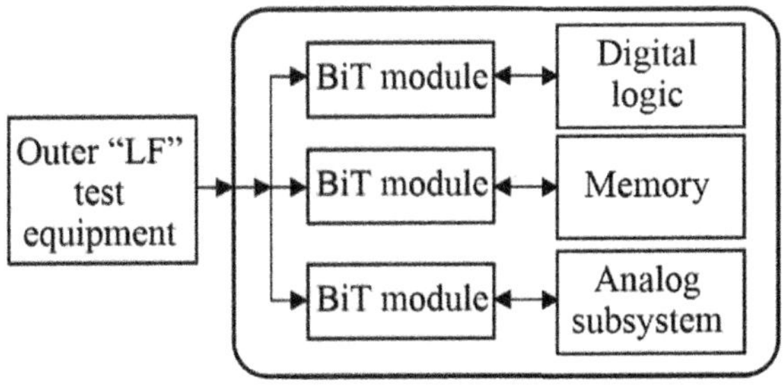

Fig. 7.8.1 Chip architecture with built-in testing

facilitate the testing of systems-on-packages (SOPs). In that, the tester (i.e., the automatic test equipment) may be programmed to behave as a BOT. This concept, being less general, will not be elaborated further in this chapter.

If some test functions are introduced into the chip, that is, if the concept of built-in testing is applied, both main obstacles can be overcome: the disproportion (number of transistors)/(number of pins) and the difference in internal and external bandwidth [3]. The architecture of the chip that would be designed on this principle would be as in Fig. 7.8.1. It can be seen that it is assumed that an analog part is also built into the chip, which is increasingly the case, especially in modern telecommunication circuits.

An external "low-frequency" signal based on the IEEE 1149.1 standard would only control the test, and special built-in units would under its control generate test signals and collect the responses to send back to the TDI.

7.8.1.1 Basic Settings of BIT Application

The concept that an electronic circuit tests itself without the need to provide a separate external test vector and a separate response evaluation circuit is very attractive especially when the complexity of the circuit being tested grows and requires more and more complex testing systems.

BIT can be done in two ways:

- by adding to the circuit some test tools that work continuously with the circuit and continuously "monitors" the response values or
- by installing structures in the circuit so that the circuit can be switched from its normal operating mode to the test operating mode as needed.

We will call the first of the mentioned possibilities *live testing*. One of the advantages of this procedure is the detection of intermittent faults. Namely, there are faults that manifest themselves only in some special conditions that, most often, are not explicitly related to the signals in the circuit. The cause of intermittent faults can be ambient conditions, for example. Bearing in mind that the circuit is continuously monitored, such faults can be detected, which is less likely when applying other concepts.

This procedure is often called concurrent or implicit testing and can be implemented as informational or hardware redundancy. The best-known way of applying informational redundancy is the application of the so-called parity bits. With this method, the number of bits in the digital word is increased so that an even (or odd) number of ones is always achieved with a correct operation. When a fault occurs, it is most likely that one bit will be wrong, which disturbs the parity of the number of ones in it and enables testing.

Hardware redundancy is reduced to replicating parts or the entire circuit and checking whether the parts produce identical results. The number of replicas can be higher than two. While informational redundancy is often implemented, hardware redundancy usually requires a larger silicon area and leads to a significant increase in the power consumption and the cost of the circuit being produced. We will not devote further attention to this type of self-testing. The reader is referred to [4] for more details.

The second way is most often recognized under the name BIT, which comes from "built-in test," and it will be given disproportionately more attention. In this approach, it is assumed that the architecture of the circuit (integrated or on the printed circuit board) is adapted similarly to the scan concept, with the fact that the test signals are not taken from the outside, and also, the comparison of the test results and the correct response is not performed from the outside. Accordingly, the main attention in the following sections will be devoted to (on chip) generation of the test signal, establishing the test conditions, and monitoring the response.

Before proceeding to the BIT of analog circuits, in order to establish some context, a concept implemented for digital circuits will be very shortly explained. Here the so-called built-in logic block observation (BILBO) concept or monitoring using a built-in block will be considered.

It starts from the fact that the circuit under test is broken down into subcircuits or subsystems. BILBO registers are installed between them, which are sequentially connected as in Fig. 7.8.2a. In normal operation, the BILBO register is part of the existing sequential logic and connects the sequential and combinational logic. In test mode, a pair of BILBO registers, one in front and one behind each block being tested, are used to test it. In doing so, the first BILBO block is configured to act as a test signal generator, while the other generates a signature based on the response of the circuit under test. The *signature* here means the summary information about the response.

The signature can later be read as serial information. For that, it is necessary to reconfigure the BILBO block as a scan chain. Accordingly, each BILBO register has the following main modes of operation (according to Fig. 7.8.2b):

- work in normal mode,
- reconfiguration in (scan-in/scan-out) scan chain,
- reconfiguration into a test signal generator, i.e., into a generator of pseudo-random test words, and

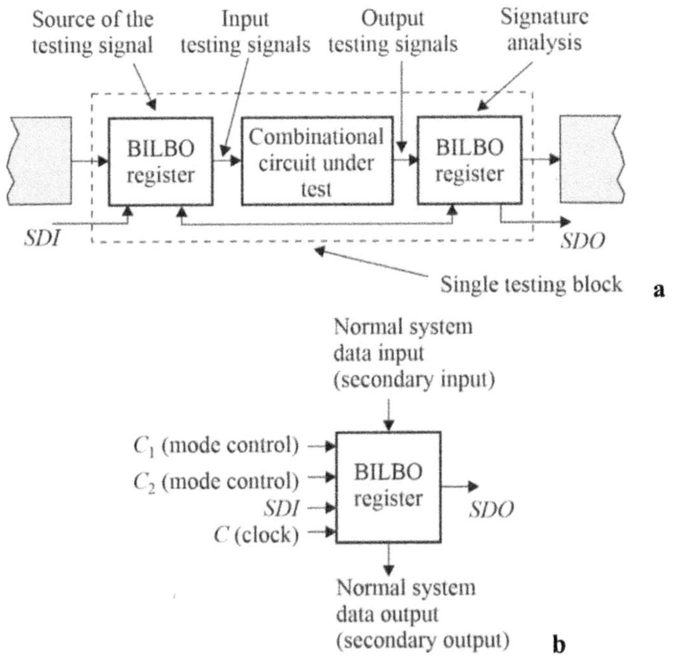

Fig. 7.8.2 **a** Application of the BILBO BIT concept to a combinational digital circuit that is broken down into smaller blocks (*SDI* scan data in, *SDO* scan data out) and **b** the BILBO functional model

- reconfiguration into the output register that generates the signature.

7.8.1.2 BIT for Analog Circuits

Testing analog and mixed-signal circuits, as we have seen in the previous chapters, is a relatively difficult job since most of the time the testing comes down to checking the functionality. In other words, testing analog and mixed-signal circuits is more expensive and requires more time, not to mention the uncertainty in measuring analog quantities either at the input or at the output of the circuit being tested. When it comes to circuits of very high complexity, which we are talking about here, that is, circuits where analog access (input) practically does not exist, things get even more complicated. That is why BIT is practically the most feasible strategy that can be used in such situations.

In order to realize the concept of BIT on a chip, when dealing with analog and mixed-signal circuits, as opposed to BIT for digital circuits, two important problems need to be solved. The first is the problem of generating analog test signals on the chip. Namely, test signals, as we have seen, can be linearly increasing voltage, sinusoid, noise, or a combination of these and other signals. Therefore, it is necessary to

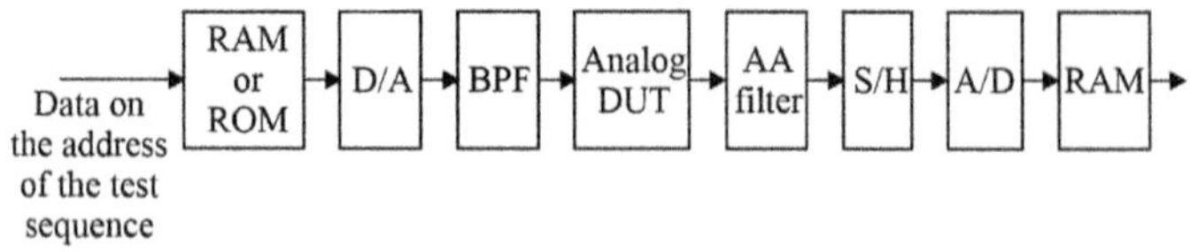

Fig. 7.8.3 A common way of testing analog subsystem (*BPF* bandpass filter)

synthesize and physically realize (on silicon) the circuits that generate those signals. On the other hand, at the output of the circuit under test, a signature should be formed, and sometimes even additional electronics for analyzing the signature so that a decision can be made inside the chip about the correct operation of the circuit (what, being similar to the concept of designing a tester (system), will also be discussed in the next chapter).

A common solution used when testing analog parts of integrated circuits using the BIT concept is shown in Fig. 7.8.3. The test is controlled by computer or digitally. The address of the data related to the test signals is communicated to the memory. The obtained data is converted into an analog test signal. If necessary, this signal can be frequency-limited by a bandpass filter (BPF). The response of the circuit under test (CUT) is first passed through an anti-aliasing (AA) filter and then converted into digital information that is stored and available for further processing.

When considering this architecture, the following should be noted. Given that the goal is testing and the output signal is given in digital form, it is possible to perform the checking of the measurement results by (digital) hardware comparing signatures. This means that it is possible to implement the standard methods used for digital circuits. The question of synthesizing the analog signal generator remains.

Within the framework of intensive research in this area in recent years, mainly two solutions are offered for the problem of generating analog test signals on a chip. According to one variant, programmable subsystems are proposed that generate an analog signal, usually a sinusoid, for a given digital description. We will call this circuit "BIT analog test generator." Alternatively, the oscillation method mentioned in the section on analog circuit testing is applied. In this method, the analog CUT becomes part of the feedback loop, under the control of the digital control circuit, and under appropriate conditions behaves like an oscillator. The properties of the signals generated in the oscillator are used to decide the correctness of the DUT. These two concepts will be described in more detail below.

7.8.1.2.1 BIT Analog Test Signal Generator

On-chip generation of analog test signals has several advantages, the most important of which is that signals that are much closer to ideal can be obtained (as compared to using the concept of Fig. 7.8.3) and that the required area on the chip can be expected to be smaller. The basic signal generation strategy for testing the analog subsystem is shown in Fig. 7.8.4. In this figure, by means of a multiplexer, the analog signals

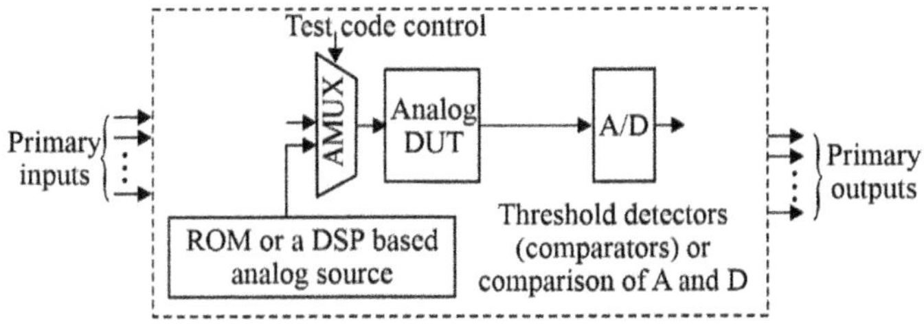

Fig. 7.8.4 Basic concept of autonomous generation of test signals on the chip

supplied during normal operation (mode) are replaced by signals generated on the chip using a suitable signal processing circuit that is able to repeat the test signal and be synchronous with other activities in the entire self-testing procedure.

Processing of the response of the analog circuit, for the sake of clarity, is not shown in Fig. 7.8.4. The response can be derived in several ways, and one of them is shown in Fig. 7.8.5. Namely, after A/D conversion, the signal is fed to the MISR (multiple input signature register) and, if necessary, read under the control of the synchronization signal.

There are several proposals for solving the problem of on-chip analog signal generation [5, 6]. Some limitations are set for the synthesis of these circuits [5], of which the following will be listed here:

- the source should be capable of generating high-quality alternating signals whose quality is higher than the analog circuits under test,
- the source should allow full digital programmability,
- ideally, it should be able to generate multiple signals at the same time,
- the source should not be sensitive to variations in the manufacturing process and should not require external calibration, and
- the source should occupy as small an area as possible on the chip, and it is preferred that its area is no larger than 10% of the area of the mixed-signal circuit chip.

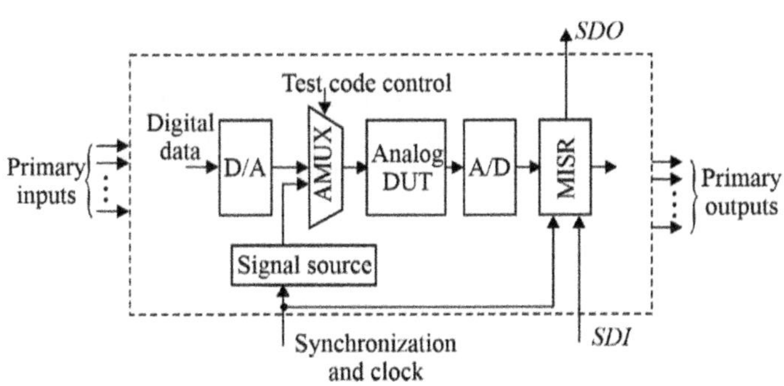

Fig. 7.8.5 Use of MISR to derive the measured digital signal

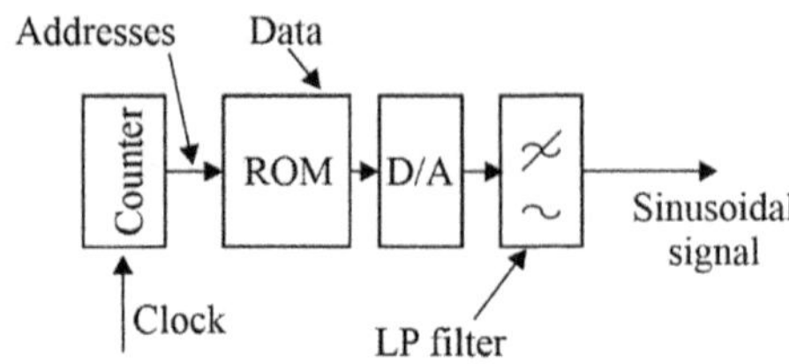

Fig. 7.8.6 Generating analog waveforms using ROM

The most natural way to generate an on-chip analog waveform is to use ROM. In this case the data that is memorized, when read and passed through a D/A converter, approximates the required analog waveform. In order to eliminate the quantization noise, a filter is usually required at the output, which is illustrated in Fig. 7.8.6. Theoretically, any waveform can be generated in this way, the limitations being the size of memory required and the value of the frequency that governs the sampling and conversion process.

However, to make the frequency of the output signal more controllable than when the signal is read from memory, it is possible to introduce some form of feedback in the system where the digital data is stored. This method of signal processing is called DDFS (from Direct Digital Frequency Synthesis) [7] Alternative solution may be found in https://www.analog.com/en/analog-dialogue/articles/all-about-direct-digital-synthesis.html. The general structure of such a subsystem is shown in Fig. 7.8.7. In this system, the accumulator periodically generates a phase angle, φ. This signal is fed to a sine function generator that generates a digital sinusoid. The signal is then converted and filtered to produce analog sinusoid. From a design point of view, the biggest problems here are related to the design of an accumulator which, in general, occupies a very large area and the design of a fast and precise sine signal generator. It is shown that the application of this concept leads to a large area on the chip, making it unpopular due to its cost, although efforts are made to achieve better results by parallelizing the hardware [8] (Fig. 7.8.7).

One of the most popular methods used today as an alternative to the DDFS procedure is the application of ΔΣ modulation (delta-sigma). It is claimed that this concept leads to a solution with less silicon consumption. The block diagram of the sine signal generator based on ΔΣ modulation is shown in Fig. 7.8.8 and is highlighted by a dashed rectangle. It can be seen that the circuit can have two outputs. The main output is an one-bit sequence that is filtered by a low-pass filter and thus a pure sinusoid is created (which depends exclusively on the selectivity of the filter). Of course, this

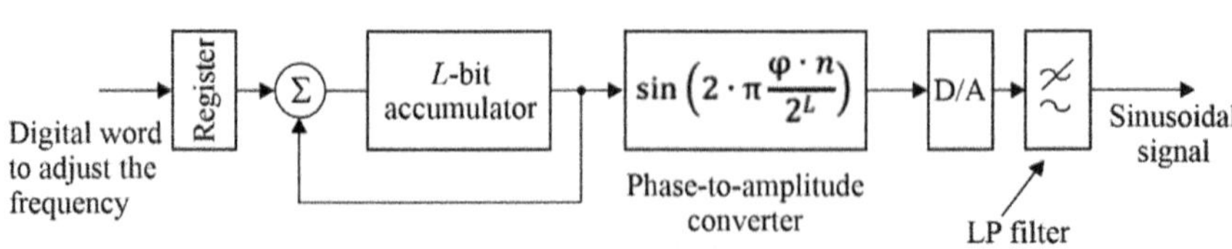

Fig. 7.8.7 Application of the DDFS concept for generation analog signals

filter also removes quantization noise. If, however, a different test signal is desired, it is possible to feed the n-bit word generated at the input of the modulator itself to a shaping circuit that generates a different waveform with the same period at its output.

The circuit of a first-order $\Delta\Sigma$ modulator is shown in Fig. 7.8.9. In this system, z^{-1} is an unity-delay circuit, and the quantizer is a threshold comparison circuit, whose output is equal to a logic 1 if its n-bit input is above a given binary number, and equal to a logic 0 in the opposite. The D/D converter gives an n-bit output 111…1 if there is logic 1 at its input, and 000…0 otherwise.

Although the flexibility of this solution is indisputable, when evaluating it, one concludes that the required number of logic operators is still considerable and that the improvement compared to DDFS is not as great as it seems at first glance. An additional problem with this solution is the fact that the frequency of the sine signal at the output is significantly lower (about three orders of magnitude) than the clock frequency, which means that the maximum frequency of the synthesized sine signal is severely limited.

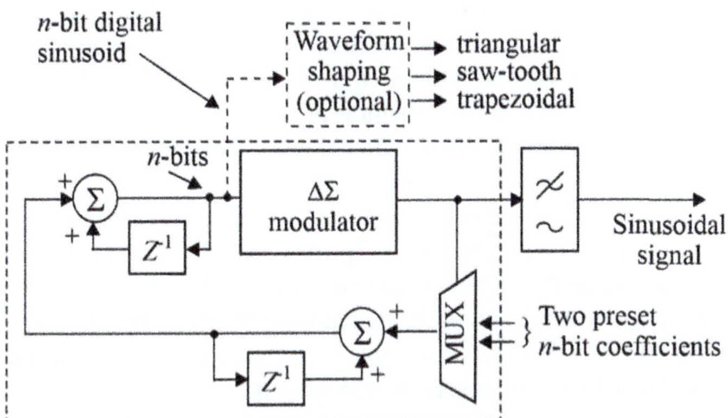

Fig. 7.8.8 Analog signal synthesis system based on $\Delta\Sigma$ modulator

Fig. 7.8.9 $\Delta\Sigma$ first-order modulator

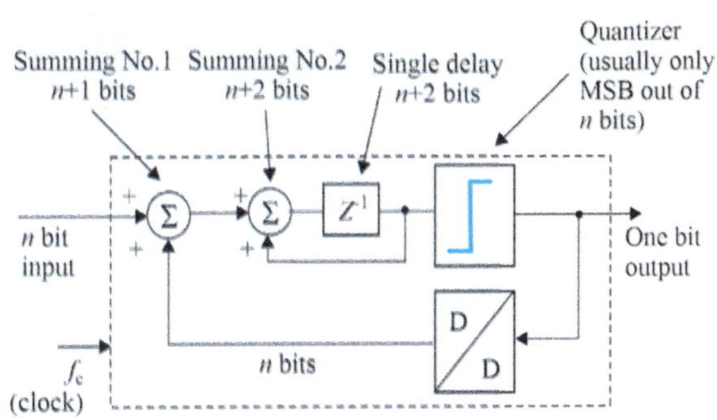

7.8.1.2.2 The Oscillation Method

Instead of applying separate subsystems for signal generation on the chip, a solution is offered that turns the circuit under test (CUT) itself into a signal generator [9–13]. This method was discussed briefly in the second chapter of this book, and here we will consider it with some more details. As said, it deserves special attention because it is conceptually simpler and does not require significant modifications to the CUT. The method is called OBT from Oscillation-Based Testing.

The procedure is to break down the complex analog system into functional blocks such as operational amplifiers, comparators, filters, A/D and D/A converters, phase-locked loops, etc., or to combinations of such blocks. Then, by adding the appropriate circuit, each block turns into an oscillator with the aim of obtaining sustainable oscillations. It remains to measure the waveform, which most often is reduced to measuring the oscillation frequency, the DC component, or harmonic distortions. Of course, the parameters of the oscillator output signal generated in this way will be different for the fault-free circuit and the faulty one. At the same time, considering that it is a signal that is universal (pulse trains or sine waves), the evaluation of the obtained signals can be standardized independently of the block being tested, which facilitates creation a processable signature and decision making related to the possible presence of faults. The general OBT strategy is shown in Fig. 7.8.10.

Although the application of this method provides a large coverage of faults, it may happen that, in addition to the mentioned ones, there is a need to observe some other response parameters or the supply current.

As an illustration of how an oscillator can be made from a given circuit, here we will take a simple CMOS inverter [12]. First, in order for the waveform to be symmetrical, the switching threshold of the inverter needs to be halfway between the supply voltages, and to achieve this, the P-channel transistors must be about 2.5 times wider than the N-channel ones. This is not a design requirement but a suggestion of how to behave in order to effectively demonstrate the method.

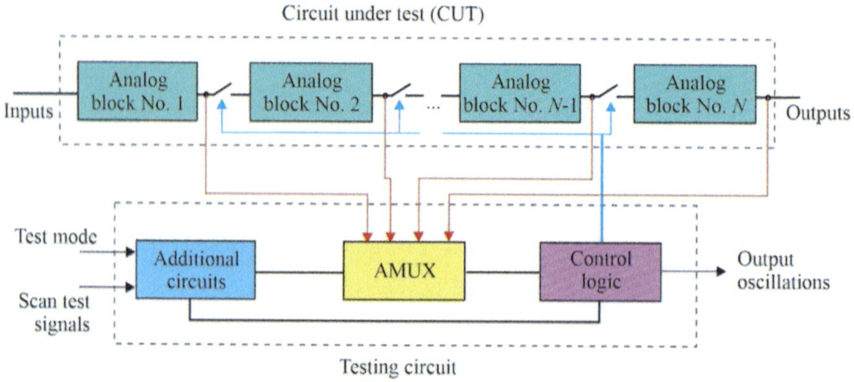

Fig. 7.8.10 OBT's fundamental framework

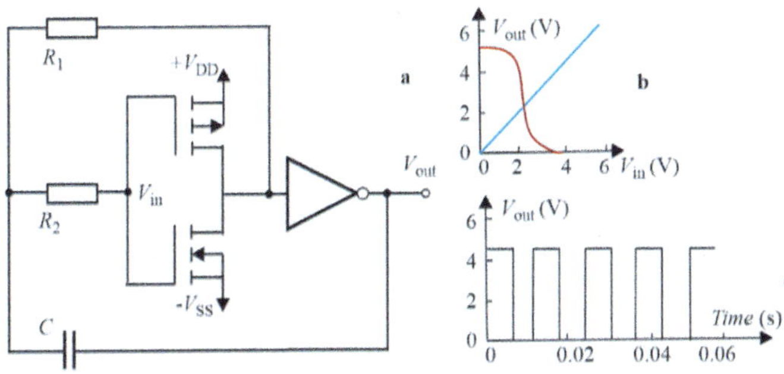

Fig. 7.8.11 CMOS inverter extended to become an oscillator. **a** Circuit schematic, **b** the transfer characteristic and the load line, and **c** output waveforms

By connecting the resistor R_1 between the input and the output, a negative feedback loop is realized that is not dominant, while a positive feedback loop with the appropriate gain and phase angle is realized by adding an inverter and a branch built of a capacitor C. R_2 serves for isolation.

The waveform of the output voltage is shown in Fig. 7.8.11c. A train of pulses is obtained the period of which is given by $T = 2.2 \cdot R_1 C_1$, and the frequency by $f = 1/T$.

The reader can familiarize her/himself with oscillator design by reading [14]. The criteria for designing oscillators for the needs of OBT methods, however, are significantly different from the general principles for designing oscillators. The essence of the difference is the following. When designing an oscillator, requirements are usually set that the amplitude and frequency of the oscillation are stable, that is, that the variations of the circuit parameters are not mapped into the amplitude and frequency of the output signal. When it comes to the OBT method, however, the requirements are just the opposite. Namely, the goal of oscillator design is to establish the presence of parameter variations by measuring *frequency* and *amplitude*, which means that the oscillator should be designed in such a way that these two response parameters are as sensitive as possible to changes in circuit parameters.

To illustrate these considerations, we will consider the principle schematic of an oscillator that is composed of a linear bandpass filter, a nonlinear element that represents the limitation of the amplitude of oscillations and in the simplest case has a simple step transfer characteristic (denoted by F), and an amplifier of finite gain (denoted by k) that is required to establish oscillations, all as on Fig. 7.8.12. For this circuit it is shown that the frequency of oscillation is

$$\omega_0 = \Omega_0 \tag{7.8.1a}$$

and the amplitude of the sinusoidal signal

$$V_0 = \frac{4 \cdot E \cdot h \cdot Q}{\pi \Omega_0}, \tag{7.8.1b}$$

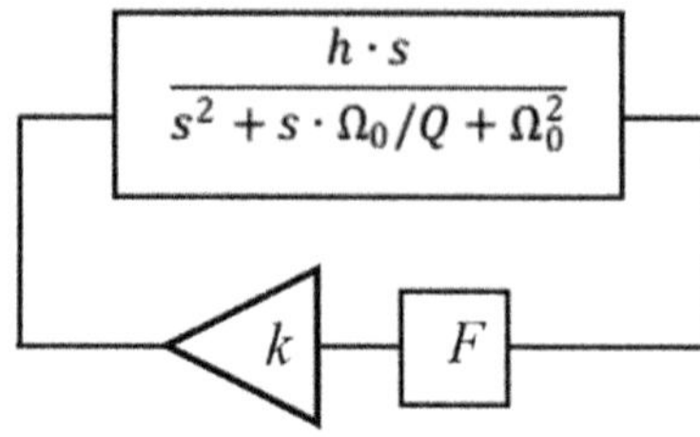

Fig. 7.8.12 Oscillator (Q quality factor of the pole, Ω_0 the center frequency of the filters passband, and $h \cdot Q/\Omega_0 =$ the gain of the filter at $s = j\Omega_0$)

where E is the saturation level, that is, the level of the flattened part of the transfer characteristic (of the nonlinear operator) F. From these expressions it can be seen that the frequency of oscillations is determined only by the value of the central frequency of the bandwidth of the filter, and that the amplitude is influenced by the quality factor of the filter poles, that is, practically all elements of the circuit affect the amplitude. Therefore, the stabilization of the oscillation amplitude would make it impossible to detect potential faults that do not affect the value of the oscillation frequency.

The filter from Fig. 7.8.13 will be used as an illustration of the method. It is about the so-called to the state-variable filter (SVF), which is transformed into an oscillator under the control of the TM (test mode) signal by connecting the output, which is a bandpass filter (BPF), to the input port. The operational amplifier shown in Fig. 7.2.2 is used in this filter. It determines the maximum amplitude of the signal, since it behaves as a nonlinear element at the peaks of the sinusoid, considering that the output transistors go into saturation or cut-off. The values of the circuit elements are as follows: $R_1 = R_2 = R_3 = R_4 = R_5 = 10\ \text{k}\Omega$, $R_6 = 1\ \text{k}\Omega$, $R_7 = 12\ \text{k}\Omega$, and $C_1 = C_2 = 20$ nF [15]. The nominal oscillation frequency of this oscillator is $f_0 =$ 419 Hz.

Table 7.8.1 shows signatures (oscillation frequencies) for different increments of circuit parameters obtained by simulation. Single fault analysis was performed. It was assumed that each element of the circuit gets two new values displaced from the nominal value: one by +20%, and the other by −20%.

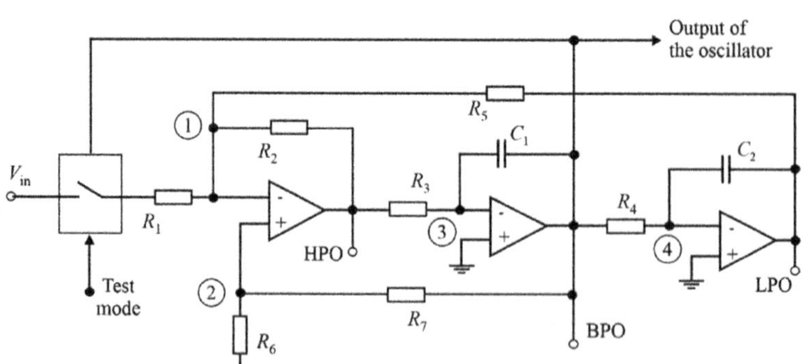

Fig. 7.8.13 Conversion of the state-variable filter (SVF) to an oscillator. *HPF* high pass, *BPF* band pass, and *LPF* low pass

Table 7.8.1 Signature for the SVF obtained by the OBT method

Element	Faulty value (kΩ) or (nF)	Oscillation frequency (Hz)
R_1	8	250.13
	12	447.78
R_2	8	406.67
	12	428.63
R_3	8	457.25
	12	389.56
R_4	8	463.39
	12	385.36
R_5	8	491.88
	12	330.25
R_6	0.8	400.64
	1.2	435.73
R_7	9.6	441.11
	14.4	402.09
C_1	16	466.64
	24	384.47
C_2	16	471.70
	24	381.24

The question of rational measurement and signature processing remains. Namely, Fig. 7.8.5 suggests that the sinusoid obtained in this way will be transposed into the digital domain and communicated to the subsystem for digital signal processing in order to determine its properties. It is shown, however, that this concept, although universal, is optimal only if there is already a DSP subsystem on the chip that can be used for OBT purposes. Otherwise, the required silicon surface area for this purpose can become a limiting factor for the application. In addition, considering that analog circuits that are converted into oscillators are generally low-frequency and considering that it is necessary to perform the measurement on several oscillation periods only after establishing a stationary oscillatory state, the duration of the test (signature generation) can become unacceptably large even for relatively small batches of components. One of the solutions offered to overcome these problems is the use of the concept of TDM (from time-division multiplexing) [15]. It is a method that integrates an A/D converter and signature registers, and at the same time exploits the slowness of the oscillator signal in order to perform the conversion more efficiently (in terms of the number of components in the circuit) with the help of time multiplex. At the same time, in order to increase the fault coverage, in accordance with (7.8.1b), the method enables the observation of signal amplitudes as well.

References

1. Zorian Y (1999) Testing the monster chip. IEEE Spectr 36(7):54–60
2. Hanai H, Yamada S, Mori H, Yamashita E, Funakura T (2001) Built-out self-test (BOST) for analog circuits in a system lsi test solution to reduce test costs. In: Proceedings 10th Asian test symposium, Kyoto, Japan
3. Lubaszewski M, Huertas ML (2004) Test and design-for-test of mixed-signal integrated circuits. In: Reis R, Lubaszewski M, Huertas ML (eds) Test and design-for-test of mixed-signal integrated circuits. Springer, Berlin
4. Hurst S (1988) VLSI testing, digital and mixed analogue/digital techniques. The Institution of Electrical Engineers, London
5. Roberts GW, Lu AK (1995) Analog signal generation for built in self-test of mixed-signal integrated circuits. Kluwer Academic Publishers, Norwell
6. Huang J-L, Ong CK, Cheng K-T (2000) A BIST scheme for on-chip ADC and DAC testing. In: Proceedings design, automation and test in Europe, Paris, France, pp 216–220
7. Tierney J, Radar CM, Gold B (1971) A digital frequency synthesizer. IEEE Trans Audio Electroacoust AU-19(1):28–57
8. Chappell M, McEwan A (2004) A low power high speed accumulator for DDFS applications. In: Proceeding of ISCAS 2004, Vancouver, Canada, pp II.797–II.800
9. Arabi K, Kaminska B (1996) Oscillation-test strategy for analog and mixed-signal integrated circuits. In: Proceedings of the 14th IEEE VLSI test symposium (VTS'96), Princeton, New Jersey, pp 476–482
10. Arabi K, Kaminska B (1997) Efficient and accurate testing of analog-to-digital converters using oscillation-test method. In: Proceedings of the European design and test conference (ED&TC 97), Paris, France, pp 348–352
11. Arabi K, Kaminska B (1997) Testing analog and mixed-signal integrated circuits using oscillation-test method. IEEE Trans Comput Aided Des Integr Circuits Syst 16(7):745–753
12. Arabi K, Kaminska B (1999) Oscillation-test methodology for low-cost testing of active filters. IEEE Trans Instrum Meas 48(4):798–806
13. Das SR et al (207) Testing analog and mixed-signal circuits with built-in hardware-a new approach. IEEE Trans Instrum Meas 56(3):840–855
14. Litovski V (2023) Lecture notes in analog electronics, electronic signal amplification and linear oscillators. Springer Nature, Singapore
15. Roh J, Abraham A (2003) A comprehensive signature analysis scheme for oscillation-test. IEEE Trans CAD Integr Circuits Syst 22(10):1409–1423

7.9 Diagnosis of Analog and Mixed-Signal Circuits

Abstract Analog and mixed-signal diagnosis is a subject rarely visited in teaching literature. Probably, the reason for that is the need for deep understanding of the functionality of the circuits in conjunction with implementation of complex algorithms frequently being very specific to this subject. In any case, in this chapter we will try to list and elaborate several deterministic algorithms that cover processing of different measured signals and implement different inference concepts. A tendency to reduction of the number of measurement points will be followed making both the process tractable and easy to understand. In that, DC, frequency, and time domain signals will be considered as well as linear and nonlinear circuits. Special attention will be paid to hierarchical diagnosis enabling to tackle complex systems as well as mixed-signal ones. In that, artificial neural networks will be promoted as an indispensable tool to capture the circuit properties and to select among hypotheses.

7.9.1 Introduction

Whenever we think about why something does not behave as it should, we are starting the process of diagnosis. Diagnosis is therefore a common activity in our everyday lives [1]. Every system is liable to faults or failures. In the most general terms, a fault is any change in a system that prevents it from operating in the proper manner. We define diagnosis as the task of identifying the cause and location of a fault manifested by some observed behavior. This is often considered to be a two-stage process: first the fact that fault has occurred must be recognized what is referred to as *fault detection*.

Secondly, the nature and location should be determined such that appropriate remedial action may be initiated. The general structure of a diagnostic system is shown in Fig. 7.9.1. Signals $u(t)$ and $y(t)$ are input and output to the system under test being denoted here as the "process," respectively. Faults and disturbances (here measurement errors and/or noise) also influence the system but there is no information about the values of these errors.

V. B. Litovski, *Lecture Notes in Analog Electronics*, Lecture Notes in Electrical Engineering 1265, https://doi.org/10.1007/978-981-97-8257-4_9

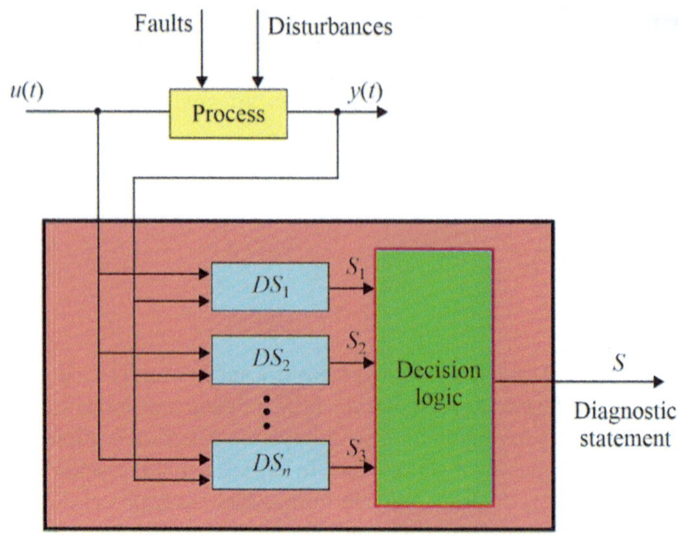

Fig. 7.9.1 General diagnostic system

The task of the diagnostic system is to generate a diagnostic statement *S,* which contains information about fault modes that can explain the behavior of the process. Note that the diagnostic system is assumed to be passive, i.e., it cannot affect the process itself. The whole diagnostic system can be divided into smaller parts here referred to as tests. These tests are also diagnostic systems, DS_i. It is assumed that each of them generates diagnostic statement S_i. The purpose of the decision logic (voting system) is then to combine this information in order to form the diagnostic statement *S.*

The number of possible faults in an electronic system may be large and they can be located everywhere in the system. To diagnose in such conditions one frequently uses hierarchical approach where successive diagnostic statements are generated as the level of description of the system is lowered, going down toward the fault itself [2, 3]. This allows for smaller sets of faults to be considered at a time for the given hierarchical level. Modern automatic test pattern generator may support such concepts [4].

7.9.2 Concepts of Diagnosis

Besides the human expert that is performing the diagnosis, one needs tools that will help, and ideally, perform the diagnosis automatically. Such tools are a great challenge to design engineers because, usually, the diagnostic problem is underspecified. In addition, it is a deductive process with one set of data creating, in general, unlimited number of hypotheses among which we try to find a solution. This is why the research community continues to be attracted by this problem [5]. Thanks to the advances in computational intelligence in the last decades new diagnostic paradigms have been applied based on: model-based concepts [1]; production rule-based artificial intelligence [6]; ANNs [7, 8]; genetic algorithms [9]; and fuzzy-reasoning [10]; all trying to create an approach that contains properties that we might consider to be "intelligent behavior."

Before proceeding with the explanations of why and how specific procedures are applied to analog electronic circuit diagnosis, the diagnostic concept (Fig. 7.9.1) will be elaborated in some detail first. It involves collaboration of design, test, and field engineers and the mutual distribution of responsibilities throughout the life cycle of an electronic product. We assume that field engineers are expected to react after a functional failure of the system. In order to diagnose such a system they need to be supplied with: testing equipment, a list of specific measurements to be done (including a set of signals and test points), and diagnostic software to process the measurement data. A similar set of data and tools would be given to a test engineer in a production-plant environment in order to evaluate the production yield and create feedback to process engineers when prototyping the circuit. We believe; however, design engineers are the most familiar with the product and the most qualified and capable to synthesize test and diagnostic signals, and procedures. This means the simulation before test (SBT) has to be applied to create fault dictionaries containing exhaustive lists of faults and corresponding responses. The *fault dictionary* is in fact a table representing the mapping from the fault list onto a list of faulty (or possibly, fault-free) responses. In that way the diagnostic process becomes a search through the fault dictionary. Alternatively, modem diagnostic techniques using traditional artificial intelligence and reasoning methods typically fall into the simulation after test (SAT) category [11]. This will increase the time spent on diagnosing systems at production time [12].

SBT systems typically require more initial computational costs but provide faster diagnosis at production (or exploitation) time being the second reason why this concept was accepted here.

We claim here that ANNs, being universal approximators [13], are the best way both to capture the mapping, and to search through the dictionary, thereby to perform diagnosis. That is the reason why most of the diagnostic concepts described in the sequel will rely on ANNs.

7.9.3 Diagnosis of Nonlinear Dynamic Analog Circuits

Analog electronic circuits are known to be difficult to test and diagnose. Apart from the huge number of possible faults, this difficulty is a consequence of the inherent nonlinearity of these circuits. Even linear circuits (having linear input–output signal interdependence) exhibit nonlinear relations between the circuit parameters and the output response. There are no linear active networks. Active networks are nonlinear with nonlinear reactive elements. They may be linearized and thought of as such in situations where signal and parameter changes are small in comparison to nominal values. When large parameter changes or even catastrophic faults occur (affecting the DC state), however, one must distinguish between (passive) linear and analog circuits. This is not the case in most research reports bringing confusion into the subject.

Several concepts were applied to diagnosis of analog networks. Among them we will first mention the ones rely on reasoning based on measured data and some measure of distance between the response of the good circuit and the faulty one. Starting with the basic research reported in [5] and [14] several ideas were reported. In [15] the fault-location phase is considered as an optimization problem where the parameter value is searched for in order to minimize the difference among the actual and simulated response. Linear circuits in the frequency domain are considered being characterized by symbolic functions. Similarly, in [16] applying SAT multiple faults may be diagnosed in linear circuits described by symbolic functions what is characterized as model-based method. SBT-based method for soft faults diagnosis in linear circuit was proposed in [17] where harmonic analysis was used for selecting the most suitable test input stimuli and nodes by means of global sensitivity approach. In [18] and [19] passive circuits were diagnosed based on graph theoretical approach, and on pass and fail regions for the circuit poles and zeroes in the complex frequency plane, respectively, while in [20] a Boolean decision scheme was proposed for the diagnosis of linear circuits described in the frequency domain. In order to diagnose multiple soft faults in the same type of circuits the Woodbury formula was applied to the modified nodal equation to construct the so-called fault equation in [21]. A decomposition method was proposed in [22] aiming to cope with circuit complexity. In one approach, small parameter changes were allowed in nonlinear circuits [23]. Soft faults were considered only when linear programming method was used for diagnostic decisions. Large parametric fault diagnosis was described in [24] using piece-wise linear models for DC analysis, and separate considerations were given for diagnosis of faults in the dynamic part of the network (considered linear) based on large change sensitivity computations. Further, in [25] the diagnostic method applied consists of injecting probable faults in a mathematical model of the linear circuit, and later comparing its output with the output of the real faulty circuit. Transfer functions transformed into the $\boldsymbol{Z}$ domain were created and fault injection was performed. In [26] methodology based on linear regression model using SBT which relates a set of measurements to the circuit's internal parameters was applied in order to solve for the circuit parameter values using iterative numerical techniques. Linear circuits in the frequency domain were diagnosed in [27] where the AC response to a set of sinusoidal input frequencies was calculated at selected test nodes. Prony's method was then utilized as a preprocessor to extract an optimal set of features representing nodal voltage waveforms. In [28] a solution to the same problem was proposed based on noise measurements.

In the next paragraph an example will be given of diagnosis linear passive RC circuit in the time domain based on the knowledge of its transfer function in symbolic form and by measuring its impulse response.

Soft and hard (shorts and opens) faults in nonlinear dynamic circuit were diagnosed in [29]. The procedure employs a statistical method of computing Mahalanobis distance (which *is the distance* of the test point from the center of mass divided by the width of the ellipsoid in the direction of the test point) to find defects in load board traces and components. Short list of defects was reported. A low-noise amplifier was diagnosed in [30] by using digital signatures suitable for built-in self-test design concepts. Hard and soft faults were diagnosed the former modeled as resistors having convenient values.

A specific aspect of diagnosis is the number and location of the test points. Simply, we can say that internal test points should be avoided and measurements on the primary inputs and outputs are preferred. This is not only related to their automatic accessibility but also to the nature of the diagnostic reasoning. Namely, one looks for functionality to diagnose something, and the function is seen at the primary terminals. Of course, in order to compensate for the small number of test points more measurements with different types of applied signals are needed to extract complete information about the system behavior. For complex analog systems, however, hierarchical approaches based on decomposition [2, 3, 5, 22, 31] are inevitable provided that no propagation of the fault effect arises between partitions what is not easy to achieve. Of course, there are circuits that may be partitioned based on functionality known a priori from the design process as mentioned in the introduction.

Another aspect of fundamental importance is related to the choice of the output quantities that are to be measured. In most cases these are voltages at the output of the circuit under test (CUT) or at selected test points. It is shown that, however, measurement of the supply current (I_{DDQ}) may be successfully used for testing of both analog and digital circuit [32–36]. This idea was used for diagnosis of analog circuits using ANN that will be discussed later.

Several results were reported where the so-called artificial intelligence concepts were applied to diagnosis of analog circuits or at least linear ones. In [37] method based on fault trajectory concept for fault diagnosis of analog linear continuous time networks, which relies on evolutionary techniques, where a genetic algorithm (GA) was coded to optimize test vector generation, was reported. GA was applied into [9] creation "transfer functions" enabling creation of a new type of fault dictionary. The classical signature dictionary has been replaced by fault decoder based on transfer functions. In order to obtain a sharpen diagnosis about the possible wrong component of the circuit; a tool based on qualitative reasoning was used in [10]. In particular, the results were refined by means of fuzzy techniques. This means that inputs, outputs, rules, and the corresponding operators to combine them were defined. A production rule-based concept was reported in [6].

ANNs have previously been applied to diagnosis several times [11, 38–48]. As in the case with the classical concepts, however, ANNs were predominantly applied to linear analog circuits. In [38] feed-forward ANNs were used for parameter identification (soft fault diagnosis) of linear circuits. In [39] linear power networks were diagnosed by feed-forward ANNs. In order to enhance the performance of the ANN applied for diagnosing of soft faults in linear active networks, in [12], new "criteria," a discriminating measure based on discrepancy of the autocorrelation function of the fault-free and the correlation function of the faulty and fault-free circuit, were introduced. The same problem was attacked in [40, 43] where the impulse response was analyzed by wavelet decomposition, principal component analysis, and data normalization preprocessors before introduced to the ANN. Soft faults were considered only. In [41] a method based on extraction of a "feature vector" from the differences between vectors of node voltages of faulty and fault-free linear circuit for every fault was described. This feature vector is then presented to the ANN as a teaching (training) session. Network tearing is applied in order to manage the circuit complexity in an 11 transistor bipolar circuit. Every partition was considered linear although catastrophic faults were present (e.g., transistor base disconnected). Two faults were diagnosed only. In [42] a linear resistive circuit was diagnosed using feed-forward ANNs. Soft and hard faults (shorts and opens) were considered. Comparably large set of faults was considered. In the schematic presented in [43] (one opamp/one capacitor, three resistors and two diodes) programmable function generator was used to generate the set of stimuli sequentially injected into the input of the CUT. Six test frequencies were chosen. For each stimulus, the frequency response of the CUT has been considered and five Fourier components were measured at the output test point with the spectrum analyzer. For the purpose of diagnosis, four neural networks were used. Euclidian distance was to be learned by the ANN in order conclusions to be created on the origin of the fault. Bipolar analog integrated circuits [44] were diagnosed and their resistances determined from the magnitudes of the Fourier harmonics in the spectrum responses to a sinusoidal input test signal using multilayer perceptron ANN. The input vector to the ANN consists of the magnitudes of the Fourier harmonics of the response waveforms owing to the input stimulus, and the class represents the type of circuit faults, while the outputs map to resistance values of the faults. Probabilistic neural network was applied in [45]. It is a four-layer feed-forward neural network that realizes the Bayes classifier. The ANN creates the probability that a circuit is faulty and points to the type of fault. In [46] a large number of circuit versions were created by introducing sets of models for every separate fault. In fact, hard faults were considered while the opens and the shorts were modeled by resistors of variable resistivity. Then statistical properties of the time domain response (in this case the supply current) to a pulse excitation were extracted in order to create knowledge on the fault to fault effect mapping. The supply current was successfully used for diagnosing gate oxide shorts in CMOS circuits by the help of ANNs in [47, 48]. After introducing a fault model of the MOS transistor built as a series connection of two MOS transistors with a common gate (i.e., considering this as a soft fault), several faults per transistor (for all transistors in an 11 transistor operational amplifier) were created by changing the possible position of the gate short relative to the source-to-drain ends of the channel. Sinusoidal and ramp signals were used for creation of a fault dictionary in an SAT method. The response, i.e., the supply current, was sampled to give a series of values used to train the feed-forward [43] and a Cohonen [44] neural network.

Here we describe the results of applying feed-forward ANNs to the diagnosis of nonlinear dynamic electronic circuits with no restriction on the number and type of faults. This method is based on fault dictionary creation and using a feed-forward

ANN for data compression by memorizing the table representing the fault dictionary. Depending on the specific case (in this and in the Chapter entitled 11 Solved Problems) DC, transient, and small signal sinusoidal excitations will be applied, so preserving the usual measurement procedure for generating the data given in a component's and/or the circuit's data-sheets. The ANN so created is, consequently, used for diagnosis by applying to it the signals obtained by field-measuring the faulty network. This process may be considered as looking-up a fault in the fault dictionary. The ANN finds the most probable *fault code* that corresponds to the measured signals.

Putting this in the general context of diagnosis we first note that the fault dictionary contains all the knowledge we need. In other words by applying the SET concept all hypotheses are memorized (within the ANN) and no further hypothesis needs to be created after the dictionary is known. This is equivalent to the structural concept of testing. The fault not conceived in advance cannot be tested nor diagnosed. Now we look among the hypotheses (by searching the dictionary, i.e., by running the ANN) to find the one most similar to the actual (faulty) circuit response. The difficulties here are the complexity of the search and the decision algorithm that finds the "most similar" entry in the dictionary. As will be shown with examples this can be an extremely difficult task. It has been successfully solved using ANNs. The network used for this diagnostic example is a feed-forward neural network structured in three layers. It has only one hidden layer, which has been proved sufficient for this kind of problem [49]. The neurons in the hidden layer are activated by a sigmoidal function, while the neurons in the output layer are activated by a linear function. The learning (training) algorithm used for training this network is a version of the steepest-descent minimization algorithm [50].

7.9.4 Fault Location in Passive Analog RC Circuits by Measuring Impulse Response—A Case Study

The method is based on comparison of circuit transfer function coefficients obtained by symbolic analysis with transfer function coefficients obtained by transformation of measured data. In the example given, simulation results were used in place of the measured ones.

The procedure contains several steps. First, to get the coefficients from the measured results one applies the Prony's method of finding the poles and the residues of the transfer function based on the sampled impulse response of the circuit. The so obtained results are then used to find the polynomial coefficients by converting its factored form into a standard form.

On the other side it is necessary to determine the transfer function of the circuit in a symbolic form. This can be done, for example, by using an existing software package such as the one described in [51]. In that way all coefficients of the transfer function are expressed in terms of circuit elements in a symbolic form. Their numerical values for the case of the fault-free circuit can be evaluated easily. Actually this means that the transfer function (in numerical form) of the fault-free circuit is determined.

To get a relation between the samples of the impulse response and the transfer function coefficient we first implement the Prony's method. As a result it produces the poles and the residues of the transfer function for a given circuit, here the faulty one. Below is a short outline of the Prony's method.

In general [52], the circuit's impulse response, as a function of the poles and residues, when the transfer function has only single poles, is given by:

$$h(t) = \sum_{i=1}^{m} K_i \cdot e^{p_i \cdot t} = \sum_{i=1}^{m} \frac{K_i}{s - p_i} \tag{7.9.1}$$

where p_i and K_i, for $i = 1,2, \ldots, m$, are single real poles and their corresponding residues, respectively. By sampling the impulse response, with sampling period T, the left equation of (7.9.1) becomes:

$$h(kT) = \sum_{i=1}^{m} K_i \cdot e^{p_i \cdot kT} \quad \text{for } k = 0, 1, 2, \ldots, 2m - 1 \tag{7.9.2}$$

which can be represented in a matrix form as:

$$[z_{ik}][K_i] = [h(kT)] \tag{7.9.3a}$$

where

$$z_{ik} = e^{p_i \cdot kT}. \tag{7.9.4}$$

The expressions presented in Eq. (7.9.2) form a system of $2m$ nonlinear equations with $2m$ unknowns (p_i and K_i, for $i = 1,2, \ldots, m$). In the case of an mth order system, the initial $2m$ samples are independent. Consequently, Prony's method can be employed to recover the poles and residues for such an mth order system. The equations also suggest that the sampled impulse response can be transformed back into a continuous signal in the time domain through the first equality in (7.9.1) and can be expressed in the frequency domain using the second equality in the same equation. This algorithm is valuable as it facilitates the conversion between the time domain and the frequency domain without relying on the Fourier transform.

The impulse response is first converted into matrix form to find the poles:

$$\begin{bmatrix} h(0) & h(1) & \cdots & h(m-1) \\ h(1) & h(2) & \cdots & h(m) \\ \vdots & \vdots & \cdots & \vdots \\ h(m-1) & h(m) & \cdots & h(2m-2) \end{bmatrix} \cdot \begin{bmatrix} a_0 \\ a_1 \\ \vdots \\ a_{m-1} \end{bmatrix} = \begin{bmatrix} h(m) \\ h(m+1) \\ \vdots \\ h(2m-1) \end{bmatrix} \tag{7.9.3b}$$

where $[a]$ is a vector of intermediate variables.

After solving for $[a]$ the characteristic equation is formed as:

$$B(\omega) = \omega^m - a_{m-1}\omega^{m-1} - \cdots - a_1\omega - a_0 = 0 \tag{7.9.5}$$

Equation (7.9.5) is then solved to get the m roots of the system (ω_m, ω_{m-1}, …, ω_1) so that the poles of the system (p_i) are obtained using

$$p_i = \frac{1}{T} \cdot \ln(\omega_i), \quad i = 1, 2, \ldots, m. \tag{7.9.6}$$

Having identified the poles, the residues of this mth order system can be determined by applying (7.9.2) in a proper manner, involving the solution of a system of linear equations:

$$h(r) = \sum_{j=1}^{m} \left(K_j \omega_j^r\right) \quad \text{for } r = 0, 1, 2, \ldots, m-1. \tag{7.9.7a}$$

or

$$\begin{bmatrix} 1 & 1 & \cdots & 1 \\ \omega_1 & \omega_2 & \cdots & \omega_m \\ \vdots & \vdots & \cdots & \vdots \\ \omega_1^{m-1} & \omega_2^{m-1} & \cdots & \omega_m^{m-1} \end{bmatrix} \cdot \begin{bmatrix} K_1 \\ K_2 \\ \vdots \\ K_m \end{bmatrix} = \begin{bmatrix} h(0) \\ h(1) \\ \vdots \\ h(m-1) \end{bmatrix}. \tag{7.9.7b}$$

So, to derive the poles and residues of an mth order system using Prony's method, a minimum of $2m$ samples is necessary. The procedure entails solving two sets of mth order linear equations—one for acquiring the poles and another for obtaining the residues. Once the poles and residues are determined, the system transfer function can be easily computed using Eq. (7.9.1). Prony's method provides an approximate transfer function based on the sampled transient response. For high-order systems, the resultant transfer function's order is typically much lower than the actual system order. However, a drawback lies in Prony's method's heavy reliance on linearization and the assumption of noiseless data, which may present challenges for nonlinear and/or noisy functions.

To complete the issue one should mention that for a given transfer function $H(s)$, with the denominator presented in factored form, one calculates the residue K_i in a pole $s = p_i \neq \infty$, of degree k, in the following way:

$$K_i = \frac{1}{(k-1)!} \lim_{s \to p_i} \left[\frac{\mathrm{d}^{(k-1)}}{\mathrm{d}s^{(k-1)}} \left((s - p_i)^k \cdot H(s) \right) \right]. \tag{7.9.8}$$

With the preliminary assumption that the transfer function has only single poles, i.e., $k = 1$, one gets the following expression for the residues in poles:

$$K_i = (s - p_i) \cdot H(s)_{|s=p_i}, \quad \text{for } i = 1, 2, \ldots, m. \tag{7.9.9}$$

As K_i and p_i, $i = 1, 2, \ldots, m$, are preliminary computed, it is possible to form system of m equations with coefficients of transfer function as unknown, and to solve it. With these coefficients one gets the transfer function of the faulty circuit.

The following notation will be used below. The coefficients of the transfer function obtained from measurement (using Prony's method) will be denoted e_i. They should represent the fault effect if any. The coefficients of the transfer function obtained by symbolic analysis of the transfer function will be denoted e_i. These will be used as reference.

Now, going to the other part of the procedure, i.e., to the coefficient obtained by symbolic analysis of the circuit, one needs to highlight a given circuit parameter within the coefficient of the transfer function. To that end one uses the result reported in [5, 53] saying that it is possible the transfer function coefficient to be expressed in the following form:

$$e_i = d_{0iF_k} + d_{1iF_k} \cdot F_k, \quad \text{for } i = 1, 2, \ldots, r, \text{ with } r \leq 2m - 1 \tag{7.9.10}$$

where e_i, are the coefficients of the transfer function obtained by measurements and F_k is a potentially faulty element value. d_{0iF_k} and d_{1iF_k} are coefficients of the binomial obtained by highlighting the circuit element value F_k within the corresponding coefficients of the transfer function (e_i) expressed in a symbolic form. The values of coefficients d_{0iF_k} and d_{1iF_k} are computed for nominal values of the rest of circuit elements using the symbolic formula for coefficients of the transfer function e_i. Relations corresponding to (7.9.10) can be written (and d_{0iF_k} and d_{1iF_k} computed) for all coefficients of the transfer function as functions of all circuit elements.

The search for the faulty circuit element may start now. It is performed in the following way. Transfer function coefficients with values determined from measured results, e_i, which deviate from their nominal values e_i, are represented in the form of (7.9.10), for all potentially faulty elements, F_k for $k = 1, 2, \ldots, m$. Then one solves (7.9.10) for F_k from two, or more, different coefficients e_i. If the obtained solutions are mutually equal, one concludes that the element F_k is a faulty one. If they are not equal, the procedure is repeated for the next circuit element, until one finds the element for which these values are equal. If there is no such element, and these values are not equal to the fault-free value of the circuit element, one assumes that there is more than one fault in the circuit.

The illustration of this method is presented by an example of a simple RC circuit, which has three poles ($m = 3$) and three zeros as depicted in Fig. 7.9.2. The circuit transfer function obtained by symbolic analysis is:

$$H(s) = \frac{a_2 s^2 + a_3 s^3}{b_0 + b_1 s + b_2 s^2 + b_3 s^3}, \tag{7.9.11}$$

where the coefficients are:

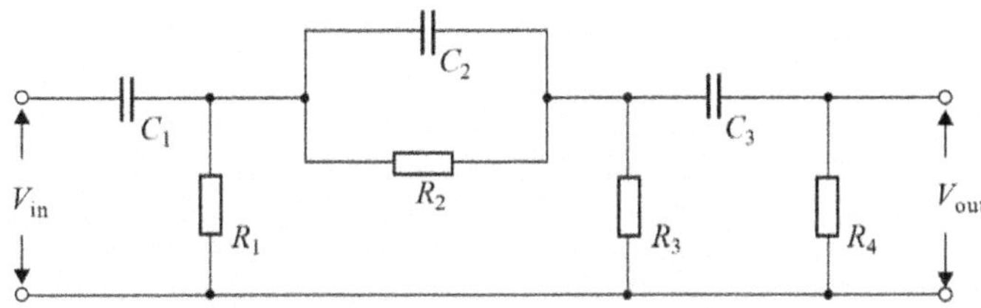

Fig. 7.9.2 Example RC circuit

$$
\begin{aligned}
a_2 &= C_1C_3R_1R_3R_4 \\
a_3 &= C_1C_2C_3R_1R_2R_3R_4 \\
b_0 &= R_1 + R_2 + R_3 \\
b_1 &= C_1R_1(R_2 + R_3) + C_2R_2(R_1 + R_3) \\
&\quad + C_3(R_1R_3 + R_1R_4 + R_2R_3 + R_2R_4 + R_3R_4) \\
b_2 &= C_1C_2R_1R_2R_3 + C_1C_3(R_1R_2R_3 + R_1R_3R_4 + R_1R_2R_4) \\
&\quad + C_2C_3(R_1R_2R_3 + R_2R_3R_4 + R_1R_2R_4) \\
b_3 &= C_1C_2C_3R_1R_2R_3R_4.
\end{aligned}
$$

The nominal values of the circuit elements are: $R_1 = 5$ kΩ, $R_2 = R_3 = 2$ kΩ, $R_4 = 10$ kΩ, $C_1 = C_3 = 10$ nF, and $C_2 = 1$ nF. Based on that one determines the values of transfer function coefficients of the fault-free circuit as: $a_2 = 10^{-5}$, $a_3 = 2.3 \times 10^{-11}$, $b_0 = 9 \times 10^3$, $b_1 = 1.254$, $b_2 = 2.38 \times 10^{-5}$, $b_3 = 2 \times 10^{-11}$. The zeros and the poles of the transfer function are: $z_1 = z_2 = 0$ rad/s, $z_3 = -5 \times 10^5$ rad/s, $p_1 = -1{,}135{,}112.430$ rad/s, $p_1 = -46{,}330.946$ rad/s, $p_3 = -8556.624$ rad/s. With the assumption that fault occurs in capacitor C_3, and if its faulty value is $C_3 = 11.5$ nF by Spice simulation one gets an impulse response as depicted in Fig. 7.9.3a. From this figure one gets the following samples of the impulse response: $h(0 \cdot T) = 1$, $h(1 \cdot T) = 0.579475$, $h(2 \cdot T) = 0.430649$, $h(3 \cdot T) = 0.369786$, $h(4 \cdot T) = 0.337801$, $h(5 \cdot T) = 0.315674$, and $h(6 \cdot T) = 0.297265$, with the sampling period $T = 1$ ms.

Having the samples, for the coefficients of the mth order polynomial (7.9.5), the following values are obtained: $a_0 = 1$, $a_1 = -5.165954233$, $a_2 = 7.448003018$, and $a_3 = -3.282796509$.

With these values one forms the polynomial (7.9.5) and computes the roots of this polynomial to be: $z_{i1} = 0.321427$, $z_{i2} = 0.954797$, and $z_{i3} = 0.992574$.

Further, by using (7.9.6), the poles of the transfer function of the faulty circuit are calculated as: $p_1 = -1{,}134{,}986$ rad/s, $p_2 = -46{,}256$ rad/s, and $p_3 = -7453$ rad/s. Residues in the poles can be obtained by substitution the impulse response sample values and the poles of the faulty circuit in the matrix relation (7.9.3). By forming relation (7.9.3) and solving for K_i one gets following values: $K_1 = -666{,}340.166$, $K_2 = -22{,}980.896$, and $K_3 = 625.410$.

Forming the system of equations, with the faulty circuit transfer function coefficients as unknown using (7.9.9) and solving them one can obtain the following values

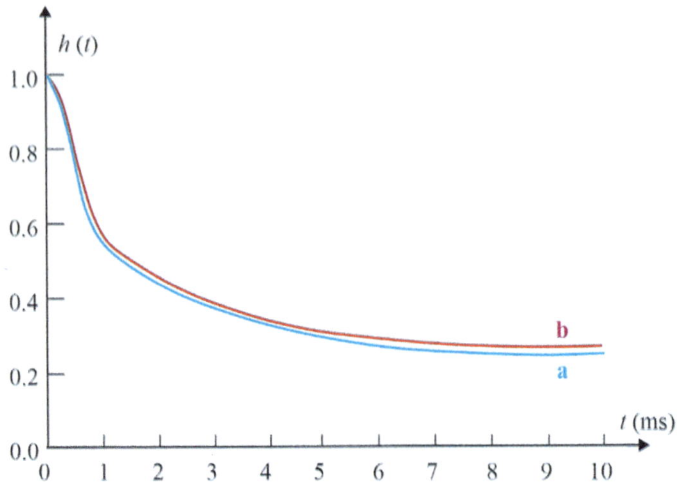

Fig. 7.9.3 Impulse response of circuit of Fig. 7.9.2. **a** With a single fault and **b** with two faults

of the coefficients of the faulty circuit transfer function: $a_{2j} = 1.1500616131 \times 10^{-5}$; $a_{3j} = 2.300132261 \times 10^{-11}$; $b_{0f} = 9 \times 10^{3}$; $b_{1j} = 1.41006389$; $b_{2j} = 2.734155718 \times 10^{-5}$; $b_{3j} = 2.300132261 \times 10^{-11}$.

By substitution, the calculated values for a_{2j}, b_{1j}, b_{2j}, and b_{3j} in (7.9.10) one gets values for circuit elements shown in Table 7.9.1. Analyzing the results from Table 7.9.1, one concludes that only one element for which these values are equal is capacitor C_3, so this element is said to be faulty one, and it is faulty value is $C_3 = 11.5$ nF.

As shown in [54], no success was obtained in the attempt to diagnose the same circuit in the presence of two faulty elements. Neither the locations nor the values were extracted. That may also be easily deduced from Fig. 7.9.3 where the response of a circuit with two faults is shown as **b**.

7.9.5 Oscillation-Based Analog Diagnosis Using Artificial Neural Networks-Based Inference Mechanism

As discussed in Sect. 7.8.1.2.2 when a suitable feedback circuit is added to an analog building block, one may create an oscillator. By measuring the output voltage (or some other responses) of that circuit one may extract quantities such as the frequency, the amplitude of the first and other harmonics, the DC value of the output voltage that may contain information on the presence of a fault in the circuit.

The main advantages of this method are avoiding the search for input stimuli and appropriate mode of operation of the circuit; independence of the method with respect to the type of faults present in the circuit (soft or catastrophic); and simplified

Table 7.9.1 Calculated values of all elements of the circuit of Fig. 7.10.2 obtained from measurements

R_1	$a_2 = 2 \times 10^{-9} \cdot R_1$	$b_1 = 0.444 + 1.62 \cdot R_1$	$b_2 = 4 \times 10^{-7} + 4.68 \times 10^{-9} \cdot R_1$	$b_2 = 4 \times 10^{-15} \cdot R_1$
	$R_1 = 5.75033$	$R_1 = 5.96336$	$R_1 = 5.75674$	$R_1 = 5.75033$
R_2	–	$b_1 = 0.9 + 1.77 \times 10^{-4} \cdot R_2$	$b_2 = 1 \times 10^{-5} + 6.9 \times 10^{-9} \cdot R_2$	$b_3 = 1 \times 10^{-14} \cdot R_2$
	–	$R_2 = 2.88172$	$R_2 = 2.51327$	$R_2 = 2.30013$
R_3	$a_2 = 5 \times 10^{-9} \cdot R_3$	$b_1 = 0.81 + 2.22 \times 10^{-4} \cdot R_3$	$b_2 = 1.1 \times 10^{-5} + 6.4 \times 10^{-9} \cdot R_3$	$b_3 = 1 \times 10^{-14} \cdot R_3$
	$R_3 = 2.30013$	$R_3 = 2.70299$	$R_3 = 2.55337$	$R_3 = 2.30013$
R_4	$a_2 = 1 \times 10^{-9} \cdot R_4$	$b_1 = 0.354 + 0.9 \times 10^{-4} \cdot R_4$	$b_2 = 2.4 \times 10^{-6} + 2.14 \times 10^{-9} \cdot R_4$	$b_3 = 2 \times 10^{-15} \cdot R_4$
	$R_4 = 11.50066$	$R_4 = 11.73404$	$R_4 = 11.65493$	$R_4 = 11.50066$
C_1	$a_2 = 1 \times 10^{3} \cdot C_1$	$b_1 = 1.054 + 2 \times 10^{7} \cdot C_1$	$b_2 = 1.6 \times 10^{-6} + 2220 \cdot C_1$	$b_3 = 2 \times 10^{-3} \cdot C_1$
	$C_1 = 11.50066$	$C_1 = 17.80319$	$C_1 = 11.59296$	$C_1 = 11.50066$
C_2	–	$b_1 = 1.24 + 1.4 \times 10^{7} \cdot C_2$	$b_2 = 2.2 \times 10^{-5} + 1800 \cdot C_2$	$b_3 = 2 \times 10^{-2} \cdot C_2$
	–	$C_2 = 12.14742$	$C_2 = 2.96753$	$C_2 = 1.15007$
C_3	$a_2 = 1 \times 10^{3} \cdot C_3$	$b_1 = 0.214 + 1.04 \times 10^{8} \cdot C_3$	$b_2 = 2 \times 10^{-7} + 2360 \cdot C_3$	$b_3 = 2 \times 10^{-3} \cdot C_3$
	$C_3 = 11.50066$	$C_3 = 11.50061$	$C_3 = 11.50066$	$C_3 = 11.50066$

One fault present. Resistances in kΩ and capacitances in nF

selection of the test points and measured quantities at the output. From the design for testability (DFT) point of view this method offers exceptional advantages. Namely, by proper structuring of the analog subsystem on the chip, as shown in Fig. 7.8.10, one can control the testing process from outside of the chip.

When diagnosis is considered, the same advantages remain and after creation of the fault dictionary by simulation one has to search it to find a match with the measured response [55, 56].

As a case study for implementation of the oscillation-based method for diagnosis of analog circuits, the Sallen-Key non-inverting second order notch filter cell will be used. The importance of this cell comes from the fact that it is the most frequently used cell within every selective analog active RC filter, implemented as a cascade of second order cells, irrespective of the position of its passband on the frequency axis. In addition, the specific shape of the transfer function of the feedback circuit makes the oscillation frequency differ significantly from the notch frequency (which is $f_0 = 1/(2\pi RC)$) of the filter cell. The schematic of the cell is based on a twin-T circuit as shown in Fig. 7.9.4, while the actual cell is depicted in Fig. 7.9.5 [57].

For testing, a positive feedback loop is created by direct connection of the output to the input of the cell. The oscillator circuit so obtained is depicted in Fig. 7.9.6.

Fig. 7.9.4 Twin-T four-pole

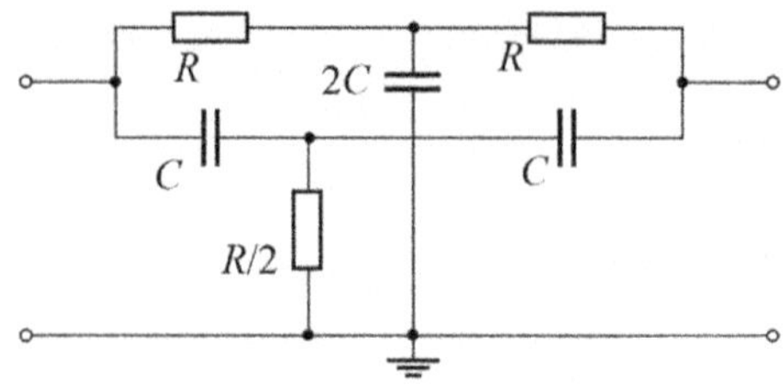

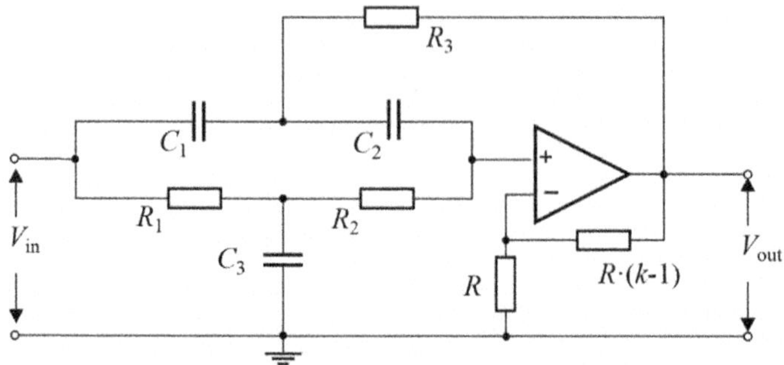

Fig. 7.9.5 Salen and Key (SK) non-inverting second order notch filter cell

The fault-free circuit will oscillate at f_N. The value of f_N will be discussed below. When faults are present one may observe three different situations: (a) the faulty circuit oscillates at f_N, (b) the faulty circuit oscillates at some other frequency $\hat{f}_N$, and (c) there are no oscillations in the circuit. The testing will be successful if there is any difference between the oscillating frequencies and if there are no oscillations. Unfortunately, that is not enough for diagnosis. One needs distinctive information in order to create a diagnostic statement which will be discussed later.

The oscillator is usually treated as a linear circuit so one may use the usual oscillator analysis method to get the expected oscillation frequencies. Namely, by writing the modified nodal equations [58] for the circuit of Fig. 7.9.6, and equating

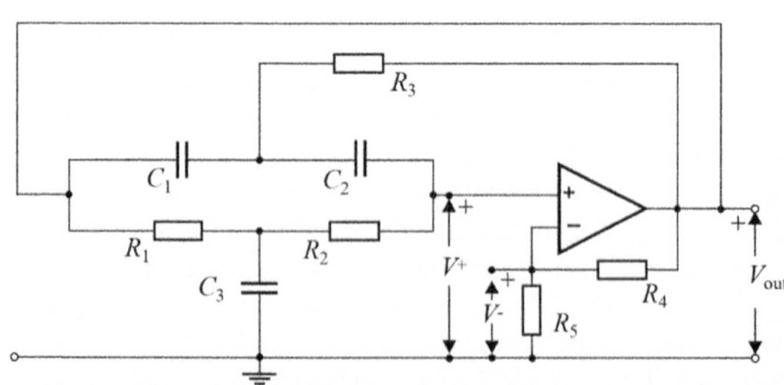

Fig. 7.9.6 Oscillator built using the RC active notch cell

the system determinant to zero, after separating the real and imaginary parts, one obtains the following two expressions for the possible oscillation frequencies

$$\omega_0^2 = (A-1)/[(A-1)C_3(R_2R_3C_2 + R_1R_3C_2 + R_1R_2C_1) - R_3C_1(R_2C_2 - R_1C_1)] \tag{7.9.12a}$$

$$\omega_0^2 = \frac{R_2C_3 + R_1C_3 + R_3C_2 + R_3C_3 - R_1C_1/(A-1)}{R_1R_2R_3C_1C_2C_3}, \tag{7.9.12b}$$

where $A = k = 1 + R_4/R_5$ with $R_4 = (k-1)R$ and $R_5 = R$.

One of these expressions is usually required for frequency calculation while the other is needed to find the necessary value of A for sustained oscillation.

The above expressions were derived assuming that the operational amplifier is ideal with infinite gain, which is not true in a real circuit. This is important since the closed-loop gain of the oscillator circuit is characterized by both modulus and phase. Both are frequency dependent and fundamentally determine the oscillation frequency. The operational amplifier's phase shift becomes of major importance in this situation and has to be considered. So, the expressions given by (7.9.12a) or (7.9.12b) may be used as reference only. This will be confirmed below after simulation experiments are shown. It is worth mentioning that the need to include the operational amplifier's phase shift has been previously noted [57].

The fault list used contains both parametric (soft) and catastrophic faults in the passive feedback circuit. When testing or diagnosing a circuit one needs to create a fault dictionary, that is, a table containing faults and fault effects. To build it, a large number (equal to the number of modeled faults) of repetitive simulations has to be performed. This is known as "simulation before test." For every simulation, a fault is inserted in the original oscillator circuit, so creating a new oscillator. Note that, what was not mentioned in Chap. 7.8, *simulation* of an oscillator is not a straightforward task since one usually uses an A-stable integration rule (such as Euler backward) for solving the differential equations of the circuit, in order to the prevent oscillations in the simulation. Here however, since the circuit is unstable, one needs to use an integration rule that is not A-stable (trapezoidal, for example) [58] (This issue is discussed in LNAE_Book 2, Chap. 7.7). For correct simulation one needs to use a complete schematic of the operational amplifier or a qualified model that performs well from the nonlinearity and phase-shift point of view. To illustrate how important the modeling of the phase difference inserted by the operational amplifier is, the simulation results for a fault-free circuit (oscillator) are depicted in Fig. 7.9.7. Two signals are shown, the operational amplifier's input (non-inverting terminal) and its output. One can deduce from the figure a phase shift of 9.65°. This is to be compared to the maximum phase shift of the feedback circuit which is 16.7° [59] meaning that a significant difference between the oscillation frequency f_N and that given by (7.9.12a) may be expected.

Faults that will be modeled in the following example are categorized into several groups:

- catastrophic faults within the RC circuit,

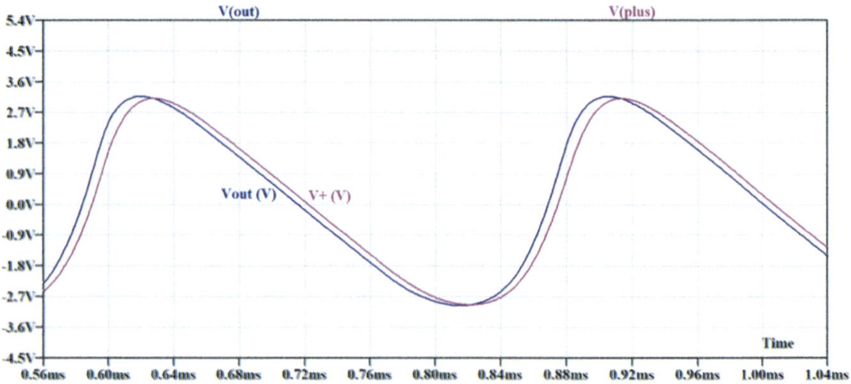

Fig. 7.9.7 Responses obtained by simulation of the circuit of Fig. 7.9.6

- parametric faults within the RC circuit,
- selected examples of multiple parametric faults within the RC circuit.

Note, since the number of possible faults in a system may be very large, when creating the fault dictionary one generally chooses a set of the most probable faults. If this approach is implemented, no faults will be diagnosed except the ones conceived in advance and included in the fault list.

Searching through the fault dictionary, in general, enables the fault coverage to be established and the test signal to be validated. In addition, by applying a specific inference mechanism, after the search, one may create a qualified diagnostic statement.

The actual process of generating the final diagnostic statement, as already mentioned, may be based on different concepts. If a large number of faults and a reduced number of outputs are to be processed at the same time, due to the resemblance of the fault effects, the search process within the fault dictionary requires highly sophisticated classification algorithms.

Let us elaborate now the creation of the fault dictionary. For every passive element within the RC circuit (Fig. 7.9.6), a short and open circuit is considered as a catastrophic fault. After insertion, the resulting schematic was simulated and the results are shown in Table 7.9.2. The resistors in the negative feedback circuit were chosen to be $R_4 = 200\ \text{k}\Omega$, and $R_5 = 20\ \text{k}\Omega$, leading to a gain $A = 11$. The operational amplifier used in the simulation was the LTC6078.

By inspection of Table 7.9.2 we find that in six out of twelve faulty cases the circuit keeps oscillating. In all the cases the fault effect is recognizable since the oscillation frequency is never the same as that of the fault-free circuit. In general, however, one may state that by implementing OBT we got almost perfect fault coverage of catastrophic faults. We also calculated the frequency increment relative to the oscillating frequency, and it is given in percentage. The minus (−) sign stands for the cases where the absolute value of the oscillating frequency is less than in the case of the fault-free circuit, and for the rest of the cases, the relative increment is positive.

Table 7.9.2 Catastrophic faults

No.	Fault type	Element value	Oscillation frequency (kHz)/Relative increment (%)
1	Fault-free		3.48/0
2	SC: R_1	$R_1 = 0\ \Omega$	No
3	SC: R_2	$R_2 = 0\ \Omega$	No
4	SC: R_3	$R_3 = 0\ \Omega$	2.12/−39.08
5	OC: R_1	$R_1 = \infty\ \Omega$	3.14/−9.77
6	OC: R_2	$R_2 = \infty\ \Omega$	No
7	OC: R_3	$R_3 = \infty\ \Omega$	3.32/−4.6
8	SC: C_1	$C_1 = \infty$ F	2.76/−20.69
9	SC: C_2	$C_2 = \infty$ F	2.12/−39.08
10	SC: C_3	$C_2 = \infty$ F	2.4/−31.03
11	OC: C_1	$C_1 = 0$ F	No
12	OC: C_2	$C_2 = 0$ F	No
13	OC: C_3	$C_3 = 0$ F	No

SC stands for short circuit, *OC* for open circuit, and *No* means no oscillations are present

When considering parametric faults one has to decide to what extent a change of the element value is to be treated as a fault having in mind that small changes may be looked upon as tolerances coming from the technology process. Here a parametric fault will be seen when the element value within the RC circuit is changed by 20% in comparison to its nominal value. Both positive and negative changes are considered. As a result Table 7.9.3 is produced. It contains the faults and the fault effects for all modeled soft faults in the RC circuit. By inspection of Table 7.9.3 one may easily conclude that in all cases there is difference between the behavior of the fault-free and the faulty circuit. In six cases the change in the frequency value is relatively low (less than 5%) so making the classification difficult. Here again we may conclude that the fault coverage is almost perfect.

The possible number of double soft faults is much larger than in the case of the single faults. This is why a reduced set of pairs of soft faults was considered as shown in Table 7.9.4. One may observe that in all cases the faulty circuit exhibits a new value for f_N. In six cases the change in the frequency value is also small.

In summary, considering the OBT approach applied to the notch filter cell, as already stated, we have to observe the following. First of all, no test signal was needed. That is a big advantage to the test engineer. Second, only one test point was observed, the circuit's output, which is the most natural point of access for measurement. Finally, only one quantity was extracted as a measure of fault coverage: the oscillation frequency. With such a simple procedure a large number of faults were covered, leading to the conclusion that the OBT method, for this example, is an excellent testing concept.

By inspection of Tables 7.9.2, 7.9.3 and 7.9.4, one may find that some of the fault effects, while different from the fault-free response, are identical to each other. We

Table 7.9.3 Parametric faults

No.	Fault type	Oscillation frequency (kHz)/Relative increment (%)
1	$1.2 \cdot R_1$	3.44/−1.15
2	$0.8 \cdot R_1$	3.54/1.72
3	$1.2 \cdot R_2$	3.1/−10.92
4	$0.8 \cdot R_2$	4.02/15.52
5	$1.2 \cdot R_3$	3.46/−0.57
6	$0.8 \cdot R_3$	3.5/0.57
7	$1.2 \cdot C_1$	3.38/−2.87
8	$0.8 \cdot C_1$	3.62/4.02
9	$1.2 \cdot C_2$	3.28/−5.75
10	$0.8 \cdot C_2$	3.8/9.2
11	$1.2 \cdot C_3$	3.32/−4.6
12	$0.8 \cdot C_3$	3.7/6.32

Table 7.9.4 Fault effects when two faults are present in the circuit simultaneously

No.	Fault type	Oscillation frequency (kHz)/Relative increment (%)
1	$1.2 \cdot C_2$; $1.2 \cdot C_3$	3.1/−10.92
2	$0.8 \cdot C_2$; $0.8 \cdot C_3$	4/14.94
3	$1.2 \cdot C_2$; $0.8 \cdot C_3$	3.52/1.15
4	$0.8 \cdot C_2$; $1.2 \cdot C_3$	3.64/4.6
5	$1.2 \cdot R_1$; $1.2 \cdot C_1$	3.34/−4.02
6	$0.8 \cdot R_1$; $0.8 \cdot C_1$	3.7/6.32
7	$1.2 \cdot R_1$; $0.8 \cdot C_1$	3.56/2.3
8	$0.8 \cdot R_1$; $1.2 \cdot C_1$	3.46/−0.57
9	$1.2 \cdot R_2$; $1.2 \cdot C_2$	2.9/−16.67
10	$0.8 \cdot R_2$; $0.8 \cdot C_2$	4.34/24.71
11	$1.2 \cdot R_2$; $0.8 \cdot C_2$	3.38/−2.87
12	$0.8 \cdot R_2$; $1.2 \cdot C_2$	3.78/8.62

refer to these types of faults as ambiguity groups or functionally equivalent faults (FEF) [24]. These groups do not affect the testing since the go-no-go concept is applied, i.e., the information that there is a fault in the system is enough to decide whether the circuit should be discarded. In cases where diagnosis is to be performed, however, FEF limits the diagnostic resolution, so one should try to keep the number of FEFs to a minimum.

Following these considerations we needed to find some additional quantity to support the creation of the diagnostic statement. Using an additional test point or even redesigning the circuit to create the possibility for monitoring the supply current, as was proposed in [60, 61], we consider to be unacceptably expensive and inconvenient for application in DFT designs. Instead we were looking for a parameter that can be extracted from the same measurement. Furthermore, we would prefer to implement the same parameter extraction procedure for the new parameter as that used for the extraction of the oscillation frequency.

So, keeping the circuit's output voltage as the only measured quantity and after performing spectral analysis to get the frequency it became natural to use the total harmonic distortion (THD) of the output signal as the additional information for diagnosis. Alternatively, as it was done in [56], the amplitude of the first harmonic ($f_{2\omega_N}$) may be used as part of the signature since the amplitude of the main harmonic (f_{ω_N}), being defined by the power supply voltage, was found to be not suitable.

For diagnostic purpose we created a merged fault dictionary (Table 7.9.5) including faults from Tables 7.9.2, 7.9.3, and 7.9.4, respectively. A column was added, however, containing the values of THD. From Table 7.9.5 we can deduce that, in the vast majority of cases, the values of THD differ from each other, so diagnosis is possible. One can find from the same Table, however, that some faults have the same values for both the oscillating frequency and THD (faults No. 1, No. 2, No. 5, No. 10, No. 11, No. 12, and also faults No. 3 and No. 8) meaning that we still have FEFs. These are extracted from Table 7.9.5 and reproduced in Table 7.9.6. The new ambiguity groups are denoted as AG1 and AG2. For the case of AG1, where there were no oscillations in the circuit, it was not possible to calculate THD (denoted as NO in the Table 7.9.5).

The presence of these ambiguity groups introduces minor changes into the fault dictionary from the Table 7.9.5, since only one representative of each ambiguity group was included in the fault dictionary. Table 7.9.7 gives the final version of the dictionary, latter used for the diagnosis. Since, even for this case, we needed to assign some concrete values that would be used in the ANN training procedure, we chose values 5 kHz for the frequency and 70% for THD, as the greatest parameter values for these groups.

The fault's coding is presented in the last column of Table 7.9.7 by: Fault code (m). This version of the fault dictionary is ready for training the ANN that will later perform the diagnosis.

The fault coding is an important issue. In fact, some faults exhibit very similar effects. Namely, input data (signatures) can have very close numerical values, and if the output values (fault codes) were also similar, the network could not always be trained successfully.

For example, we notice that faults Nos. 9 and 19, Nos. 4 and 17, Nos. 13 and 29, Nos. 11 and 26, Nos. 18 and 24 have the same values for the oscillating frequency and very close values for THD, and hence we assigned codes that are not close, (1 and -15, 5 and -13, 6 and -10, 2 and -14, -3 and 13, respectively), in order to differentiate them. Fault code $m = 0$ denotes the fault-free circuit, and codes are in the interval $[-15, 15]$. In general, faults are coded randomly, so that faults with similar effects are unlikely to have similar codes, and if that happens, the codes are reordered in order to make a bigger distinction. This approach has proven to be good, because the coding influenced the training time of the ANN, and also, the final training error.

Table 7.9.5 Fault dictionary

No.	Fault type	Oscillation frequency (kHz)/Relative increment (%)	THD (%)
0	Fault-free	3.48	44.25
1	SC: R_1	NO	NO
2	SC: R_2	NO	NO
3	SC: R_3	2.12	67.09
4	OP: R_1	3.14	9.826
5	OP: R_2	NO	NO
6	OP: R_3	3.32	26.85
7	SC: C_1	2.76	12.443
8	SC: C_2	2.12	67.09
9	SC: C_3	2.4	46.44
10	OP: C_1	NO	NO
11	OP: C_2	NO	NO
12	OP: C_3	NO	NO
13	$1.2 \cdot R_1$	3.44	33.555
14	$0.8 \cdot R_1$	3.54	40.877
15	$1.2 \cdot R_2$	3.1	17.701
16	$0.8 \cdot R_2$	4.02	23.108
17	$1.2 \cdot R_3$	3.46	26.902
18	$0.8 \cdot R_3$	3.5	48.13
19	$1.2 \cdot C_1$	3.38	30.56
20	$0.8 \cdot C_1$	3.62	53.35
21	$1.2 \cdot C_2$	3.28	30.256
22	$0.8 \cdot C_2$	3.8	24.3
23	$1.2 \cdot C_3$	3.32	34.23
24	$0.8 \cdot C_3$	3.7	26.93
25	$1.2 \cdot C_2; 1.2 \cdot C_3$	3.1	23.59
26	$0.8 \cdot C_2; 0.8 \cdot C_3$	4	28.812
27	$1.2 \cdot C_2; 0.8 \cdot C_3$	3.52	39.144
28	$0.8 \cdot C_2; 1.2 \cdot C_3$	3.64	28.7
29	$1.2 \cdot R_1; 1.2 \cdot C_1$	3.34	19.28
30	$0.8 \cdot R_1; 0.8 \cdot C_1$	3.7	51.97
31	$1.2 \cdot R_1; 0.8 \cdot C_1$	3.56	26.45
32	$0.8 \cdot R_1; 1.2 \cdot C_1$	3.46	23.75
33	$1.2 \cdot R_2; 1.2 \cdot C_2$	2.9	25.2
34	$0.8 \cdot R_2; 0.8 \cdot C_2$	4.34	17.476
35	$1.2 \cdot R_2; 0.8 \cdot C_2$	3.38	28.65
36	$0.8 \cdot R_2; 1.2 \cdot C_2$	3.78	28.461

Table 7.9.6 Ambiguity groups

Ambiguity group	Faults included	Oscillation frequency (kHz)	THD (%)
AG1	SC: R_1 SC: R_2 OP: R_2 OP: C_1 OP: C_2 OP: C_3	NO	NO
AG2	SC: C_2 SC: R_3	2.12	67.09

For diagnosis, fully connected feed-forward ANNs with sigmoidal activation function, for the neurons in the hidden layer, and with a linear activation function, for the neurons in the output layer, were used. With two pieces of data for each fault, the neural network input structure was restricted to two input terminals. The relative increment (%) was the first ANN input. We chose the increment instead of the oscillating frequency because these data have a wider range, $[-39.08, 43.68]$, so being numerically easier to discern while the oscillating frequency is in the range $[2.12, 5]$. This allows the ANN to be trained to learn the fault code more successfully with a smaller number of neurons in the hidden layer. The THD is the second input. The ANN learns the fault code.

In the exploitation phase, when excited by the measured relative increment of the oscillation frequency and the THD values, the ANN diagnoses the fault by outputting the fault code as a signal level, so we needed only one output neuron. The number of hidden neurons n, was found by trial and error after several iterations starting with an estimation based on that in [26, 62]. The goal was to find the optimum n that leads to a satisfactory classification. Using too many neurons would increase the training time, but using too few would starve the network of the resources needed to solve the problem. Also, an excessive number of hidden neurons may cause the over-fitting problem [63]. In practice, $n = 20$ hidden neurons were used.

The effectiveness of the training process of the obtained ANN was verified by exciting the ANN with faulty inputs. The responses of the ANN show that there were no errors in classifying the faults as is presented in Table 7.9.8. The ANN response was considered to be correct when its value was in the range $[(m - 0.5), (m + 0.5)]$. After successful training, no mistakes were observed for all 30 faults.

Further validation of the model was needed since noise is always present when measurement is performed. This situation was simulated by adding white noise, which plays the role of the voltmeter measurement error, in series to the circuit output. The white noise amplitude was chosen to be 1% of the signal amplitude. The simulation results show that white noise does not change the oscillating frequency at all, but it changes the THD. These changes are presented in Table 7.9.9. THD in the third column of the table is for the ideal case, with no error (as was presented earlier in the text), and THD_N in the fourth column is calculated after the simulation with white noise.

In the last column of the table the relative THD increment is presented. These data show that this increment is in all cases smaller than 1%. One can conclude now that the measurement error does not affect the diagnostics capacity of the ANN,

Table 7.9.7 Fault dictionary used for ANN training

No.	Fault type	Oscillation frequency (kHz)/Relative increment (%)	THD (%)	Fault code
0	Fault-free	3.48	44.25	0
1	AG1	5/43.68	70	3
2	AG2	2.12/−39.08	67.09	−11
3	OP: R_1	3.14/−9.77	9.826	12
4	OP: R_3	3.32/−4.6	26.85	5
5	SC: C_1	2.76/−20.69	12.443	−8
6	SC: C_3	2.4/−31.03	46.44	8
7	$1.2 \cdot R_1$	3.44/−1.15	33.555	9
8	$0.8 \cdot R_1$	3.54/1.72	40.877	−2
9	$1.2 \cdot R_2$	3.1/−10.92	17.701	1
10	$0.8 \cdot R_2$	4.02/15.52	23.108	−12
11	$1.2 \cdot R_3$	3.46/−0.57	26.902	2
12	$0.8 \cdot R_3$	3.5/0.57	48.13	11
13	$1.2 \cdot C_1$	3.38/−2.87	30.56	6
14	$0.8 \cdot C_1$	3.62/4.02	53.35	−9
15	$1.2 \cdot C_2$	3.28/−5.75	30.256	−5
16	$0.8 \cdot C_2$	3.8/9.2	24.3	14
17	$1.2 \cdot C_3$	3.32/−4.6	34.23	−13
18	$0.8 \cdot C_3$	3.7/6.32	26.93	−3
19	$1.2 \cdot C_2; 1.2 \cdot C_3$	3.1/−10.92	23.59	−1
20	$0.8 \cdot C_2; 0.8 \cdot C_3$	4/14.94	28.812	7
21	$1.2 \cdot C_2; 0.8 \cdot C_3$	3.52/1.15	39.144	−4
22	$0.8 \cdot C_2; 1.2 \cdot C_3$	3.64/4.6	28.7	−15
23	$1.2 \cdot R_1; 1.2 \cdot C_1$	3.34/−4.02	19.28	−7
24	$0.8 \cdot R_1; 0.8 \cdot C_1$	3.7/6.32	51.97	13
25	$1.2 \cdot R_1; 0.8 \cdot C_1$	3.56/2.3	26.45	15
26	$0.8 \cdot R_1; 1.2 \cdot C_1$	3.46/−0.57	23.75	−14
27	$1.2 \cdot R_2; 1.2 \cdot C_2$	2.9/−16.67	25.2	10
28	$0.8 \cdot R_2; 0.8 \cdot C_2$	4.34/24.71	17.476	4
29	$1.2 \cdot R_2; 0.8 \cdot C_2$	3.38/−2.87	28.65	−10
30	$0.8 \cdot R_2; 1.2 \cdot C_2$	3.78/8.62	28.461	−6

which is one supplementary advantage of this method, and one can also conclude that extracting THD as the second parameter was a good choice.

Table 7.9.8 ANN responses

No.	Fault type	Fault code expected	ANN response
0	Fault-free	0	0.00023
1	SC: R_1	3	2.98743
2	SC: R_3	−11	−11.0297
3	OP: R_1	12	12.0047
4	OP: R_3	5	5.00297
5	SC: C_1	−8	−7.92505
6	SC: C_3	8	7.99517
7	$1.2 \cdot R_1$	9	9.00308
8	$0.8 \cdot R_1$	−2	−1.78515
9	$1.2 \cdot R_2$	1	1.01037
10	$0.8 \cdot R_2$	−12	−11.9991
11	$1.2 \cdot R_3$	2	2.0044
12	$0.8 \cdot R_3$	11	11
13	$1.2 \cdot C_1$	6	6.00198
14	$0.8 \cdot C_1$	−9	−9.00061
15	$1.2 \cdot C_2$	−5	−4.9262
16	$0.8 \cdot C_2$	14	14.0097
17	$1.2 \cdot C_3$	−13	−12.9816
18	$0.8 \cdot C_3$	−3	−2.98707
19	$1.2 \cdot C_2$; $1.2 \cdot C_3$	−1	−0.970082
20	$0.8 \cdot C_2$; $0.8 \cdot C_3$	7	7.00478
21	$1.2 \cdot C_2$; $0.8 \cdot C_3$	−4	−3.74645
22	$0.8 \cdot C_2$; $1.2 \cdot C_3$	−15	−14.9913
23	$1.2 \cdot R_1$; $1.2 \cdot C_1$	−7	−6.98762
24	$0.8 \cdot R_1$; $0.8 \cdot C_1$	13	13.0002
25	$1.2 \cdot R_1$; $0.8 \cdot C_1$	15	15.008
26	$0.8 \cdot R_1$; $1.2 \cdot C_1$	−14	−13.9907
27	$1.2 \cdot R_2$; $1.2 \cdot C_2$	10	10.0079
28	$0.8 \cdot R_2$; $0.8 \cdot C_2$	4	3.9985
29	$1.2 \cdot R_2$; $0.8 \cdot C_2$	−10	−9.99773
30	$0.8 \cdot R_2$; $1.2 \cdot C_2$	−6	−5.98843

7.9.4 Fault Location in a CMOS Operational Amplifier Using ANNs—A Case Study

For a universal procedure for fault diagnosis in nonlinear dynamic electronic circuits a method for creation and search of hypotheses is necessary. It is the experience reported in [8] that use of artificial neural networks (ANNs), as shortly introduced

Table 7.9.9 THD changes due to the measurement error

No.	Fault type	THD (%)	THD_N (%)	Relative increment (%)
0	Fault-free	44.25	44.16	0.2
1	AG1	NO	NO	/
2	AG2	67.09	67.06	0.04
3	OP: R_1	9.826	9.829	0.03
4	OP: R_3	26.85	26.63	0.82
5	SC: C_1	12.443	12.438	0.04
6	SC: C_3	46.44	46.64	0.43
7	$1.2 \cdot R_1$	33.555	33.438	0.35
8	$0.8 \cdot R_1$	40.877	40.813	0.16
9	$1.2 \cdot R_2$	17.701	17.669	0.18
10	$0.8 \cdot R_2$	23.108	23.236	0.55
11	$1.2 \cdot R_3$	26.902	26.893	0.03
12	$0.8 \cdot R_3$	48.13	48.06	0.15
13	$1.2 \cdot C_1$	30.56	30.64	0.26
14	$0.8 \cdot C_1$	53.35	53.33	0.04
15	$1.2 \cdot C_2$	30.256	30.425	0.56
16	$0.8 \cdot C_2$	24.3	24.095	0.84
17	$1.2 \cdot C_3$	34.23	34.182	0.14
18	$0.8 \cdot C_3$	26.93	26.83	0.37
19	$1.2 \cdot C_2$; $1.2 \cdot C_3$	23.59	23.55	0.17
20	$0.8 \cdot C_2$; $0.8 \cdot C_3$	28.812	29.09	0.96
21	$1.2 \cdot C_2$; $0.8 \cdot C_3$	39.144	39.2	0.14
22	$0.8 \cdot C_2$; $1.2 \cdot C_3$	28.7	28.88	0.63
23	$1.2 \cdot R_1$; $1.2 \cdot C_1$	19.28	19.43	0.78
24	$0.8 \cdot R_1$; $0.8 \cdot C_1$	51.97	51.87	0.19
25	$1.2 \cdot R_1$; $0.8 \cdot C_1$	26.45	26.43	0.08
26	$0.8 \cdot R_1$; $1.2 \cdot C_1$	23.75	23.765	0.06
27	$1.2 \cdot R_2$; $1.2 \cdot C_2$	25.2	25.082	0.47
28	$0.8 \cdot R_2$; $0.8 \cdot C_2$	17.476	17.55	0.42
29	$1.2 \cdot R_2$; $0.8 \cdot C_2$	28.65	28.73	0.28
30	$0.8 \cdot R_2$; $1.2 \cdot C_2$	28.461	28.478	0.06

above, is the most successful one. Thus, here, we will outline the corresponding procedure which consists of several parts.

First one needs to create the fault dictionary. It is created by simulation and, when related to Fig. 7.9.1 it should be considered as a list of hypotheses among which, after search, one may find the one representing the faulty version of the circuit.

In the second step, the fault dictionary is mapped into the parameters of a neural network. This process is usually referred to as a training of the neural network. Here the response parameters obtained by simulation are used as input signals while the fault effect is enumerated (coded) and that code is used as the output to be learned.

Now, having the measured results which, of course, are contaminated by noise and measurement error, one should search the dictionary to find the fault that produces a fault effect which is the most similar to the measured one. For this purpose we use the ANN already created but now in "running" mode. Now, the measured parameters of the circuit response are used as inputs and the ANN's output is calculated. The resulting (output) code should represent the fault present in the circuit if any.

In order to describe the way in which the fault dictionary was created, the circuit in Fig. 7.9.8 is used as an application example. This is a CMOS operational amplifier consisting of seven transistors. According to the authors [8] this example belongs to the category of the most complex ones reported, both from the number of circuit elements point of view and the number of faults inserted. Note that three (nonlinear) capacitors are associated with every transistor totaling the number of nonlinear circuit elements to 28 but, for the sake of simplicity, are not shown in the figure. In order to emphasize the method as such, while not offering a full solution of the diagnostic problem for this circuit, having in mind the abundance of possible faults, a reduced set of faults was considered. To this end only single transistor faults are sought. That, of course, does not affect the generality of the ideas implemented in the next. There was no intention to diagnose simultaneous presence of several faults.

Ten faults per transistor, six catastrophic and four parametric were added to the dictionary. As shown in the figure there exist three open-circuit faults (OC) and three short-circuit faults (SC) per transistor. For example, in Table 7.9.10, OC3G stands for open gate of transistor T_3, and SCdg1 stands for drain and gate shorted in transistor T_1, and so on.

As opposed to [46] and some others, the shorts (some of them behaving as bridging fault) and opens, were really implemented as such, instead of using resistors for modeling them. To effectively simulate perfect short and opens the model of the ideal switch [47] was used (which is not implemented in any commercially available Spice type simulator). In addition, two faulty values for every channel length (±20%) (Denoted as $L+$ and $L-$ in Table 7.9.2), and two for every channel width (±20%) (Denoted as $W+$ and $W-$ in Table 7.9.2) were introduced, totaling 10 faults per transistor. The soft faults considered here are expected to model design errors and, in a specific way, gate oxide short having in mind the fault model reported in [64]. For the whole circuit this gives a set of 70 faults observed.

The DC output values (V_{oDCm}) were first obtained by simulation. Here $m = 0, 1, 2, \ldots, 69$ stands for the fault code. In addition, the frequency response of the circuit (the non-inverting input terminal was excited by a signal of amplitude 1 mV) was obtained by simulation over a fixed frequency range in order to extract two response parameters: the nominal gain (A_m) and the 3 dB cut-off frequency ($f_{3\,\mathrm{dB}m}$). For the example given, it was considered this signature to be satisfactory complex. If

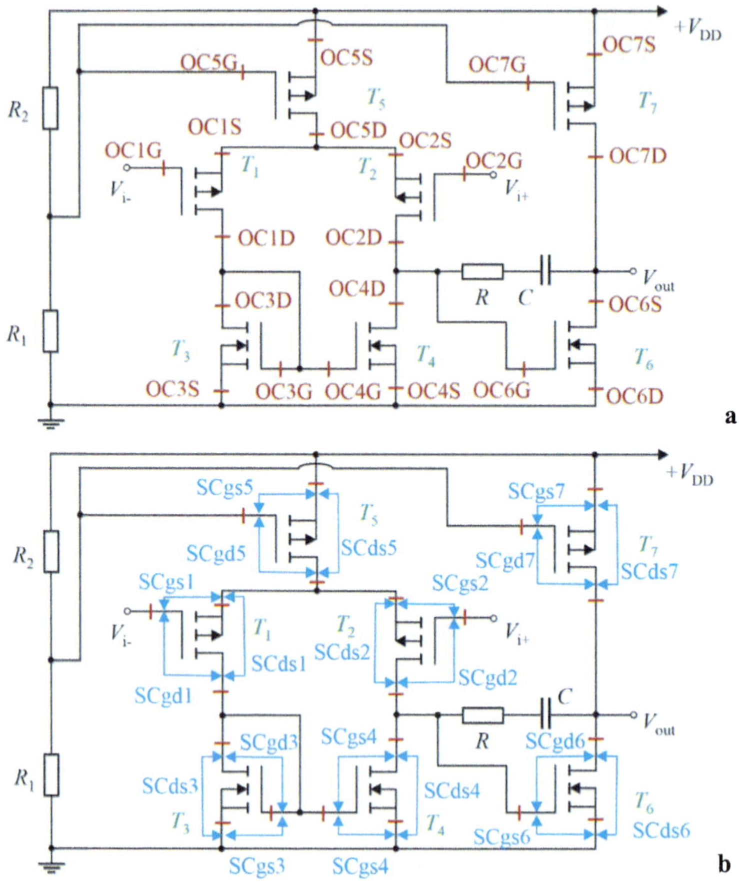

Fig. 7.9.8 Operational amplifier circuit. **a** Open-circuit (OC) faults shown and **b** short-circuit (SC) faults shown

additional fault were to be used one might think on additional measurements such as supply current. Note that, the DC open faults at sources and drains in series connected transistors may have equivalent signatures.

Because of the nonlinearity of the circuit, every fault is expected to change the quiescent operating points of all or some of the transistors. Consequently, new linear transistor models are created by Spice-like program and used for frequency domain performance extraction for each fault. In order to find the new quiescent operating point for every fault, we have to insert the fault, i.e., to create a faulty model of the circuit for DC analysis. This procedure is described elsewhere [65, 66] and will not be elaborated here.

Fault SCdg3 is untestable because of the existing connection between the gate and drain of T_3. This reduces the fault dictionary to 69 elements. Therefore, the fault dictionary created here has four columns containing the set of circuit performances,

Table 7.9.10 Part of the fault dictionary for the circuit of Fig. 7.9.8 (the fault codes are created as random integers)

Type	A_m	$f_{3\,\text{dB}m}$ (MHz)	$V_{\text{oDC}m}$ (V)	Code (m)
FF	419	0.01527	0.127	0
1L+	0.0053	6.791	0.0497	37
OC1G	0.047	501.187	0.127	49
OC3G	0.049	544.042	0.093	47
SCdg1	0.042	320.440	0.0458	6
SCds2	0.071	312.071	3.3	27
SC5ds	0.656	0.57	0.0186	55
6W−	5770	0.0018	0.2146	13
OC5D	0.056	507.298	3.3	25
SCgs5	0.109	0.036	0	2

i.e., the signatures: $\{V_{\text{oDC}m}, A_m, f_{3\,\text{dBm}})$ and the fault code. In the ANN training phase the three items of the signature will be considered inputs to the neural network, while the fault code will be learnt as an output.

As mentioned in the previous paragraph, the fault coding is an important issue. In fact, some faults exhibit very similar effects. So, input data (signatures) can have very close numerical values, and if the output values (fault codes) were also similar, the network could not always be trained successfully. Such a situation is illustrated by Fig. 7.9.9. Here the signatures of three faults are compared. By careful inspection we can see that only the values of $f_{3\,\text{dB}}$ suggest a difference between the fault effects. Having that in mind, to help the final diagnostic phase, the faults are coded randomly so that faults with similar effects are unlikely to have similar codes. This approach is proven to be good, because the way of coding influenced the training time, and also, the final training error of the ANN. Part of the fault dictionary for the circuit in Fig. 7.9.8, is given in Table 7.9.10, where $m = 0$ denotes the fault-free circuit.

Here we come to an additional issue concerning the applicability of rule-based approaches to the diagnosis of systems of this kind. Because of the similarity of the circuit performance in the presence of different faults, no set of rules can be established to distinguish between the three faults depicted in Fig. 7.9.9. Furthermore, by inspection of Table 7.9.8, we can see that the performance parameter values span over a broad range. For example, for the voltage gain, the smallest value in the table is 0.0053, and the largest is 5770. So, we cannot establish a rule defining the difference that occurs as a consequence of the presence of a certain fault. Thus rule-based approaches are impractical for systems exhibiting responses as continuous functions.

Note, in addition, that we expect noisy data to be obtained when field measurements are performed for diagnosis which further complicates the creation of any rules. We may claim that similar is related to use fuzzyfication in order to boost the difference among signatures.

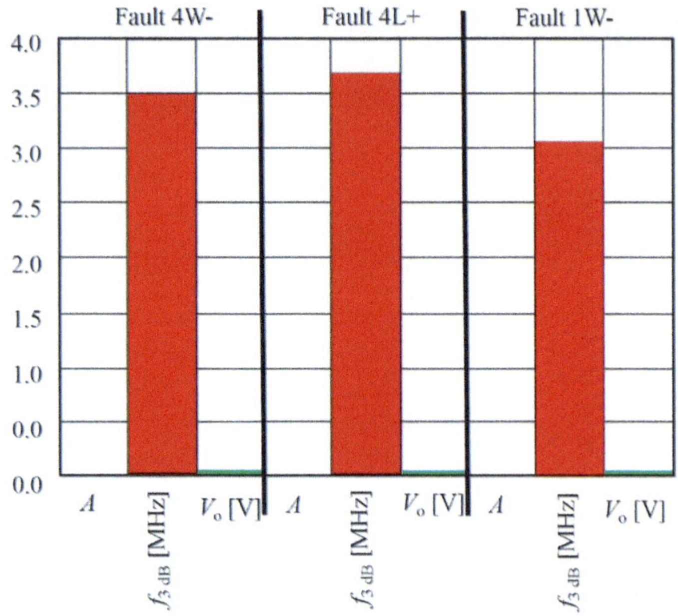

Fig. 7.9.9 Fault effects of faults $4W-$, $4L+$, and $1W-$

To go further, this puts in a similar prospective the simulation-after-test, i.e., the model based concept (which was not elaborated in detail here). Namely, in this concept one is supposed to create a set of hypotheses that will be checked against the measurement data by successive simulations of the circuit under test at the repairing site. Having in mind Fig. 7.9.9, however, we do not believe the creation of a qualified set of hypotheses is an achievable task.

The fault dictionary can be further reduced in size by processing the *ambiguity groups* or the groups of equivalent faults. According to [26, 62] "an ambiguity group is, essentially, a group of components where, in case of fault, it is not possible to uniquely identify the faulty one." Here, we can say that an ambiguity group consists of a set of *faults* that propagate identical signatures to the output, making the faults detectable and the circuit testable, but no distinction between the individual faults is possible making them undiagnosable. Table 7.9.11 shows all ten ambiguity groups for this example, systematically collected after simulation.

A specific ambiguity group is the case when the gain (A), although small, ever rises within the given frequency range. This is denoted as $A \rightarrow \infty$. The 3 dB cut-off frequency is in these cases indeterminate. Some of faults exhibiting this property, however, have different V_{oDC} values, so they can be distinguished. In these cases, to avoid use of infinite large numbers during the training of the ANN we assigned a value of 1000 to the gain. This is the case when simulating faults OCS7 and OCD7, but since these faults produce completely the same effect, they form ambiguity group number 6.

Table 7.9.11 Ambiguity groups and fault effects

Ambiguity group	Faults included	A	$f_{3\,\mathrm{dB}}$ (MHz)	V_{oDC} (V)
1	OC1D OC1S	0.31	20,000	0.0179
2	OC3D OC3S	0.041	365.8	3.3
3	OC4D OC4S SCgs4 SCgs3 SCds3	0.303	20,000	0.0458
4	OC5D OC5S	0.056	507.298	3.3
5	OC6D OC6S	0.063	0.039	3.3
6	OC7D OC7S	$A \rightarrow \infty$	Indeterminate	0
7	SCgs1 SCgs2	0.055	515.993	0
8	SCgs5 SCgs7	0.109	0.036	3.3
9	SCds4 SCgs6 SCds7	$A = 0$	Indeterminate	3.3
10	$3L$+ $4W$+	0.05	2.37	3.3

A similar situation occurs when the gain is almost zero, $A = 0$. The 3 dB cut-off frequency is then again indeterminate. Ambiguity group number 9 covers three such faults, with the same V_{oDC}.

Only one representative of each ambiguity group was included in the fault dictionary. From Table 7.9.12, we find that the complete fault dictionary in this case has $70 - 1 - 24 + 10 = 55$ elements.

With three pieces of data for each fault, the neural network input structure was restricted to three input terminals. The ANN diagnoses the fault by outputting the fault code (m) as a signal level, so we needed only one output neuron. The number of hidden neurons, n, was found by trial and error after several iterations starting with an estimation based on that in [66]. The goal was to find the optimum n that leads to a satisfactory classification even with noisy excitations. Using too many neurons would increase the training time, but using too few would starve the network of the resources needed to solve the problem. Also, an excessive number of hidden neurons may cause the *over-fitting* problem [48], when a network has so much information capability that it learns insignificant aspects of the training sets, irrelevant to the general population. In practice, 30 hidden neurons were used. After successful training, no mistakes were observed for all 55 faults.

Table 7.9.12 Noisy inputs and ANN response

Code	A_j	$f_{3\,\text{dB}}$ (MHz)	$V_{\text{oDC}j}$ (V)	ANN response
0	419	**0.0145**	0.127	−0.02128
1	129.6	**0.0248**	0.079	1.09057
2	0.109	0.036	**−0.05**	2.01405
3	6028	**0.001575**	0.1712	2.93868
5	4453	**0.002415**	1.0255	5.03203
6	**0.0441**	320.44	0.0458	6.03224
9	1000	1000	−0.05	9.0707
10	**0.043**	365.8	3.3	10.0278
12	1000	1000	**3.39**	12.1771
13	5770	**0.00171**	0.2146	13.2376
16	8220	**0.00197**	0.4876	16.031
18	0.32	1000	**0.133**	17.8458
20	0	1000	**3.46**	20.4409
21	0.83	1000	**3.46**	20.6497
25	**0.0588**	507.298	3.3	25.0605
26	11.739	**0.114**	0.127	26.0098
27	0.071	312.071	**3.46**	27.0091
34	5809	**0.00169**	0.1811	33.7541
35	209	0.0237	**0.115**	35.47
36	**0.05**	1000	0.8824	36.3514
37	**0.00556**	6.791	0.0497	37.2652
43	**0.004**	17.191	0.0509	43.0008
46	**0.0523**	515.993	3.3	45.99
47	**0.0514**	544.042	0.093	47.0133
49	**0.04935**	501.19	0.127	49.042
50	6030	**0.001425**	0.2466	49.9284
52	0.005	133.757	**3.46**	52.0044
53	119.4	**0.0258**	0.0843	53.0205
54	0.041	**428**	3.3	53.5346
55	0.688	0.57	0.0186	54.861

The generalization property of the network was verified by supplying noisy data to its inputs. This is presented in Table 7.9.12. Thirty samples were examined. For each sample, one input (boldfaced in the Table 7.9.12) is incremented by +5% or −5%, representing noise generated during the measurement process. The responses of the network are given in the last column of the table. The ANN response was considered to be correct (i.e., acceptable) when its value was in the range $[(m -$

0.5), $(m + 0.5)$]. We can see that all faults can still be diagnosed though some with difficulties (for $m = 20$, 35, and 54).

The method proposed above may be summarized as follows:

A set of test points and a set of measurements need to be determined in accordance with some general principles. The number of test points should be minimized while at the same time using the system primary inputs and outputs as much as possible. In addition, the number of measurements required for successful diagnosis must be kept small while, at the same time, no measurement based on "extraordinary" signals should be applied. Keeping in mind maintainability, which is the basic purpose of diagnosis, measurements that are already prescribed in the data-sheets of the product (i.e., measurements that are used for functional verification) should be used.

Careful analysis of the list of faults may be performed. Generally speaking, the number of faults that may occur within a circuit is enormous. The number of shorts between lines, for example, is a combinatorial problem that, even for moderate circuit leads to the "curse of dimensionality." Accordingly, a list of the most probable faults should be established before the project starts.

From the list of faults one creates a fault dictionary by inserting fault models into the circuit netlist and by repetitive simulation of the system under test. Related to that, a small digression will be made now. Namely, very frequently, as an alternative to the netlist the designers use a graphical editor to create the circuit schematic. We find, it will be more difficult to insert open-circuit catastrophic faults in that case.

The fault dictionary representing the mapping of the faults into a set of fault effect is captured by an artificial neural network (ANN). The measured (or simulated) performances of the system under test are used as the input signals to the ANN, while the output of the ANN is the fault code. It is up to the designer how the fault code is expressed. Here we propose the output neuron of the ANN to be linear, so the signal level expresses the code. Alternatively, several output neurons could code the signal. For example, six output neurons could create a six bit binary word at the output (this way of representing the fault code will be exemplified in Problem 7.11.3.2).

After learning the fault dictionary, and given the actual system performances, the ANN is able to diagnose the system even with noisy data. In some cases the ANN might be implemented in hardware and used on-line, even in real time. Such application requires that the measurements should be simple especially when additional (pre)processing of measurement data is recommended.

7.9.7 Hierarchical Analog and Mixed-Signal Diagnosis

The explosion of integrated circuit technology has brought with it some difficult testing problems. The recent growth of mixed analog and digital circuits complicates the testing problem even further. It becomes more complicated to determine a set of input test signals and output measurements that will provide high degree of fault coverage. There is also a timing problem when testing such circuits even on the fastest automated equipment. Analog electronic circuits are known to be difficult to test and diagnose. Apart from the huge number of possible faults, this difficulty is a consequence of the inherent nonlinearity of this circuit category. Even linear circuits (having linear input–output signal interdependence) exhibit nonlinear relations between circuit parameter values and the output response. There are no linear active networks. Active networks are nonlinear with nonlinear reactive elements. They may be linearized and thought of as such in situations where signal and parameter changes are small in comparison to nominal values. When large parameter changes or even catastrophic faults occur (affecting the DC quiescent state), however, one must distinguish between linear and analog circuits. This, unfortunately, is not the case in most research reports bringing confusion into the subject [67, 68]. A specific aspect of diagnosis is the number and location of the test points. Simply, we can say that internal test points should be avoided and measurements on the primary inputs and outputs are preferred. This is not only related to their automatic accessibility but also to the nature of the diagnostic reasoning. Namely, one looks for functionality in order to start diagnosing, the function being seen at the primary terminals. Of course, in order to compensate for the reduced number of test points additional measurements with different types of applied signals may be needed to extract complete information about the system behavior. For complex analog systems, however, hierarchical approaches based on decomposition [2, 3, 5, 20, 21] are inevitable provided that no propagation of the fault effect arises between partitions. That is not easy to achieve. Of course, there are circuits that may be partitioned based on functionality known a priori from the design process as mentioned in the introduction.

In this chapter we describe the results of applying feed-forward ANNs to the diagnosis of nonlinear dynamic electronic circuits that are mixed with digital ones with no restriction to the number and type of faults. This method is based on fault dictionary creation and using an ANN for data compression by memorizing the table representing the fault dictionary. The ANN created in this way is, consequently, used for diagnosis by applying to its inputs the signals obtained by measurement of the faulty network. The procedure was earlier applied to analog circuits and illustrated in [8] and outlined in the previous paragraph. A specific experiment was described in related to the robustness of the ANN used as a fault code classifier for analog circuit diagnosis. Namely, the measured data were corrupted by disturbances as shown in Fig. 7.9.1. That was random addition of noise (up to $\pm 5\%$). Not in one case false response was obtained.

Putting the SBT approach in the general context of diagnosis we first note that the fault dictionary contains all the knowledge we need. In other words all hypotheses are memorized (within the ANN) and no further hypothesis needs to be created after the dictionary is known. For a mixed-signal system such the one depicted in Fig. 7.9.10, we are faced with additional difficulties related to the different nature of the responses sought at different nodes. In order to tackle that problem the fault dictionary created at the system level was partitioned in two parts enabling implementation of the concept described in Fig. 7.9.1.

The ANNs used for this diagnostic example are the well-known feed-forward neural networks structured in three layers. The ANN topologies will be graphically depicted later on while presenting the example. They have only one hidden layer

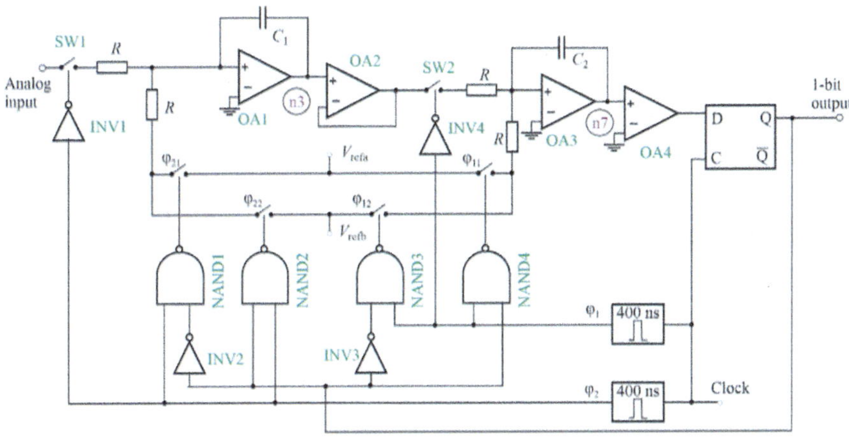

Fig. 7.9.10 Sigma-delta modulator structure

which has been proved sufficient for this kind of applications, i.e., approximation [43]. The neurons in the hidden layer are activated by a sigmoid (logistic) function, while the neurons in the output layer use linear activation function. Linear activation function at the output is of crucial importance for the implementation of the method since the response value of the output neuron is limited by the number of faults conceived to be diagnosed only. The learning algorithm used for training this network is a version of the steepest-descent minimization algorithm [44].

7.9.7.1 Fault Dictionary Creation and Application Example

In order to describe the way in which the fault dictionary was created, the sigma-delta modulator circuit depicted in Fig. 7.9.10 was used. It is a mixed-signal system with representative functional complexity having analog, digital, and switching elements. The switches in the circuit are modeled as truly ideal, exhibiting zero and infinite resistance for closed and open state, respectively.

The integrator's charging time is invariable with respect to the clock rate in order to keep the gain constant. This means that the analog switch must be turned on for fixed time duration regardless of clock rate. This is achieved by using monostable multivibrator as a fixed-width pulse generator in the circuit. The monostable multivibrator between the clock-input and the switch-control block operates as a pulse generator to produce control signals of fixed time duration.

We consider here faults in the whole circuit, meaning in analog, digital, and the switching part. We do not intend to diagnose multiple faults.

There are two types of faults in the digital part of the system observed: catastrophic (stuck-at) and delay faults (delays of rising and falling edge of digital signals). In the system of Fig. 7.9.10, the analog switches are controlled by digital signals, so there are pairs with the same fault effects, i.e., the effect is the same when the switch is stuck at ON (OFF) and when the logic circuit's output is "stuck-at-1" ("stuck-at-0"). So, we will consider hard faults (that are associated to the analog part of the circuit) as stucked switches [69].

Having in mind that the clock period in the system is 1.2 μs (half period is 600 ns), we examined effects of delays not greater than 400 ns. In fact, effects of rising-edge delay are simulated for delay values of: 100, 250, 400 ns, while for the falling edge, we inserted smaller values: 50, 100, 150 ns. The goal was to determine the mapping of the delay faults onto the output digital signal. All digital gates were examined (4 inverters and 4 NAND circuits). Simulations were performed using Alecsis [70–72] simulator. The first conclusion was that delays in the circuit of inverter 2 (INV2, Fig. 7.9.10) do not influence the output signal, meaning that the output was not changed. The fault dictionaries for both analog and digital part of the system were created using the response of the circuit to an input ramp signal (Fig. 7.9.11a). The system output value was registered after every clock period (Fig. 7.9.11b), so these output digital values form the output signature (Fig. 7.9.11c). These are then represented in more compact hexadecimal form (Fig. 7.9.11d). As an illustration, part of the fault dictionary for the digital part of the system is created as shown in Table 7.9.13.

The fault type coding in the Table 7.9.13 refers to the notation given in Fig. 7.9.11. For example, na1(tf=100ns) stands for the falling-edge delay of NAND1 being 100 ns, na3(tr=250ns) stands for the rising-edge delay of NAND3 being 250 ns, φ_{12}OFF is coding the switch φ_{12} stuck-at OFF, while sw2OFF is coding the switch sw2 stuck-at OFF. FF stands for the fault-free circuit.

There exist groups of delay values causing the same fault effect. Such groups are referred to as ambiguity groups or the groups of functionally equivalent faults (FEF).

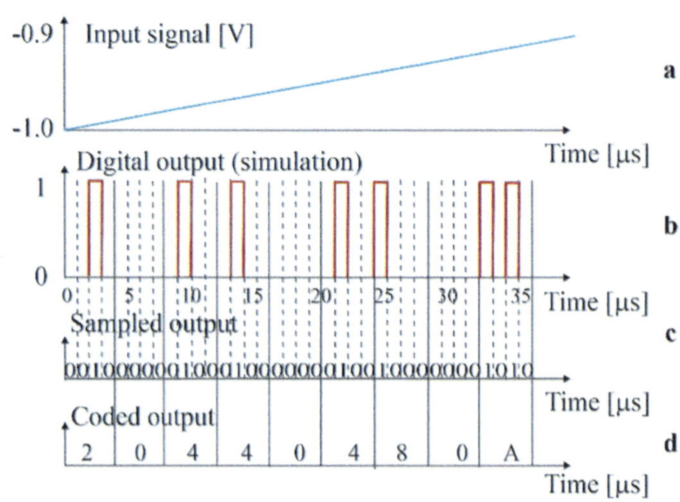

Fig. 7.9.11 Fault dictionary creation: response of the fault-free system

Table 7.9.13 Part of the fault dictionary for the digital part of the circuit

Fault code	Fault type	Signature
0	FF	20440480A
2	inv1(tr=150ns)	000010010
5	na1(tf=100ns)	008000010
8	na3(tf=50ns)	104108210
15	inv3(tr=100ns)	202404410
22	φ_{11}OFF	001C00038
24	sw2ON	018018030
27	na3(tr=250ns)	021041084
30	na3(tf=100ns)	082202208
33	sw2OFF	996696699
36	inv1(tr=50ns)	040810108
42	φ_{12}OFF	300038003
45	na4(tf=150ns)	802408811

No external testing can distinguish among FEF. The partition of all faults into distinct subsets of FEF defines the maximal fault resolution. A test that achieves the maximal fault resolution is said to be a *complete fault-location test*. For the subset of faults belonging to the digital part of the system FEF are listed in Table 7.9.14. The first four FEF represent cases where delay increments in different circuits cause the same effect at the system output, i.e., the same delay. In the second column of the Table 7.9.14, groups of faults giving rise to the same effects are listed, and accordingly, the third column presents that common fault effect (signature). The fifth ambiguity group is in a way different. The members of that group are both catastrophic and delay faults. Note that only one representative of each ambiguity group is introduced in the fault dictionary as depicted in Table 7.9.13.

In the analog part of the system, we have considered both parametric and catastrophic faults [73, 74]. As parametric faults we considered variations of resistance and capacitance values. The capacitances of both capacitors are changed. The first stage of the system is more sensitive to parameter variations, while the changes in the second stage have reduced effect on the performance, due to noise shaping. That is best seen when the fault effect of the capacitance in the first stage is observed. However, changes of capacitance in the second stage cause exactly the same effect as the fault-free circuit.

Catastrophic (hard) faults in an analog system change the circuit topology. In order to illustrate this, we have observed the situation when the feedback capacitor of the operational amplifier is disconnected, and also the situation when there is an open circuit at the operational amplifier's output (in the example given here, the output of the third operational amplifier-OA3 is disconnected, node n7 in Fig. 7.9.10). Part of the fault dictionary for the analog part of the system is depicted in Table 7.9.15, where fault code 0 denotes the fault-free system as it does in Table 7.9.13.

Table 7.9.14 Functionally equivalent faults

Ambiguity group	Fault type	Signature
1	na3(tf=50ns) na4(tr=250ns)	104108210
2	na3(tr=400ns) na4(tr=400ns)	102104208
3	na4(tf=100ns) inv3(tf=100ns)	404210240
4	FF inv2(tr=50ns) inv2(tr=100ns) inv2(tr=150ns) inv2(tf=50ns) inv2(tf=100ns) inv2(tf=150ns)	20440480A
5	φ_{21}OFF na1(tf=150ns) sw1ON	000000000

Table 7.9.15 Part of the fault dictionary for the analog part of the circuit

Fault code	Fault type	Signature
0	FF	20440480A
1	C_1 disconnected	E38E38E38
3	$0.8 \cdot C_1$	102204210
6	$1.2 \cdot R_1$	822211110
10	$0.8 \cdot R_4$	804809011
12	OA3 output disconnected	805005012

Here, again, the faults are coded randomly (column 1 in Tables 7.9.13 and 7.9.15), so that faults with similar effects are unlikely to have similar codes. This approach is proven to be good, because the way of coding influences the training time and error. The second column of Tables 7.9.13 and 7.9.15, describes the type of the fault. The third column contains the signature seen at the output. Note that, for obtaining the nine-digit hexadecimal number coding the binary output, one had to get 36 samples of the output waveform.

7.9.7.2 Synthesis of a Hierarchical Diagnostic System

A two-level hierarchical system is depicted in Fig. 7.9.12. The idea is to look at both the system and the subsystem (or circuit or component) level. Accordingly, one diagnostic system is to be created at the topmost level the task of which is to locate the faulty subsystem (component) and to deliver enough information (fault code and, possibly, type of hypothesis) to the lower diagnostic level to locate the fault within the subsystem (component). For that, of course, one needs as many diagnostic subsystems at the subsystem level as many parts are conceived within the system. Generally, such subsystems in the analog part may be operational amplifiers, circuits built of passive components and operational amplifiers (filters, for example), etc. while in the digital part one meets basic logic gates, flip/flops or even registers, and memory blocks. It is assumed that fault dictionaries and corresponding ANNs for diagnostic purposes are at disposal and incorporated into the diagnostic software at the time of diagnosis of the whole system.

At the system level, the modular approach was implemented first making the search for the diagnostic statement easier. The digital and analog part of the system were considered as modules and two artificial neural networks were trained for capturing the look-up tables, one for diagnosis in the digital part and another for diagnosis in the analog part of the system. Note, the partition is natural from the point of view of creation of fault dictionaries being obtained by simulation successively: firstly for the digital and later for the analog part. Both ANNs are feed-forward with one hidden layer. The signatures are inputs to the ANNs, and the fault code is ANN's output to be learned. It means that both neural networks have nine inputs (one input per hexadecimal digit) and one output terminal. After learning was completed, the number of hidden neurons in the resulting ANN was 10 (Fig. 7.9.13), for the network implementing the fault dictionary related to the digital part, and 3, for the other, what was found by trial and error after several iterations starting with an estimation based on [66].

The effectiveness of the training process of the obtained ANNs was verified by exciting the ANNs with faulty inputs. Responses of the ANNs show that there were no errors in identifying the faults what is partly presented in Table 7.9.16, where, results for the faults listed in Table 7.9.13 are shown. Only negligible discrepancies may be observed.

To mention again, two groups of faults were considered in order to reduce the number of faults per ANN so enabling easier learning and reduced complexity of the

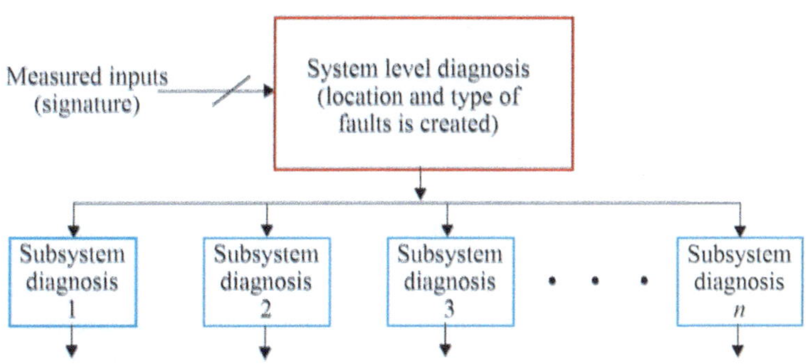

Fig. 7.9.12 Two-level hierarchical system

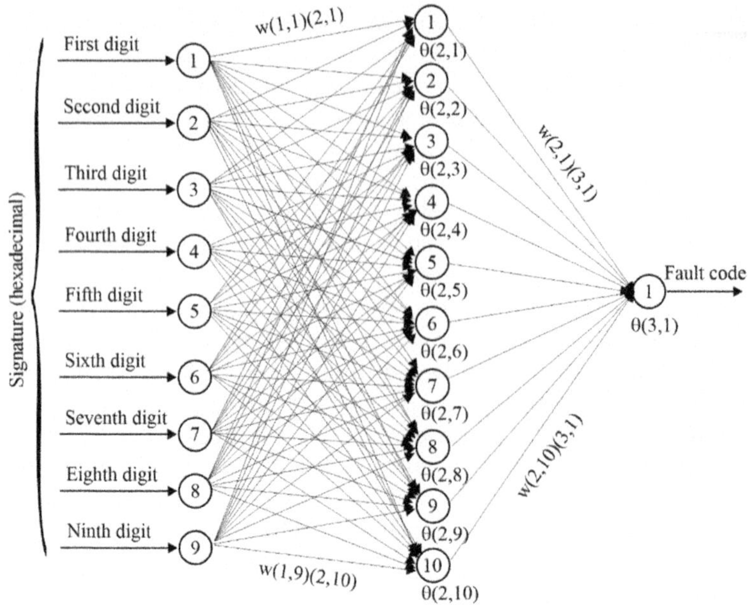

Fig. 7.9.13 Structure of one of the diagnostic ANNs

Table 7.9.16 ANN output results

Fault type	Fault code	ANN output
FF	0	−0.000215
inv1(tr=150ns)	2	1.99969
na1(tf=100ns)	5	4.99988
na3(tf=50ns)	8	7.99998
inv3(tr=100ns)	15	15
φ_{11}OFF	22	22
sw2ON	24	23.9997
na3(tr=250ns)	27	27
na3(tf=100ns)	30	30
sw2OFF	33	33.0021
inv1(tr=50ns)	36	36
φ_{12}OFF	42	41.9996
na4(tf=150ns)	45	41.9996

ANNs. Now, the task is to have a complete diagnosis at system level responding to every signature. The practical implementation of the concept of Fig. 7.9.1 is depicted in Fig. 7.9.14. ANN1 diagnoses faults in the digital part of the system and fault codes

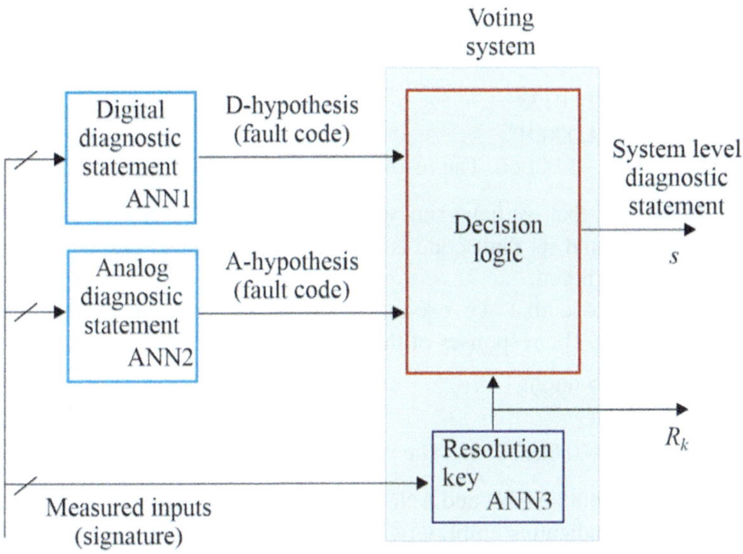

Fig. 7.9.14 ANN-based hierarchical diagnostic system

are in the range from 0 to 45. ANN2 diagnoses faults in the analog part of the system and fault codes are in the range from 0 to 12.

We can notice that we use numbers starting from 0 in both cases in order to denote fault codes. With that notation, when both diagnostic ANNs work in parallel, one cannot distinguish whether the fault code refers to analog or digital fault. So, we provided ANN3 in order to help distinguishing if certain fault is digital or analog. ANN3 also has 9 inputs and it gets the measured signature as an input as ANN1 and ANN2 do. It gets trained so that its output code takes values from the set {−1, 0, 1}. We refer to these values as to *resolution keys*. Namely, if the fault comes from the digital part, the output code is set to 1, while if it comes from the analog, the output code is set to −1. In the special cases when ambiguity arises, that is when one has the same signature coming from faults belonging to the digital and analog part, we assign 0 to the output of ANN3. We will give a few examples now, in order to illustrate the previous explanation. Suppose that we excite our ANNs with the input signature: $\{0\ 8\ 2\ 2\ 0\ 2\ 2\ 0\ 8\}$. The responses of the three networks are as follows:

ANN1 response: 30
ANN2 response: −0.0800663
ANN3 response: 0.99934. The resolution key is 1.

The decision logic (Fig. 7.9.14) decides that we have digital fault (because the ANN3 output value is approximately 1), and its code is 30 (because the ANN1 output value is 30). ANN2 response is ignored.

Next, we suppose that we excite our diagnostic system with the input signature: $\{8\ 0\ 4\ 4\ 1\ 0\ 4\ 2\ 1\}$. The responses of the three networks are as follows:

ANN1 response: 29.0138
ANN2 response: 4.00001
ANN3 response: -1.00066. The resolution key is -1.

The conclusion is that we have analog fault (because the ANN3 output value is approximately -1) and its fault code is 4 (because the ANN2 output value is 4). ANN1 response is ignored.

Finally, we suppose that we excite our 3 ANNs with the input signature: $\{1\ 0\ 4\ 1\ 0\ 8\ 2\ 1\ 0\}$. The responses of the three networks are as follows:

ANN1 response: 7.99998
ANN2 response: 11
ANN3 response: –0.00172622. The resolution key is 0.

We consider now both ANN1 and ANN2 responses because the response of ANN3 is approximately 0, indicating ambiguity. The conclusion is that we have analog fault with fault code 11, or digital fault coded with 8. We cannot decide which one of them really happened in the system because they have exactly the same response, and this is a problem that may be resolved by increasing the number of sampling intervals or by introducing additional signals for fault dictionary creation. That will be not discussed here anymore.

7.9.7.3 Hierarchy Application

The diagnostic statement obtained so far at the system level may be handed down to, say, a component level where the final diagnostic act may be performed. To do that one needs to perform diagnosis for the subsystems.

In the A/D converter depicted in Fig. 7.9.10 the role of the subsystems may be played by the digital gates and operational amplifiers. Identical process of fault dictionary creation and ANN synthesis is expected to be done for every single subsystem in order to perform diagnosis at the lower level.

An example of such a solution for an operational amplifier is given in [8] and will be used here for illustration of the rest of the diagnostic process. Earlier in this chapter we took into consideration a fault within the operational amplifier OA3, in the analog part of the system of Fig. 7.9.10, diagnosed by ANN2. It is the fault coded by 12: "*OA3 output disconnected*." Let that be the result of diagnosis at the system level: the location and the fault code related to that location. The structure of this operational amplifier is shown in Fig. 7.9.8. At the lower level, after searching the fault dictionary of OA3, being created separately for this subsystem as described in Sect. 7.9.7, we come to a final conclusion: the fault affecting the output of the whole system is that the drain of transistor T_6 is disconnected.

Table 7.9.17 Activity-flow of a diagnostic system

A. *Fault dictionary creation phase* ***at the subsystem level***. This activity is to be performed once for a given technology and the results are kept as intellectual property (i.e., library)
For every subsystem {
Choose a signal(s) expected to activate all the faults from the fault list (of length m);
For $j = 1, m$ (i.e., all the faults from the fault list) {
Create a signature by simulation of the faulty subsystem;
Create a random number k_j with uniform distribution such that $1 \leq k_j \leq m$, and $k_j \neq k_l, l = 1, 2, \ldots, j - 1$. This will be considered fault-code
Add the fault-code and signature to the fault dictionary
}
Perform ANN training to memorize the local fault dictionary. This includes not only determination of the numerical values of the weights and thresholds but the complexity of the ANN, too
}

B. *Fault dictionary creation phase* ***at the system level***
Choose a signal(s) expected to activate all the faults from the fault list (of length r);
For $j = 1, r$ (i.e., all the faults from the fault list) {
Create a signature by simulation of the faulty system;
Create a random number k_j with uniform distribution such that $1 \leq k_j \leq r$, and $k_j \neq k_l, l = 1, 2, \ldots, j - 1$. This will be considered fault-code
Add the fault code and signature to the fault dictionary
}
Perform ANN training to memorize the system's fault dictionary

C. ***Diagnosis*** Create signature by processing the measured response of the faulty system Run the appropriate ANN to get the diagnostic statement at the system level Run the appropriate ANN to get the diagnostic statement at subsystem level

Similar activities may be performed for every other fault that is diagnosed at the topmost level to belong to a fault located within one of the subsystems.

To summarize, in Table 7.9.17, we explain the complete algorithm representing the activity-flow for a diagnostic system to be developed and implemented.

It is well known that complex approximation activities are implemented whenever new ANN is created. That involves initial solution and convergence problem solving, not to mention handling of local minima. All that, however, was already discussed by many authors and will not be elaborated here.

References

1. Benjamins R, Jansweijer W (1994) Toward a competence theory of diagnosis. IEEE Expert 9(5):43–52
2. Ho CK, Eberhardt F, Tenten W (2001) Hierarchical fault diagnosis of analog integrated circuits. IEEE Trans CAS II Analog Digit Signal Process 48(8):921–929
3. Sheu H-T, Chang Y-H (1997) Robust fault diagnosis for large-scale analog circuits with measurement noises. IEEE Trans CAS-I 44:198–209

4. Soma M, Huynh S, Zhang J (2001) Hierarchical ATPG for analog circuits and systems. IEEE Design Test Comput 72–81
5. Bandler J, Salama A (1985) Fault diagnosis of analog circuits. Proc IEEE 73(8):1279–1325
6. Pipitone F, Dejong K, Spears W (1991) An artificial intelligence approach to analogue system diagnosis. In: Liu R-W (ed) Testing and diagnosis of analog circuits and systems. Van Nostrand Reinhold, New York, pp 187–215
7. Hayashi S, Asakura T, Zhang S (2002) Study of machine fault diagnosis system using neural networks. In: Proceedings of the international joint conference on neural networks, Honolulu, Hawaii, pp 233–238
8. Litovski V, Andrejević M, Zwolinski M (2006) Analogue electronic circuit diagnosis based on ANNs. Microelectron Reliab 46(8):1382–1391
9. Golonek T, Rutkowski J (2002) Use of genetic programming to analog fault decoder design. In: Proceedings of the international conference on ICSES'2002, Wroclaw-Swieradow Zdroj
10. Pous C, Colomer J, Melendez J, de la Rosa JL (2002) Introducing qualitative reasoning in fault dictionaries techniques for analog circuits analysis. In: Proceedings of the sixteenth international workshop on qualitative reasoning, Barcelona, Spain
11. Stratigopoulos H-G (2015) Test techniques for analog circuits and systems. Micro- and nanotechnologies/microelectronics. Université de Grenoble
12. Spina R, Upadhyaya S (1997) Linear circuit fault diagnosis using neuromorphic analyzers. IEEE Trans CAS II Analog Digit Signal Process 44(3):188–196
13. Scarselli F, Tsoi AC (1998) Universal approximation using feed forward neural networks: a survey of some existing methods and some new results. Neural Netw 11(1):15–37
14. Milor L, Visvanathan V (1989) Detection of catastrophic faults in analog integrated circuits. IEEE Trans Comput Aided Design 8(2):114–130
15. Luehetta A, Manetti S, Piccirilli MC (2002) Critical comparison among some analog fault diagnosis procedures based on symbolic techniques. In: Proceedings of the 2002 design, automation and test in Europe conference and exhibition, Paris, France, 1105
16. Catelani M, Giraldi S (1998) Fault diagnosis of analog circuits with model based techniques. In: Proceedings of the IEEE instrumentation and measurement technology conference, vol VI, pp 501–504
17. Alippi C, Catelani M, Mugnaini M (2002) SBT soft fault diagnosis in analog electronic circuits: a sensitivity-based approach by randomized algorithms. IEEE Trans Instrum Meas 51(5):1116–1125
18. Huang J, Cheng KT (2000) Test point selection for analog fault diagnosis of unpowered circuit boards. IEEE Trans CAS II Analog Digit Signal Process 47(10):977–987
19. Yoon H, Hou J, Bhattacharya SK, Chatterjee A, Swaminathan M (1999) Fault detection and automated fault diagnosis for embedded integrated electrical passives. J Signal Process Syst 21(3):265–276
20. Chang YH (2002) Frequency-domain grouping robust fault diagnosis for analog circuits with uncertainties. Int J Circuit Theory Appl 30:65–86
21. Liu D, Starzyk A (2002) A generalized fault diagnosis method in dynamic analogue circuits. Int J Circuit Theory Appl 30:487–510
22. Starzyk JA, Liu D (2002) A decomposition method for analog fault location. In: Proceedings of IEEE ISCAS'02, vol III, pp 157–160
23. Tadeusuewicz M, Halgas S, Korzybski M (2002) An algorithm for soft-fault diagnosis of linear and nonlinear circuits. IEEE Trans CAS II Analog Digit Signal Process 49(11):1648–1653
24. Worsman M, Wong MWT (2000) Non-linear analog circuit fault diagnosis with large change sensitivity. Int J Circuit Theory Appl 28:281–303
25. Cota EF, Carro L, Lubaszewski M (1999) A method to diagnose faults in linear analog circuits using an adaptive tester. In: Proceedings of the design, automation and test in Europe conference, DATE'99, Munich, Germany, pp 184–188
26. Cherubal S, Chatterjee A (1999) Parametric fault diagnosis for analog systems using functional mapping. In: Proceedings of the design, automation and test in Europe conference, DATE'99, Munich, Germany, pp 195–200

27. El-Yazeed MF, Mohsen A, Adel AK (2003) A preprocessor for analog circuit fault diagnosis based on Prony's method. Int J Electron Commun 1:16–22
28. Dai Y, Xu J (1999) Analog circuit fault diagnosis based on noise measurements. Microelectron Reliab 39(8):1293–1298
29. Pinjala KK, Kim BC, Varuyam P (2003) Automatic diagnostic program generation for mixed signal load board. In: Proceedings of the international test conference, Charlotte, NC, USA, pp 403–409
30. Liobe J, Margala M (2004) Fault diagnosis of a GHz CMOS LNA using high-speed ADC-based BIST. In: Proceedings of the IEEE defect-based testing workshop, Napa, CA, pp 85–89
31. Starzyk JA, Liu D (2002) A decomposition method for analog fault location. In: IEEE International symposium on circuits and systems, Scottsdale, Arizona, vol III, pp 157–160
32. Dragic S, Margala MA (2002) 1.2 V Built-in architecture for high frequency on-line Iddq/ delta Iddq test. In: Proceedings of the IEEE computer society annual symposium on VLSI, pp 148–153
33. Margala M, Dragic S, El-Abasiry A, Ekpe S, Stopjakova V (2002) 1 V Fast IDDQ current sensor for on-line mixed-signal/analog test. In: Proceedings of the IEEE computer society annual symposium on VLSI, Pittsburgh, PA, USA, pp 165–170
34. Papakostas DK, Hatzopoulos AA (1991) Supply current testing in linear bipolar ICs. Electron Lett 30(2):128–130
35. Bell IM, Caraplin DA, Taylor GE, Bannister BR (1991) Supply current testing of mixed analogue and digital ICs. Electron Lett 27(17):1583–1591
36. Zwolinski M, Bartt A, Wilkins BR, Suparjo BS (1996) Analogue circuit test using RMS supply current monitoring. In: Proceedings of the IEEE international mixed signal testing workshop
37. Savioli CE, Calvano JV, de Mesquita Filho AC (2005) Fault trajectory approach for fault diagnosis on analog circuits. In: Proceedings of the design automation and test in Europe conference, DATE'05, Munich, Germany, pp 174–177
38. Materka A (1994) Neural network for parametric testing of mixed-signal circuits. Electron Lett 31(3):183
39. Rodrigez C et al (1996) A modular neural network approach to fault diagnosis. IEEE Trans Neural Networks 7(2):326–340
40. Aminian M, Aminian F (2000) Neural network based analog circuit fault diagnosis using wavelet transform as preprocessor. IEEE Trans CAS II Analog Digit Signal Process 47(2):151–156
41. He YG, Tan YH, Sun Y (2002) A neural network approach for fault diagnosis of large-scale analog circuits. In: Proceedings of ISCAS'02, vol 1, pp 153–156
42. Andrejević M, Litovski V (2004) ANN application in electronic diagnosis—preliminary results. In: Proceedings of IEEE 24th international conference on microelectronics (MIEL2004), Niš, Serbia, pp 597–600
43. Catelani M, Gori M (1996) On the application of neural networks to fault diagnosis of electronic analog circuits. Measurement 17:73–80
44. Maidon Y, Jervis BW, Dutton N, Lesage S (1997) Diagnosis of multifaults in analogue circuits using multilayer perceptrons. IEE Proc Circuits Devices Syst 144(3):149–154
45. Yang ZR, Zwolinski M, Chalk CD, Williams AC (2000) Applying a robust heteroscedastic probabilistic neural network to analog fault detection and classification. IEEE Trans CAS Int Circuits Syst 19(1):142–151
46. Stopjakova V, Malosek P, Misucik D, Matej M, Margala M (2004) Classification of defective analog integrated circuits using artificial neural networks. J Electron Test Theory Appl 20:25–37
47. Yu S, Jervis BW, Eckersall KR, Bell IM, Hall AG, Taylor GE (1994) Neural network approach to fault diagnosis in CMOS opamps with gate oxide short faults. Electron Lett 30(9):695–696
48. Collins P, Yu S, Eckersaal KR, Jervis BW, Bell IM, Taylor GE (1994) Application of Cohonen and supervised forced organization maps to fault diagnosis in CMOS opamps. Electron Lett 30(22):1846–1847
49. Masters T (1993) Practical neural network recipes in C++. Academic Press, San Diego

50. Zografski Z (1991) A novel machine learning algorithm and its use in modeling and simulation of dynamical systems. In: Proceedings of 5th annual European computer conference, COMPEURO'91, Bologna, Italy, p 860. (Published in: Advanced computer technology, reliable systems and applications, 1992)
51. Petkovic P, Stevanović R, Litovski V (1993) Symbolic approach to linear circuit analysis. In: Proceedings of 2nd Serbian conference on microelectronics and optoelectronics, Niš, Yugoslavia, pp 437–442
52. Duhamel P, Rault J (1979) Automatic test generation techniques for analog circuits and systems: a review. IEEE Trans Circuits Syst CAS 26(7):411–440
53. Khilari SS (2014) Transfer function and impulse response synthesis using classical techniques. Ms thesis, University of Massachusetts, Amherst, USA
54. Panić V, Milovanović D, Petković P, Litovski V (1995) Fault location in passive analog RC circuits by measuring impulse response. In: 20th International conference on microelectronics MIEL'95, Niš, Serbia, vol 2, pp 517–520
55. Andrejević Stošović M, Milić M, Litovski V (2012) Analog filter diagnosis using the oscillation based method. J Electr Eng 63(6):349–356
56. Andrejević Stošović M, Milić M, Zwolinski M, Litovski V (2013) Oscillation-based analog diagnosis using artificial neural networks based inference mechanism. Comput Electr Eng 39:190–201
57. Litovski VB (2019) Electronic filters, theory, numerical receipts, and design practice using the RM software. Springer, Berlin
58. Litovski VB, Zwoliński M (1997) VLSI circuit simulation and optimization. Chapman and Hall, London
59. Milić M, Andrejević Stošović M, Litovski V (2011) Oscillation based analog testing—a case study. In: 34th International convention on information and communication technology, electronics and microelectronics, MIPRO 2011, Opatija, Croatia
60. Alli PK (2004) Testing a CMOS operational amplifier circuit using a combination of oscillation and Iddq test methods. Ms thesis, Louisiana State University, USA
61. Hu G, Wang H, Hu M, Yang S (2007) Oscillation test strategy for analog filters by monitoring output voltage and supply current. Tsinghua Sci Technol 12(S1):78–82
62. Manetti S, Piccirilli C (2003) A singular-value decomposition approach for ambiguity determination in analog circuits. IEEE Trans Circuits Syst I Fundam Theory Appl 50(4):477–487
63. Mrčarica Ž, Ilić T, Litovski VB (1999) Time domain analysis of nonlinear switched networks with internally controlled switches. IEEE Trans Circuits Syst I Fundam Theory Appl 46:373–378
64. Milovanovic D, Litovski V (1991) Fault models of CMOS transmission gate. Int J Electron 71(4):675–683
65. Milovanovic D, Litovski V (1994) Fault models of CMOS circuits. Microelectron Reliab 34(5):883–896
66. Baum EB, Haussler D (1989) What size net gives valid generalization. Neural Comput 1:151–160
67. Mismar D, Abu Baker A (2010) Neural network based algorithm of soft fault diagnosis in analog electronic circuits. Int J Comput Sci Network Secur IJCSNS 10(1):107–111
68. Manikandan V, Devarajan N (2007) SBT approach towards analog electronic circuit fault diagnosis. Active Passive Electron Compon, Article ID 59856
69. Andrejević M, Litovski V, Zwolinski M (2006) Fault diagnosis in digital part of mixed-mode circuit. In: Proceedings of IEEE 24th international conference on microelectronics (MIEL2006), Niš, Serbia, pp 437–440
70. Glozić D (1994) Alecsis 2.1: an object-oriented hybrid simulator. PhD thesis, University of Niš, Serbia. (in Serbian)
71. Litovski V, Dimić Ž, Damnjanović M, Mrčarica Ž (1998) Electronic circuit simulation in a mixed-language environment. Microelectron J 29(8):553–558
72. Litovski V, Maksimović D, Mrčarica Ž (2001) Mixed-signal modelling with AleC++: specific features of the HDL. Simul Pract Theory 8(6–7):433–449

73. Andrejević M, Litovski V (2006) Fault diagnosis in analog part of mixed-mode circuit. In: VI simpozijum Industrijska Elektronika, INDEL 2006, Banja Luka, Bosnia and Herzegovina, pp 117–120
74. Andrejević M, Litovski V (2006) Fault diagnosis in digital part of sigma-delta converter. In: Proceedings of the 8th seminar on neural network applications in electrical engineering, NEUREL 2006, Belgrade, Serbia

7.10 Testing as an Industrial Activity

Abstract In the previous chapters the synthesis of test signals and algorithms for diagnosis were emphasized. To complete the subject, in this chapter the very technology of testing will be discussed. We start with the overview of failure modes of electronic components and circuits which, we expect, will familiarize the reader with the vast number of possible faults that are lurking in an electronic system. Then we will go through the life cycle of the electronic system and discuss the most frequent failures and testing methods related to that part of the life cycle. Having an overview of the problem from industrial point of view, we then address the hardware and the software needed to perform testing including the automatic test pattern generator (ATPG). The tester himself and its interface to the device under test will be studied including testing bare die (chips), encapsulated components, and printed circuit boards. A study of the testing economy will be offered and challenges of the future will be discussed.

7.10.1 A Short Overview of Failure Modes of Electronic Components and Circuits

In [1] and [2] the physics, construction, and function mechanisms of most discrete and integrated electronic components were discussed. That, we think, allows the reader to comprehend the failure mechanisms which may happen in them to which this paragraph is devoted. What was not given within the LNAE series were the construction and physics of passive components, including the printed circuit boards. These will be discussed briefly below with emphasis to the failure mechanisms.

Electronic components have a wide range of failure modes. Various classifications exist for these, including categorizations based on time or cause. Failures may stem from factors like elevated temperature, excessive current or voltage, ionizing radiation, mechanical shock, internal stress, impact, and various other triggers. Semiconductor devices, for instance, might encounter issues caused by contamination, mechanical stress on the device, or the occurrence of open or short circuits within the device package.

V. B. Litovski, *Lecture Notes in Analog Electronics*, Lecture Notes in Electrical Engineering 1265, https://doi.org/10.1007/978-981-97-8257-4_10

The primary cause of electronic parts failures is often associated with the packaging. The packaging, acting as a protective barrier between electronic components and their surroundings, proves highly vulnerable to environmental influences. Mechanical stresses induced by thermal expansion, particularly when materials with differing coefficients of thermal expansion are involved, can lead to material fatigue. Environmental factors such as humidity and corrosive chemicals contribute to the corrosion of packaging materials and leads. This corrosion has the potential to break the packaging and damage internal components, resulting in electrical failures.

Additionally, surpassing the allowable temperature range can lead to overstressing of wire bonds, causing them to break loose, semiconductor dies to crack, or packaging to develop cracks. Cracking can also occur due to humidity, elevated temperatures, mechanical damage, or shock.

The contact resistance in electrical contacts is widespread, with its intensity influenced by surface structure and the composition of surface layers. Ideally, contact resistance should remain low and stable. However, factors such as insufficient contact pressure, mechanical vibrations, corrosion, and the development of passivating oxide layers on contacts can significantly alter contact resistance. This alteration may result in resistance heating and eventual circuit failure.

Soldered joints are susceptible to various failure modes, including electromigration and the formation of brittle inter-metallic layers. Some failures only become apparent at extreme joint temperatures, making troubleshooting challenging. The thermal expansion mismatch between the printed circuit board material and its packaging puts strain on the part-to-board bonds. Leaded parts can absorb the strain by bending, but lead-less part rely on the solder to absorb stresses. Thermal cycling can induce fatigue cracking in solder joints, especially those with elastic solders.

Printed circuit boards (PCBs) are susceptible to environmental factors. For instance, the conducting stripes are prone to corrosion, and improper etching may leave them partially shorted. Vias might be inadequately plated through or filled with solder. Mechanical loads can lead to the cracking of conducting stripes, resulting in unreliable PCB performance. Solder flux residues can promote corrosion, while residues of other materials on PCBs can induce electrical leaks. Ionic compounds like chlorides can enhance corrosion. Alkali metal ions have the potential to migrate through plastic packaging and impact semiconductor functionality. Over time, chlorinated hydrocarbon residues may undergo hydrolysis, releasing corrosive chlorides. These issues often manifest after several years. Additionally, polar molecules can dissipate high-frequency energy, causing parasitic dielectric losses.

Examples of semiconductor failures related to semiconductor crystals include:

- formation of dislocations,
- buildup of trapped charge carriers in the gate oxide of MOSFETs,
- movement of charge carriers from floating gates,
- inadequate passivation.

Vias frequently contribute to undesirable serial resistance in chips. Faulty vias exhibit unacceptably high resistance, leading to increased propagation delays. Alternatively, the drop in their resistivity with rising temperature serves as an indicator of a fault, manifested by the degradation of the chip's maximum operating frequency.

Metallization failures, some of which are illustrated in Fig. 7.1.1, are more common and serious causes of integrated circuit degradation than material processes. Examples of such failures include:

- Electromigration involves the movement of atoms out of active regions, leading to dislocations and point defects that act as non-radiative recombination centers, generating heat. This phenomenon is particularly observed in MESFETs with aluminum gates under RF signals, resulting in erratic drain current. Dislocated aluminum may also lead to shorts in neighboring structures. In every instance, electromigration has the potential to induce alterations in the dimensions and parameters of transistor gates and semiconductor junctions.
- Ohmic contact degradation between metallization and semiconductor layers.

The majority of semiconductor failures attributed to stress are fundamentally electro-thermal at the microscopic level. Elevated temperatures at localized points can result in immediate failures through processes such as melting or evaporating metallization layers, semiconductor melting, or structural changes. High temperatures tend to accelerate diffusion and electromigration, thereby reducing the device's lifespan. Damage to junctions that does not cause immediate failure may be evident in altered current–voltage characteristics of the junctions. Failures due to electrical overstress can be categorized as thermally induced, electromigration-related, or electric field-related.

Electrostatic discharge (ESD) falls within the category of electrical overstress and can result in instantaneous device failure, lasting alterations in parameters (as depicted in Fig. 7.1.1), and latent damage leading to an elevated degradation rate. Contemporary VLSI circuits exhibit heightened sensitivity to ESD, due to smaller dimensions which leads to smaller capacitances exhibiting larger voltages for the same charge loaded.

7.10.2 The Life Cycle of an Electronic Product and Related Testing

There are four phases in the life of an electronic system or component:

- design and development,
- production,
- exploitation, and
- end-of-life.

Faults can occur in each of these phases (excluding, naturally, the last one), be it production, design or exploitation. Therefore, it is necessary to perform testing in each of these phases in order to remove faults before proceeding to the next phase. At the same time, although a large number of faults may occur in each phase, it is known that some of the faults appear most often. That is why fault detection procedures can be adapted to individual components and to individual phases of component life. In the following text, we will consider some such characteristic situations.

In the design phase, we distinguish three stages: specification of requirements, synthesis, and realization. In the first stage, we formalize the user's requirements into the final order document; in the synthesis stage, we convert the order into a functional schematic; and in the realization stage, we convert the functional schematic into an electrical schematic using a set of available components. All this is followed by layout design be it an integrated circuit or a printed circuit board. Each of these stages is subject to human or design errors. In order to ensure correctness, tests are needed that check the agreement between the input requirements and the performance of the synthesized system.

Figure 7.10.1 shows the testing stages of the type of dominant faults and the tests that are usually applied. The specification stage is dominated by faults related to incorrect translation of ideas into formal requirements, logical inconsistency among requirements, omission of parts of specifications or requirements due to carelessness, and imperfection of the algorithms used to perform the conversion process of specifications into formal requirements. In the synthesis stage, the most common faults refer to the omission of part of the circuit due to carelessness and problems arising from the imperfection of the algorithms used for synthesis. Finally, in the stage of realization, faults arise related to errors in the conversion of the functional schematic into a schematic composed of available components, violations of design rules, installation of faulty components, faulty wiring and assembly, and too long connections that disrupt system synchronization (Table 7.10.1).

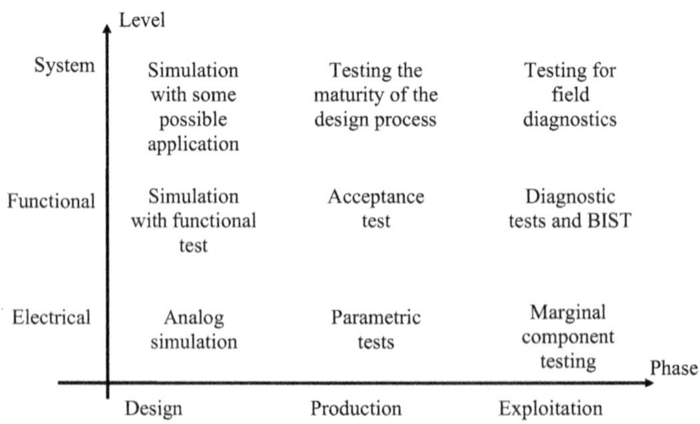

Fig. 7.10.1 Testing as a function of system level and product life stage

Table 7.10.1 Design stages, dominant types of faults in each stage and type of testing used

Activity	Result	Source of failure	Testing method
Marketing specifications	Functional specifications	• Translation error • Inconsistencies • Omission of design requirement • Algorithmic error	• Examination by a specialist • Simulation • Ad hoc review
Synthesis	Functional schematic	• Omitting part of the scheme • Bad interpretation • Algorithmic error	Simulation that performs functional testing
Realization	Prototype	• Violation of design rules • Malfunction of the installed component • Bad connections and installation • Delay inconsistencies	• Non-functional tests on prototypes • Timing testing

During the production phase, two testing approaches are required. One of them refers to the products themselves and the other to the production process as such. Starting from the raw materials and ending with the final product, it is necessary to constantly perform quality control and testing in order to minimize the number of samples that contain undetected faults, thus transferring them to higher stages of production. The types of testing required, of course, depend largely on the product itself and the processes involved. When dealing with electronic systems, however, three levels of testing are typically required: component level, circuit board level, and system level. At this stage, it is considered that the faults due to the design have been removed, so the testing is related to checking the "black boxes" that were created by mass production. The faults are mostly structural as stated above.

The production process can, in itself, manifest problems that can be reflected in an unacceptably large number of faults. In this case, fault diagnostics should not only detect and locate the fault but should also find the phase in the production process that led to that fault. For the sake of illustration Fig. 7.10.2 shows the quality control testing process and Fig. 7.10.3 the process of evaluation of the production process.

After installing the component or system, in the exploitation phase, routine control and preventive maintenance is performed. The most common source of failures in this phase is component failures. When it is shown that the system is not working, testing is applied for the purpose of repair, which also includes diagnostics as we mentioned above, in order to carry out the repair on the spot. Since it is assumed that the system design is in order and that the number of possible physical faults is small, this form of testing is relatively easy to implement (Tables 7.10.2 and 7.10.3).

There are two types of testing for repair. One is done in the field and it tries to reduce the loss of the user of the system by getting the repair done quickly. The goal of this testing is to find a defective unit that is replaceable and to replace it on the spot. If it turns out that this method of testing for the purpose of repair does not give

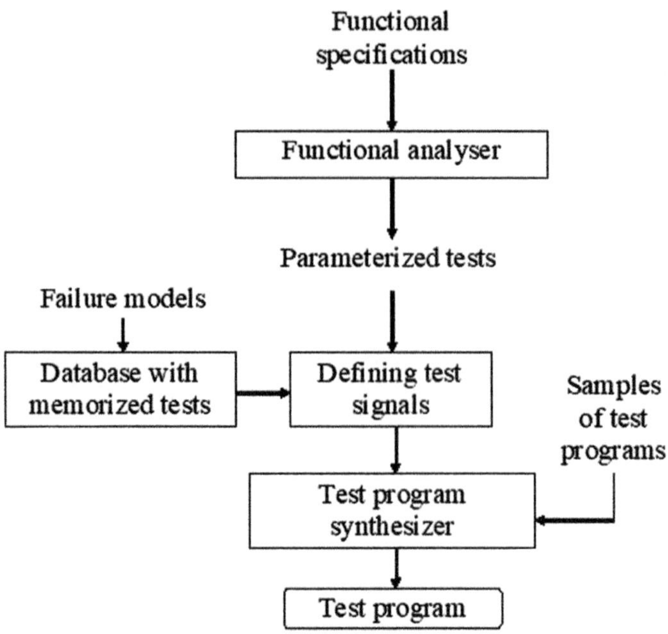

Fig. 7.10.2 Automatic generation of functional tests

Fig. 7.10.3 Diagram to demonstrate testing progress

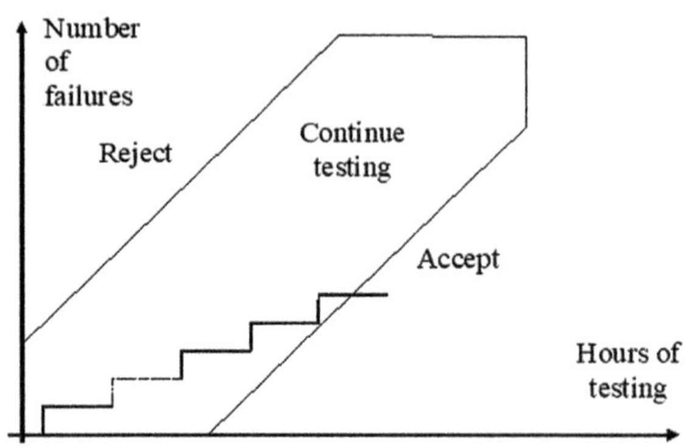

Table 7.10.2 Quality control in the production phase

Level	Nature of the fault	Testing method	Resolution
Chip	• Initial fault • Component failure	Burn-in	Go/No go
Printed circuit board	• Fault in assembly • Component failure	• Connection test • Diagnostics	• Level of connections • Chip level
System	• Connection • Assembly • Component fault	• System test • Diagnostics	• Connections • Boards

Table 7.10.3 Testing in the production phase

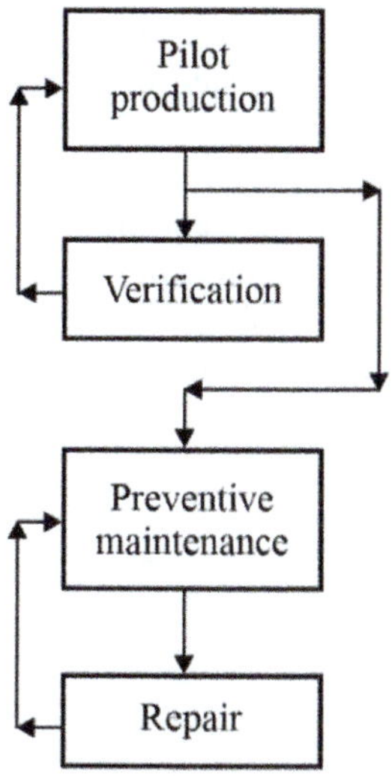

Testing undertaken	Resolution
Testing of the maturity of the project	Go/No go
Testing of the maturity of the process as a whole	Steps in the production to be improved

results, testing in the factory is approached, whereby the cost of the repair itself is minimized by replacing the simplest component that has failed. The stages of system operation and testing related to each stage are shown in Fig. 7.10.4.

Component failures can be gradual, causing intermittent faults. Even permanent faults are sometimes undetectable during normal operation. That is why preventive maintenance is needed, which enables the correctness of the system to be achieved for a longer period of time. During preventive maintenance, the system is sometimes

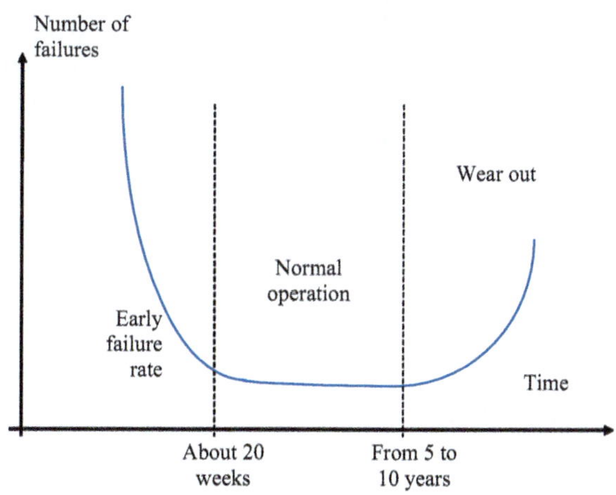

Fig. 7.10.4 Number of failures of semiconductor components as a function of time of use

subjected to additional stress. Such a stress would be running at a higher clock speed in the hope that the marginal components would exhibit a fault under that stress.

Summarizing the considerations in this paragraph, we can conclude that at any stage of a system's life cycle, testing can be done at every level in the system hierarchy. Figure 7.10.5 provides a classification of the types of testing that are typical for each phase.

Before proceeding with a more detailed study of testing in different life cycle phases and different levels of the system we will shortly pay attention to the subject of end-of-life of an electronic product being the last phase of existence of a product. We expect this subject to be considered in detail in a separate book of the LNAE series while here we are giving a short glimpse only (Table 7.10.4).

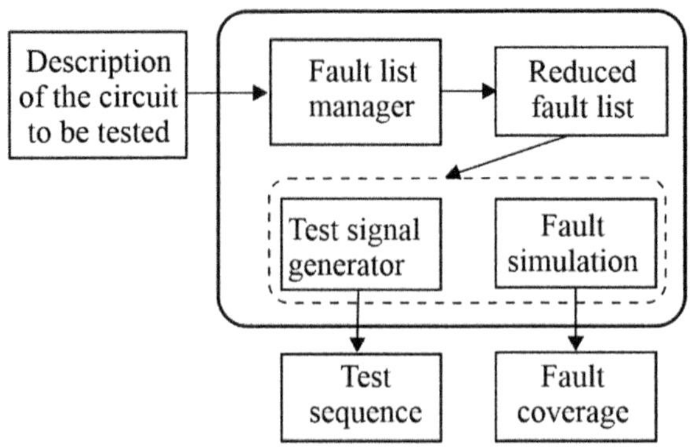

Fig. 7.10.5 ATPG system architecture

Table 7.10.4 Testing in the exploitation phase

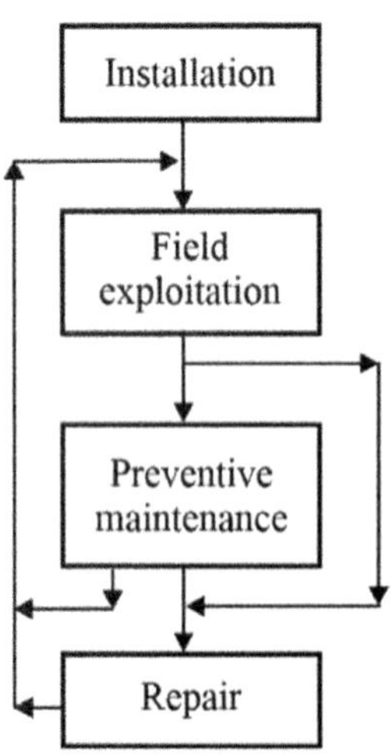

Source of the fault	Testing method	Resolution
• Wiring • Mounting	• System level test • Diagnostics	• Connection
• Component failure • Handling error	An integrity test such as a parity check	Subsystems that can be observed off-line
Marginal components	Checking of marginal components	Units that can be replaced in the field

According to a 2019 UN report, a staggering 53.6 million tons of e-waste were globally disposed of, with only 17.4% of discarded electronics being recycled. The rapid advancements in software and hardware development are accelerating the obsolescence of computers, leading to an alarming increase in discarded electronic equipment. Consequently, the substantial volume of electronic waste poses a dual challenge—both in terms of waste management and recycling. This presents a critical issue not only for environmental preservation but also for the conservation of energy and raw materials.

Quantity, in itself, is not the biggest problem. Electronic products (as described in [1]) contain a large amount of hazardous substances that can endanger the environment if they are disposed in an inappropriate manner. It has been established that although electronic waste makes up only 1–3% of land filled waste, it brings between 50 and 70% of heavy metals to the landfill. Furthermore, it is estimated that 22% of the annual consumption of mercury (which is associated with brain damage) goes to the production of electrical appliances. Mercury is found in flat-panel liquid crystal monitors and printed circuit boards. And just one more example, the plastic (PVC) used to insulate the cables contains highly toxic dioxin and furans that are released during incineration.

The responsibility for proper end-of-life of electronic products, in our opinion, is not only related to the system of take-back of obsolete electronics and proper end-of-life management but to the design engineers, too. Namely, the design engineer is the one defining the use of materials, the power consumption, the life longevity, and the facilitation of disassembly at the end-of-life. Accordingly, he/she is in charge to ensure compliance with standards and directives prescribing the use of materials, energy and similar. These thoughts frequently are not in the front row of considerations when new products are developed which is leading to a difficult end-of-life management.

7.10.3 Introduction to Testing as Industrial Activity

In this paragraph a brief overview of the testing technique will be given. In fact we will first elaborate the claims from the preceding paragraph which will allow us to escape from the test signal synthesis and simulations and to enter the physical world of testing.

We initially require a testing strategy, which is a detailed and overarching document outlining the approach to software testing. This document includes elements such as scope, objectives, and communication strategy with customers, documentation formats, test processes, and more.

The test strategy outlines how risks should be mitigated at the testing stage, specifies the criteria for application, and identifies the required types of testing. This information communicates critical process issues to the project leader, software developers, and quality assurance (QA) experts.

The primary goal of a test strategy is to provide a clear understanding of the overall approach, tools, objectives, and schedule for the planned test activities. It aims to elucidate the fundamental tasks and challenges inherent to the specific test project. An effective test strategy should be tailored to the project's requirements, easy to comprehend, practically suitable, and have realistic and attainable objectives.

Next, we will need a test plan. The test plan, commonly referred to as a QA test plan, serves as the foundation for software testing. It is a detailed document that encompasses various testing elements, including test scope, objectives, template, test schedule, necessary resources (software, hardware, and personnel), test environment, test deliverables, test estimation, risk assessment, defect tracking, mitigation plan, entry and exit criteria, etc.

The main objective of a test plan is to create documentation outlining how the QA tester will verify that the system functions according to its intended design. This document primarily outlines the testing approach, specifies what needs to be tested, and identifies the individuals responsible for carrying out the tests.

The test plan also ensures that all functional and design requirements are implemented according to the specifications in the documentation. By creating a test plan, team members can collaborate effectively, clearly defining their responsibilities and roles to facilitate seamless communication.

7.10.3.1 Testing in the Design Phase

Bearing in mind the three stages of design which are given in Fig. 7.10.1, three different tests should be carried out. They will refer to the test results obtained in those phases.

1. *Specification phase*: In this phase, ad hoc verification is mainly performed since most of the design requirements cannot be formalized. Usually, the verification is entrusted to experienced designers or parts of the obtained results are verified by simulation.
2. *Synthesis*: The result of the synthesis is a functional schematic that can be drawn or expressed in some hardware description language (HDL). Simulation is usually used for limited design verification. Due to the high complexity of the schematic and the large number of tests that need to be simulated, verification is usually performed by simulating a part of the network rather than the entire network.

In principle, the test program can be developed immediately after receiving the specifications. In real life, however, preference is usually given to synthesis so that verification is done after synthesis. It should be emphasized here that the practical realization can affect the tests that will be used in the verification for the simple reason that it determines the most probable faults.

Considering the total number of faults, it is practically impossible to test all faults and all signals of the system. That is why the so-called functional testing is in use. In this case, the correct operation is checked, which is related to a given functional specification. In the ideal case, such tests are generated only on the basis of the functional requirement (that is because they could be obtained even before the synthesis phase), so it is expected that they can be used independently of the results of the synthesis, that is, of the functional schematic obtained by the synthesis. This, of course, differs from other test philosophies where the test is developed for a given fault and for a given schematic. Functional testing that is independent of the synthesis procedure is considered theoretically attractive but difficult to apply in practice. However, if some limitations are set and thus reduce the generality of the methods and get closer to the physical realization, according to some authors [3], functional testing can be successfully applied. One example of the concept for generating a functional test is shown in Fig. 7.10.2. It consists of four main components:

- functional analyzer,
- test synthesizer (test signal generator),
- the database contains test samples for functional faults,
- test program synthesizing.

In the following text, we will briefly consider this concept.

Before proceeding with the development of the test, it is necessary to functionally describe the system in a rigorous way using some HDL. After generating such a description, a functional analysis can be started. The verification itself is usually based on the generation of a graph in which nodes represent control or processing operators, and branches represent data distribution paths. Various types of faults can be entered into such a graph and parametric faults can be generated for which tests are developed. It is usually assumed that only one fault exists at a time, and the basic test generation algorithm is based on path sensitization (through the graph). The result of these activities are parameterized tests that are general in the sense that they are valid for several systems that are functionally equal but differ in the values of the structural parameters.

The definition of specific test signals is achieved using parameterized solutions and memorized tests from the database. In other words, specific tests are required for the given parameters. Thus, ready-made tests for functional primitives are obtained. The database is generated in several ways, among which previous knowledge and experience is an important category. When the required test is found in the database, it is downloaded and the test program is generated. Programming itself is a complex and laborious job, and therefore its automation is of great importance. Therefore, it is recommended to use the existing experience through the use of advanced programming techniques and a hierarchical approach.

3. *Realization phase*: As already mentioned, the testing of the functional schematic is difficult due to the shortcomings of the simulation tool itself, that is, the impossibility of generating realistic models for the simulation. This especially applies to timing testing where real models can only be assembled after the layout is known. This means that all technological data should be available. Therefore, in order to be able to claim that the prototype will be suitable for production,

in addition to the technologically independent tests, it is necessary to perform a series of tests determined by specific solutions used in the application.

7.10.3.2 Testing in the Manufacturing Phase

In the production phase, the design is completed and the problem becomes how to ensure the mass production of a fault-free product. Before the system is recommended for production, however, it must be established that design and production will not introduce an unacceptably high level of faults. Once the production line is checked, routine testing is necessary to eliminate faultive components at each production step. As part of these activities, the following types of testing are performed.

Design maturity testing. This procedure estimates the mean failure time for new products before going into production. Design maturity testing will allow systematic design problems to be corrected. If such problems were ignored, it would lead to high maintenance costs and, of course, customer dissatisfaction.

This method of testing is performed by running several (five to six) samples (components or devices) for a certain longer period of time (two to three months). During work, failures are observed and remembered, then categorization of them is approached based on the nature of the failures, the time when the failure occurred and the conditions in which it occurred. Mutually similar failures are now grouped, with groups being prioritized based on frequency of occurrence.

Based on this, the priority of eliminating the cause of faults is established. When the fault is found and corrective actions are taken, the effectiveness of the steps taken is checked by observing the repaired device. In this way, an iterative procedure is established which stops when the required mean failure time is reached.

Testing progress can be tracked using a preprepared chart for each product being tested. Such a diagram is shown in Fig. 7.10.3 [4]. It provides an objective criterion for estimating the average failure time with a predetermined statistical risk. The construction of the diagram is determined by four parameters:

- a predetermined mean time to failure τ_0,
- minimum acceptable mean time to failure τ_1,
- the customer's risk *a*, which in fact represents the probability that a product with a mean time to failure less than τ_0 will be accepted,
- by the risk of the manufacturer *b*, which represents the probability of rejecting a product whose mean failure time is greater than τ_0.

The quotient of τ_0 and τ_1 of the order of 1.5–2 is usually used. The risks of the producer and the buyer are usually taken to be around 20%. On the abscissa is the test time (time of operation until failure), and on the ordinate is the cumulative number of failures. The line representing the behavior of the product is expressed as stepped and moves up (with the occurrence of failure) and to the right (over time). The diagram is divided into three parts: accept, reject, and continue testing. When the stepped response line intersects the diagram and enters the "accept" region (as shown in Fig. 7.10.3), the mean time to failure of the sample under test is at least equal to the minimum acceptable mean time to failure (with a predetermined risk), and the product can be accepted.

If the response line crosses into the "reject" region, it is likely that the mean time to failure is lower than the minimum acceptable one, with the appropriate tolerance as stated above. In such a case, the testing is stopped and the repair of the project is approached until the necessary improvement is achieved.

Process maturity testing. The term process includes all steps in production that are applied in the manufacture of parts, assembly, inspection, and testing during mass production. The purpose of this testing is to check for hidden faults that arise during the production of a new (unknown) product. In order to carry out this test, it is necessary to encompass a large number of produced units for a long period of time.

Within a test lot, the characteristics of initial failures can vary considerably. The stacking of initial failures of individual built-in components generates the "normal flow of initial failures" of the device. Here, the manufacturer's goal should be to deliver products of good quality, while keeping the initial failure testing time to a minimum with an acceptable mean time to failure. The usual testing time for initial failures, the so-called burn-in testing, lasts between 20 and 40 h.

Based on the cause, faults can be classified into the following categories: initial, production, design, and others.

Initial faults usually refer to components that were installed and were defective at the time of installation. Such faults can be encountered in highly complex semiconductor components that have been improperly stored or transported.

Production faults refer to the imperfection of the production process in terms of assembly, connection of components, and installation in boxes. These faults are usually repaired as opposed to faultive components being discarded.

Design faults refer to cases when, due to inexperience in the design phase, the cause of the fault was not eliminated.

Other faults would refer to unspecified (unknown) causes that should be corrected during the operation of the device.

Experience has shown that the three most probable faults cause about seventy-five percent of the failures. Therefore, it is believed that if the first six most probable faults are removed, the mean time to failure will increase tenfold.

Let us now consider in more detail the impact of faults in semiconductor components. All semiconductor processes generate some faultive components. Even when the component manufacturer discovers and removes these faulty components, failures will still occur in what is known as the "child mortality period." This period lasts about twenty weeks and the number of failures gradually decreases over time. At the end of this period, the number of failures stabilizes at a very low value that tends to decrease. This period, known as "normal working time," lasts a very long time, sometimes over ten years. However, at the end of this period, the number of failures begins to increase and a period of "wear out" occurs. The dependence curve of the average failure time on the time of use of one series of products is shown in Fig. 7.10.4. This curve is said to have the shape of a bathtub.

The cost of repairing a failure depends on the production step at which the failure was detected. The later the fault is discovered, the more expensive the repair is. Fault detection at the level of semiconductor components minimizes cost. Based on the considerations in [5], the following can be said. At the first next repair level, the circuit board level, if the cost is estimated at five dollars, then at the system repair level it would be at fifty dollars, and at the field system repair (after delivery) at five hundred dollars. Of course, the level at which the fault will be detected depends on the test program that has been applied and hence its importance.

Even a small percentage of faults at the semiconductor components level can cause a significant percentage of failures at the printed circuit board level. It is affected by the density of the mounted components on the board. For example, consider a board with forty semiconductor components that have an initial failure rate of one percent. The percentage of printed circuit boards that use these components is equal to $(0.99)^{40} = 0.67$, which means that even thirty percent of the printed circuit boards will have at least one fault. In order to handle such a low yield, it is necessary to produce a significantly larger number of printed circuit boards for testing.

At the circuit level, testing can include measurements in a wide spectrum starting with simple electrical parameters and functional tests and ending with stress measurements to eliminate faults related to child mortality period. Stress means mechanical shock (vibration), voltage overload, burn-in, and thermal shocks. As much as the testing is more expensive, so is the check that follows. During mass production, not every sample is tested, but representatives are selected from among them to be processed. For example, below is a set of tests that are applied in MOS technology and that relate to the values of the so-called parameters:

- gate-to-oxide voltage analysis,
- analysis of drain-to-substrate voltage and leakage current,
- drain-to-source breakdown voltage and breakdown resistance,
- threshold voltage and leakage current,
- drain current for zero gate voltage and for a predefined voltage,
- transconductance at a predefined voltage.

These are parametric tests that determine whether a component meets electrical requirements.

7.10.3.3 Testing in the Exploitation Phase

When the system is installed in the user's premises and in the conditions set by the user, it is necessary to perform testing in order to check whether there was any damage during transportation, bad connections, bad coupling with other devices, and the power supply. For this purpose, tests that were used in the final stage of testing during production are usually applied.

Field testing must address both actual faults and those declared by the user. Due to the complexity of the system, it is not unusual for a large number of false fault reports to appear and for this number to be significantly higher than the actual number of faults. Permanent failures are, of course, easy to diagnose and fix. More difficult problems, however, can be related to the interaction between system components or be determined by the way the system is loaded. It is a very common situation that such faults cannot be predicted and therefore tests are compiled for that purpose.

Therefore, when it comes to field testing, the first thing to consider is whether the failure actually exists. That is why testing should start with a small set of function-verifying logic tests. Other functions are checked by gradually increasing the complexity of the test. In order to minimize the time that the user is not using their faulty device, it is necessary to design the devices in such a way that they can be tested as stand-alone devices. For example, some communication device should have a test mode where the output port will send a signal to its own receiving port so that the logic used to send and receive the signal can be tested with an external tester.

Fault recording and reporting can often be of great benefit to maintenance. Very often the test signals do not stress the system under test as much as actual operation does. Fault notes, however, also contain information about the conditions under which the system was operating, so that it is easier to find the correct signals for testing and to support diagnostics.

The use of external test equipment should be minimized or even eliminated. The reason for this is that the test equipment is complex and sensitive, so it is difficult to transport, its fault would most likely mask the fault of the system under test.

An important factor in maintenance is determining which component to replace in the field. These are typically printed circuit boards or VLSI chips. The physical layout of the system should provide conditions for easy replacement of components. There are simple and more complex replacement procedures. In the first case, only one component (board) is changed, and in the second, based on sophisticated and preprepared instructions, several components are changed simultaneously. Of course, it is cheaper and simpler to change one component, which, however, requires a high-quality and qualified diagnostic procedure.

7.10.4 Testing Systems

Having looked at certain technical aspects of electronic circuit testing, in this chapter we will consider this design activity in the wider context of its application. First, the so-called testing device will be briefly described. It is usually called tester. Then testing will be viewed in the context of the entire economy of designing and manufacturing electronic circuits and systems, and at the end, additional views of advanced test planning methods and appropriate integrated circuit design procedures will be given.

The device for testing electronic components or integrated circuits can be viewed from several aspects. First, it should be a machine that generates test signals and sends them to the component under test. This part of the tester is usually known as automatic test pattern generation (ATPG). The structure of this subsystem is given in Fig. 7.10.5. As one can see, it starts from the description of the circuit being tested, which includes a list of faults. This description is processed in the fault list manager in order to perform the activities described in Chap. 7.9 of this book such as FEF, for example. The reduced list of faults together with the description of the circuit is accepted by the program for the generation of the test signal, which is based on one of the already described algorithms. A comprehensive overview of the automatic test generation algorithms for analog systems may be found in [6]. At the end, the verification of the obtained test signals is carried out with the help of simulation in order to establish the coverage of faults.

The second part of the system is the so-called probe. It is a sophisticated device that physically carries the component under test and transmits the signals coming from the ATPG to it through a set of contacts. Depending on whether components are being tested in packages or bare wafers with chips (that are not physically separated from each other), the part of the probe that accepts the component has a different shape. Thus, in Fig. 7.10.6a shows a part of the probe for testing bare chips. It can be seen that a series of pins (cantilevers) are used that come down to the chip. A part of them communicates signals to the circuit; a part is used for deducting the measured responses, and a part for biasing or grounding. It is important to keep in mind that integrated components are produced in very large series and that due to their complexity the testing time is long. This sets very strict requirements regarding the mechanical and electrical properties of the probe legs (pins) and the entire system for moving and positioning (aligning) the wafer (which is under these legs).

Looking at today's achievements of standard testing devices, i.e., their technological level, it is safe to say that the testing speed is over 2.5 Gbps with over 144 channels, with further improvements that will raise the testing speed to 5 Gbps and higher.

Considering the large dissipation that is created during such high testing speeds, when the probe and the component are in contact, it is necessary to provide intensive cooling. In some situations, liquid nitrogen is also used.

There is an alternative to the use a probe based on cantilevers. Namely, a membrane is designed for the proper chip on which contacts are created to touch the testing points on it. The contacts are then connected (using four metal layers) to signals, ground, and power supply in within the probe. One such probe is depicted in Fig. 7.10.6b. Figure 7.10.6c illustrates the intimate contact between the bare die (the chip) and the membrane of the probe.

Utilizing mechanical probes for examining bare integrated circuits (ICs) presents several disadvantages:

- Application of mechanical probing has the potential to harm the circuits beneath the probe pad on the IC.
- Repeated probing may result in damage to the probe pad on the IC, rendering further probing of that IC unfeasible.
- The probe card is susceptible to damage through repetitive contact or contamination with debris generated during contact with the wafer.

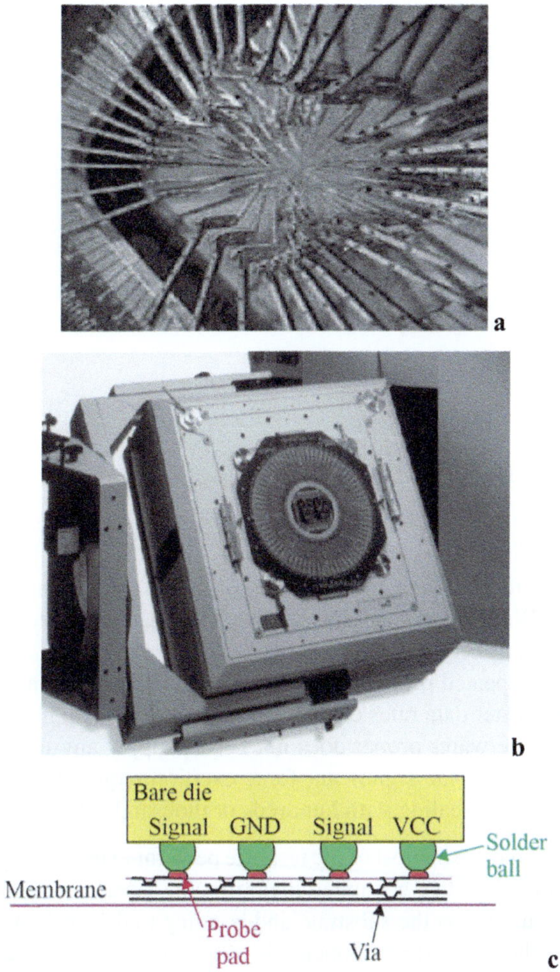

Fig. 7.10.6 **a** Probe pins (cantilevers) contacting the bare chip, **b** probe with exposed contact pads, and **c** cross-sectional detail of membrane and chip contacted (after [8])

- The probe assumes the role of a circuit, influencing the test results. Consequently, tests conducted during wafer sort may not always be identical or as comprehensive as those performed during the final device test after the packaging is concluded.
- As the probe pads are typically situated on the perimeter of the IC, the IC can quickly become pad-limited. Reducing pad sizes poses a challenge in the design and manufacturing of smaller and more precise probes.

Instead of employing mechanical probing techniques on integrated circuits (ICs), one can utilize miniature RF antennae as an alternative to both the mechanical probes and metal probe pads, as proposed by [7]. When the antennae on the probe card and IC are appropriately aligned, a transmitter on the probe card can wirelessly transmit data to the IC's receiver through RF communication.

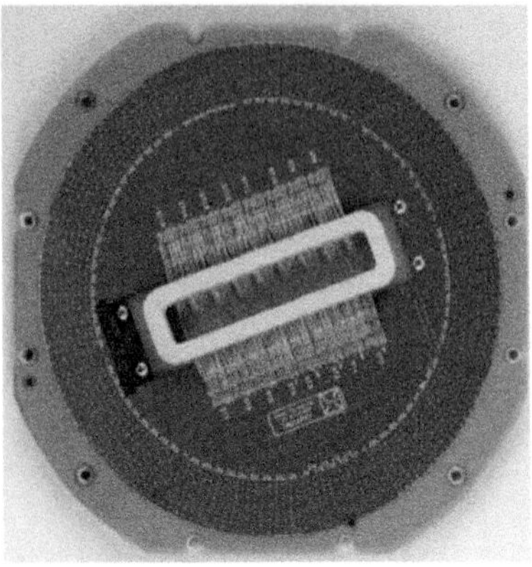

Fig. 7.10.7 Printed circuit board used to fit the given socket (integrated circuit)

This approach offers several advantages:

- No harm is inflicted on circuits, pads, or probe cards.
- No debris is generated.
- Probe pads on the periphery of the IC are no longer necessary.
- Wireless probe points can be positioned anywhere on the IC, not restricted to the periphery.
- Repeated probing is feasible without causing damage to the probe points.
- Faster data rates can be achieved compared to mechanical probes.
- The wafer prober does not need to apply any force to the probe location, eliminating the significant force exertion required in traditional probing, especially when dealing with hundreds or thousands of probes.

Of course, most of the tests are performed in this way, but considering the series of operations that are undertaken from the separation of the chips on the wafer, through mounting on the substrate and bonding to the final encapsulation, additional testing of the finished components is necessary. In such cases, the end of the probe has a completely different shape and is mostly different for different packages (Fig. 7.10.7).

The measured signals are returned to the computer, which stores them, gives a report on the success of the testing and, depending on the sophistication of the tester, performs additional actions such as statistical processing, DSP, etc. When bare wafers are tested, the malfunction of the circuit is usually marked with a dot on the chip, and when the components are finished, they are usually mechanically separated (discarded) into a defective group.

Here it is of great importance to keep in mind the methods that are applied when deciding on the correctness of the circuit. Namely, in order to establish the malfunction, it is necessary to have memorized properties of the component being tested and to compare them with the obtained measurement results.

In other words, it is necessary to have memorized the correct response and to carry out the comparison. Graphically, this activity is shown in Fig. 7.10.8b. This seems simple at first glance, but in the case of very complex circuits with a large number of

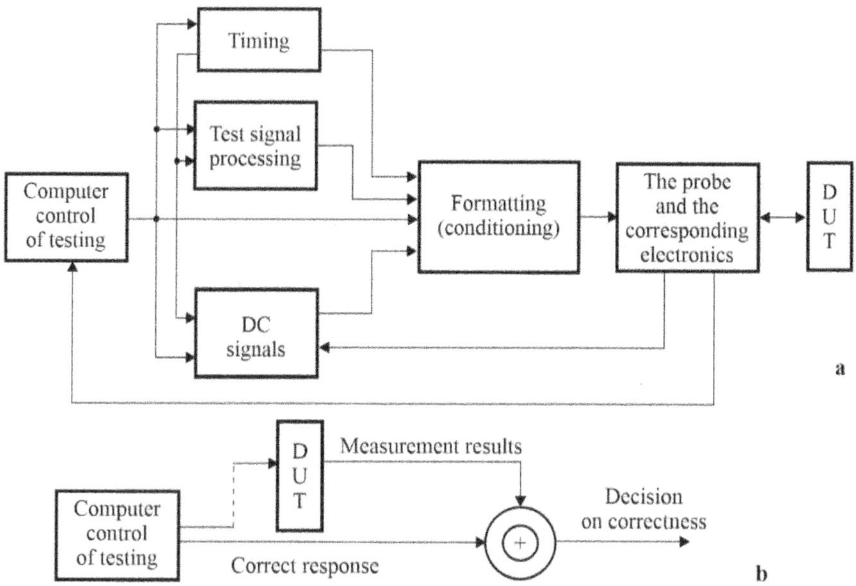

Fig. 7.10.8 Architecture of a tester. **a** Complete tester and **b** implementation of the EXOR operation within the decision system

test signals, it becomes a time-consuming step in the entire testing process and that is why such data are often condensed and the comparison shortened.

Based on this, it can be easily concluded that the tester is a very complex and expensive machine that must have significantly better performance than the component being tested, both mechanically and electrically (speed, power, temperature). Fig. 7.10.8a shows the architecture of a tester.

Its components are:

- a powerful computer,
- a powerful signal processor that includes a DSP for mixed-signal circuit testing,
- ATPG,
- Probe including adapters for different packages.

The testers that are used in mass production are equipped with an additional electromechanical system that transports the integrated circuits to the probe and, after measurement, takes them away from the tester.

The purchase price of one tester is calculated according to the formula

$$C = b + \sum (m \cdot x), \tag{7.10.1}$$

where C is the price while the meaning of other variables is given in Table 7.10.5 [9]. To this should be added the life cycle costs of the tester such as depreciation and obsolescence, maintenance, operating costs (energy, replaceable components, etc.)

Table 7.10.5 Price of the tester

Tester capabilities	Base price (b) k$	Price per pin (m) $	Pin number (x)
ASIC and high-performance μ-processor	250–400	2700–6000	512
Mixed-signal systems	250–350	3000–18,000	128–192
DfT tester	100–350	150–650	512–2500
Low performance ASIC and μ-controller	200–350	1200–2500	256–1024
Memory chips	200+	800–1000	1024
RF	200+	~50,000	32

The above numbers are taken in one time instant and should be considered informative only (Inflation and improved performances would rise them). Nevertheless, it is easy to conclude that the tester is a very expensive machine.

7.10.5 The Economics of Testing

One measure of production quality is the level of faults. If company A sells 100,000 components of one type and 10 of them are defective (We should not forget that it is considered that structural and not a functional test was done, so there is always a possibility that some sample will be defective.), then we say that the fault level is 0.1% or 100 ppm (parts per million). The mean quality level (MQL) is the percentage of correct components. For company A it would be MQL = 99.9%.

Suppose that Company A manufactures custom-made circuits (ASICs) κ. Let the price of each κ series circuit, as tested by the component manufacturer, be $10. Let Company B now buy 100,000 circuit components from the κ series, and let it manufactures motherboards by assembling them among other things. The total investment in this type of components is, therefore, one million dollars. Company C manufactures personal computers into which it installs motherboards manufactured by B. Now Company C is testing the motherboards it received from B, which requires $100 per motherboard to repair due to a faulty circuit κ. Table 7.10.6 provides an overview of repair costs if testing is performed at the motherboard level. It can be seen that a fault level of 5% leads to a repair cost of $500,000. This means that practically the purchase price has been increased by 50%. At the system level, things get even worse.

Let Company C now sells its computers for $2500, with a profit of $250. Suppose, however, that 10% of motherboards containing defective κ-series components pass the test. This percentage seems high, but it is realistic that hard-to-detect faults are even more difficult to detect at the system level than at the component level, so testing here is very difficult, which justifies the percentage. The cost of repair at the system level as a function of the level of faults within the κ series is given in Table 7.10.7. It can be seen that a 5% fault level at the component level leads to a cost of $2,500,000

Table 7.10.6 Repair costs of a series of printed circuit boards

Fault level in the series κ (%)	Number of purchased defective circuits	Cost of repair printed circuit boards
5	5000	500,000$
1	1000	100,000$
0.1	100	10,000$
0.01	10	1000$

Table 7.10.7 System-level repair costs

Fault level in the series κ (%)	Number of defectives chips in κ	Number of defectives printed circuit boards	Total repair costs at the system level
5	5000	500	2,500,000$
1	1000	100	500,000$
0.1	100	10	50,000$
0.01	10	1	5000$

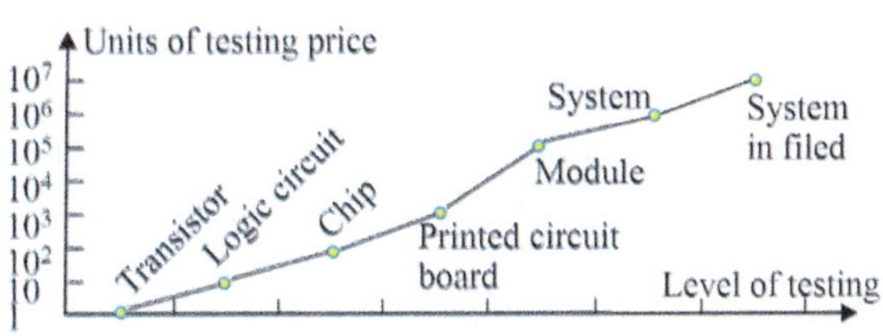

Fig. 7.10.9 Dependence of testing costs on the level

to repair at the system level. It also shows that it pays to invest all $2,000,000 in the manufacturing phase of the component in order to reduce the level of faults in the κ series, even if it was only in the testing phase.

This discussion is summarized in the diagram in Fig. 7.10.9.

7.10.6 Challenges to the Testing Technology

The beginning of the development of integrated circuits is characterized by the term small-scale integration (SSI). At the same time, ten transistors were integrated on a 3 × 3 mm chip. In the early seventies of the last century, the development of large-scale integration (LSI) integrated circuits made it possible to place thousands of transistors and other components on an approximately equal surface. This technology made it possible to build microprocessors on one chip that integrate all the basic functions of a computer. During the eighties, VLSI technology (Very LSI) was created, which led to a significant increase in the packing density of microprocessors, memory components, and other chips that surround them. Up to 20 million transistors could be placed on a surface of less than 2 cm^2. Today, the terms VVLSI (Very-Very), UVLSI (Ultra VLSI), and ASIC (application-specific integrated circuits) are in use. This dramatic development of technology and dense packaging has led to an equally dramatic reduction in the cost of logic functions and an increase in the reliability of components and systems. Table 7.10.8 (Originating from the data of the Semiconductor Industry Association) gives the status and Fig. 7.10.10 (Originating from the data of "The International Roadmap for Devices and Systems: 2022," published by IEEE) some prediction that best describe this technology. The table tells about the dimensions, the number of transistors and the maximum clock frequency of chips during time. To this should be added the reading from Fig. 7.10.10 that claims that in the year 2020 the most complex chips had 21 million equivalent NAND gates per squared millimeter. The market is estimated to growth and to reach about USD 601.49 billion by 2032.

As a consequence of the high degree of integration, the circuits have become very complex. In order to verify the function of such circuits, exhaustive testing is necessary. Even when dealing with LSI chips, the cost of testing becomes an important part of the total cost. Moreover, the speed with which the degree of integration is rising is greater than the speed of advancement of testing methods and equipment. As a direct consequence of that, we have the fact that the test methods that were not

Table 7.10.8 Status in the development of microelectronics

Year	Minimum dimension (nm)	Internal clock (GHz)	Scale of integration ($10^6/cm^2$)
1993	500	0.20	2
1995	350	0.30	4
1999	180	0.75	6.6
2001	130	1.68	13.0
2003	100	2.31	24.0
2005	80	5.17	44.0
2008	70	6.74	109
2011	50	11.5	269
2014	34	19.3	664

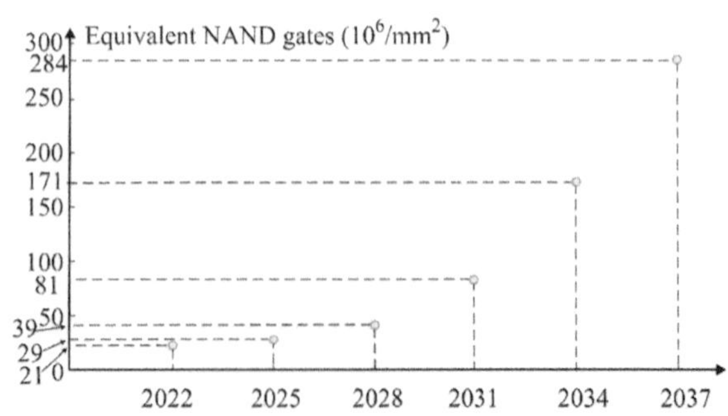

Fig. 7.10.10 Forecasting the chip complexity. *Source* IEEE

Fig. 7.10.11 Costs in cents ($/100) per transistor, for silicon, and for testing

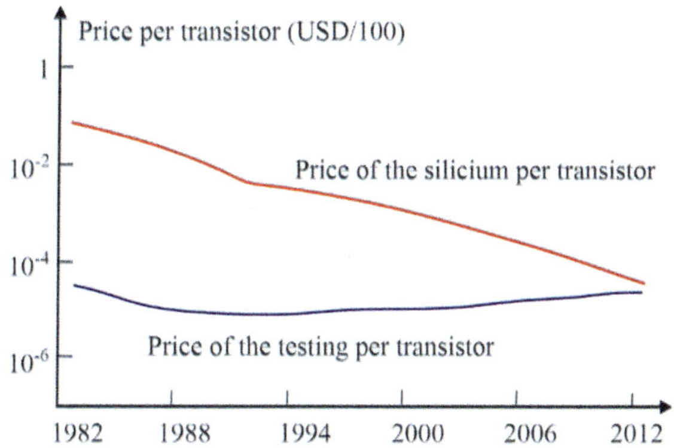

adequate even for technology with a lower degree of integration become incapable of solving the problem of testing modern circuits. Therefore, testing has become one of the most important aspects of research and development in the field of designing integrated electronic circuits [10].

As confirmation of this claim Fig. 7.10.11 [11] shows the dependence of costs as a function of time for silicon costs per transistor and testing costs per transistor. It can be seen that the former decreases, while the latter remain practically constant.

The main challenges faced by the modern development of testing are related to the shortening of testing time and fault coverage on the one hand, as well as the method of testing increasingly complex functions, which includes modern components with micro-electromechanical and micro-electromagnetic functions. For example, test time is a quadratic function of the complexity of a digital circuit, and the introduction of faults increases it by a further three powers. Table 7.10.9 illustrates the time needed to perform a functional test of a combinational circuit with a tester whose operating frequency is 10 MHz.

In order to successfully overcome the problems we have discussed so far, in addition to the usual test generation algorithms, the concepts of design for testability (DFT = design for testability) must be applied. This, as described in the previous chapters, leads to completely new chip architectures, especially if one insists on a large coverage of faults and the possibility that testers can access as many faults as possible from the outside.

The price to be paid in this case is expressed in the larger surface area of the chip, which, given the large series, is not insignificant. Nevertheless, bearing in mind

Table 7.10.9 Testing time

Number of primary input terminals n	Time required for 2^n test vectors
32	7 min
40	30 h
64	50.8×10^8 centuries

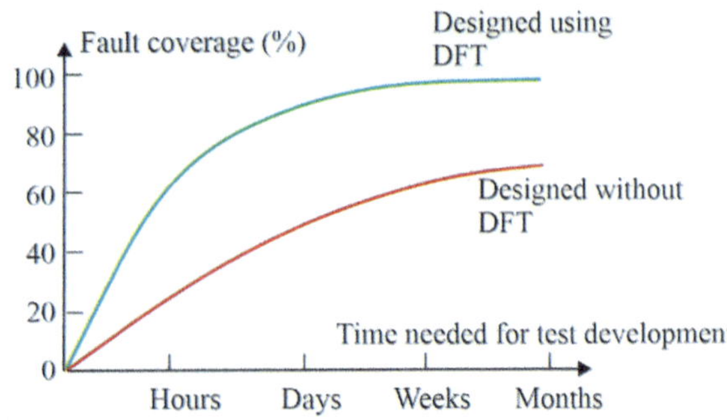

Fig. 7.10.12 Defect coverage depending on the time required to develop the test

the considerations about the economics of testing from this section, we can easily conclude that the application of DFT pays off. To that should be added the fact that for circuits whose architecture is adapted to DFT, the test is easier to synthesize. This effect is illustrated in Fig. 7.10.12. The dependences of fault coverage as a function of the time required to develop the test are given, for the case when DFT was not applied and for the case when DFT was applied. It is not difficult to conclude that the efficiency in the second case is incomparably higher. Moreover, it can be concluded that without DFT we will not be able to test many of the faults, which can have unforeseeable consequences for the final product.

The growth in the number of transistors per integrated circuit is significantly faster than the growth in the number of pins so that the possibility of accessing individual transistors and groups of transistors from the outside is decreasing more and more. If the number of transistors per pin is called an *index of testing complexity*, then in the year 2002 its value reached about 75,000 [12].

7.10.6.1 Testing Advanced Packaged Systems

Modern integration is undergoing a redefinition facilitated by systems-on-packages (SOPs). These sophisticated systems not only amalgamate analog and digital subsystems but also incorporate radio frequency (RF) components for high-bandwidth communication, optics as a conduit for multigigahertz data transfer, and micro-electro-mechanical-systems (MEMS) as an interface with the external environment. The design of such systems leverages numerous advantages not found in traditional designs. The compact and uniform nature of the medium enveloping these systems mandates shorter data paths and reduced parasitics, thereby enhancing the speed and bandwidth of subsystem communication well beyond the capabilities of classical data buses [13, 14].

Additionally, designers are now unrestricted by the discrete attributes of passive components; instead, they have the flexibility to employ embedded passive elements custom-tailored for the specific application. Moreover, systems-on-package (SOP) iterations of sensor-processor, actuator-driver, and electrical-optical subsystems experience advantages such as heightened bandwidth, the absence of package-to-package interference, and reduced power consumption in contrast to conventional interfaces.

The primary issue when testing this type of systems arises from the fact that the external automatic test equipment (ATE) lacks direct access to all internal embedded functions of the SOP. While it might be feasible to route certain internal electrical signals out of the package to the external tester, these signals operate at frequencies that the external tester cannot directly observe due to the frequency limitations of the encapsulating package and the lower speed of external input/output.

Similar concerns about speed and integrity also apply when validating the system's subcomponents. Unlike traditional systems with test nodes for individually verifying subsystem operations, the SOP faces challenges in controllability and observability of its subsystems through a conventional testing approach. Additionally, the system specifications, deemed "guaranteed by design," rely on the validation of associated subsystem specifications, which may no longer be accessible in an SOP configuration.

As a result, a conflict may be observed between the integration level facilitated by the design process and the attainable level of testability by an external tester. One practical resolution involves situating the functionalities of the ATE in close proximity to the SOP module undergoing testing. This approach enhances the test access speed, reduces test signal degradation, and enhances the controllability and observability of signals within the DUT. A potential solution is integrating ATE functions into the load board itself, where the test functions shift from the external tester to additional circuitry constructed around the system under test. This additional circuitry maintains the capability to apply high-speed stimuli to the system under test and capture the corresponding high-speed test responses, preventing degradation caused by cable parasitics from the low-bandwidth external ATE [15].

The resulting solution, termed built-off test (BOT), provides a cost-effective alternative to the steep expenses associated with a traditional ATE. The already discussed alternative, built-in test (BIT), takes a more aggressive approach by incorporating external tester functionality directly into the package and, where feasible, into the bare dies. It is worth noting that without BOT and BIT, economically testing high-performance standard testing may be impractical. This is due to the exorbitant cost of external test equipment required for test signal speeds exceeding 1 GHz. However, as multigigahertz system designs become more commonplace, especially in high-bandwidth communications, the economic feasibility of testing is significantly enhanced. This improvement stems from integrating high-speed test functions either on the load board (BOT) or within the SOP itself (BIT), complemented by low-bandwidth communication with an external tester operating at lower speeds. Consequently, this approach enables the testing of very high-speed systems using a cost-effective external tester without compromising the quality of the testing process.

References

1. Litovski VB (2023) Lecture notes in analog electronics, low voltage electronic components. Springer Nature, Singapore
2. Litovski VB (2023) Lecture notes in analog electronics, discrete and integrated large signal amplifiers. Springer Nature, Singapore
3. Siewiorek DP, Bell CG, Newell A (1982) Computer structures: principles and examples. McGraw Hill Book Company
4. von Alven WH (ed) (1964) Reliability engineering. Prentice Hall, NJ
5. DerMarderosian A Jr, Deptola M (2015) Electronics inspection and test. Raytheon Company, 4/22–23/15
6. Soma M (1996) Automatic test generation algorithms for analogue circuits. In: IEE Proceedings—Circuits devices and systems, vol 143, no 6, Dec 1996, pp 366–373
7. Shunichi I et al (2007) System debug method, system debug apparatus, data processor, wireless communication interface IC, and interface method. Patent No. US20100040123
8. Barber A, Lee K, Obermaier H (1994) A bare-chip probe for high I/O, high speed testing. HPL-94-18
9. Barber A, Lee K, Obermaier H (2023) International technology roadmap for semiconductors (ITRS™). https://irds.ieee.org/editions
10. Afaq A (2005) Testing of complex integrated circuits (ICs)—the bottlenecks and solutions. Asian J Inf Technol 4(9):816–822
11. Semiconductor Industry Association (1997) The national technology roadmap for semiconductors. http://www.semichips.org and https://go.gale.com/ps/i.do?id=GALE%7CA21188528&v=2.1&it=r&linkaccess=abs&issn=0038111X&p=AONE&sw=w&userGroupName=anon%7Ed563be85&aty=open-web-entry
12. Zorian Y (1999) Testing the monster chip. IEEE Spectr 36(7):54–60
13. Akbay SS, Halder A, Chatterjee A, Keezer D (2004) Low-cost test of embedded RF/analog/mixed-signal circuits in SOPs. IEEE Trans Adv Packag 27(2):352–363
14. Hanai H, Yamada S, Mori H, Yamashita E, Funakura T (2001) Built-out self-test (BOST) for analog circuits in a system LSI test solution to reduce test costs. In: Proceedings of the 10th Asian test symposium, Kyoto, Japan
15. Burns M, Roberts GW (2001) An introduction to mixed-signal IC test and measurement. Oxford University Press, Oxford

7.11 Solved Problems

Abstract We will try in this chapter to help the reader to do some experiments with his knowledge by her/himself. The example problems are so selected to lead the reader in "all" parts of the subject of testing and diagnosis of analog circuits including application of different concepts. Examples of diagnostics will be given including use of ANNs.

7.11.1 Fault Modeling and Testing Signal Synthesis for Analog Circuits

Problem 7.11.1.1

For the circuit of Fig. 7.11.1 compile a list of catastrophic faults in two parts. Place permanently open (or stack-at-open)-type faults in the first part and short-circuit (or stack-at-short)-type faults in the second part.

If the output voltage, V, is the only output quantity (the only testing point) to which the effect of the fault is mapped, analyze the equivalence of the faults without generating the circuit function. For the reduced set of faults from both parts, compile the fault dictionary if the numerical values of the circuit elements are: $I = 1$ mA, $R_1 = 2$ kΩ, $R_2 = 4$ kΩ, and $R_3 = 8$ kΩ.

Solution to Problem 7.11.1.1

In the second row of Table 7.11.1 the permanently open faults are listed while the third contains the permanently short ones. Here, the disconnection at point α is denoted by α-s-o. "**o**" is a common notation for **open** circuit. When it comes to short-circuit faults, according to Fig. 7.11.3, they are listed in the third row of Table 7.11.1. For two nodes in the circuit, α and β, which are **short**-circuited, the designation α-s-β is used here (Fig. 7.11.2).

Let us consider the fault equivalence by inspection of the circuit. When considering the equivalence of the α-s-o type of faults, first of all, it is easy to recognize the following: E-s-o ↔ G-s-o, B-s-o ↔ C-s-o, and F-s-o ↔ H-s-o. Here the symbol

V. B. Litovski, *Lecture Notes in Analog Electronics*, Lecture Notes in Electrical Engineering 1265, https://doi.org/10.1007/978-981-97-8257-4_11

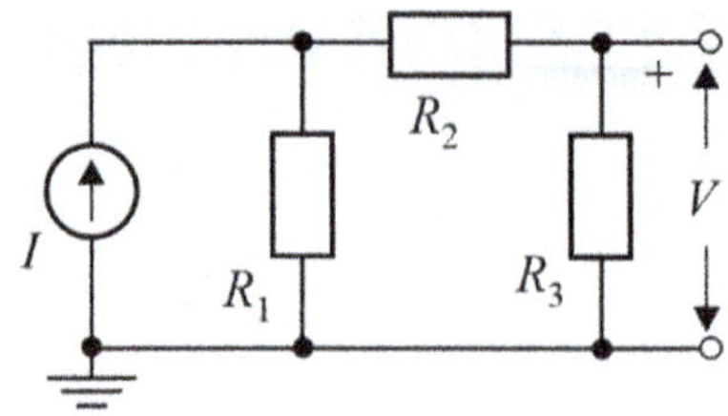

Fig. 7.11.1 Example resistive circuit

Table 7.11.1 Stack-at-open and stack-at-short faults

Fault no.	1	2	3	4	5	6	7	8
α-s-o	*A*-s-o	*B*-s-o	*C*-s-o	*D*-s-o	*E*-s-o	*F*-s-o	*G*-s-o	*H*-s-o
α-s-β	*X*-s-*Y*	*X*-s-*Z*	*Y*-s-*Z*					

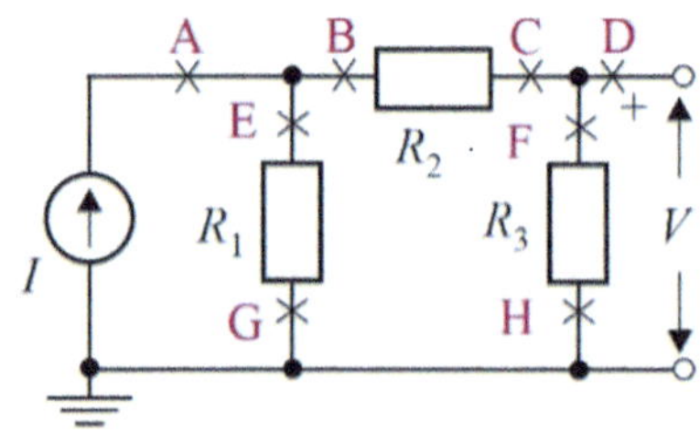

Fig. 7.11.2 Potential locations of stack-at-open faults

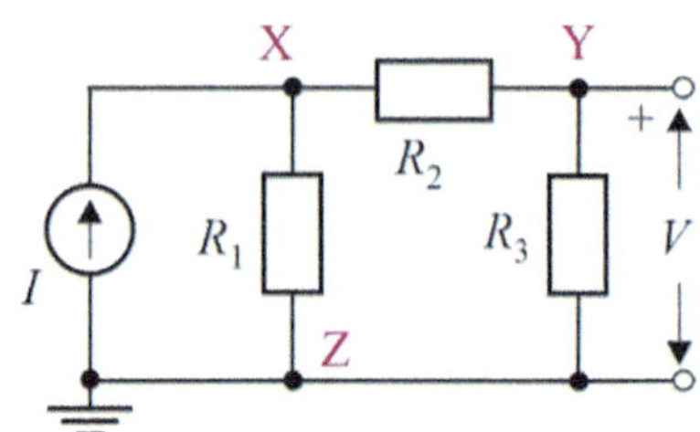

Fig. 7.11.3 Nodes potentially subject of stack-at-short faults

"↔" was used to denote the equivalence. Based on that the number of α-s-o type faults, which may be considered to have specific fault effect, is reduced to 5.

Going deeper, if we look at the faults that separate the input from the output, we recognize two categories. In the former there are faults that result in the output voltage being equal to zero, and in the latter those that make it undetermined. Thus, faults *A*-s-o, *B*-s-o, *C*-s-o separate the output from the input, but do not break the connection to the ground, that is, reduce it to zero. That is why they are also equivalent. The *D*-s-o fault, however, isolates the output terminal so that the output potential is indeterminate. That is why there is no equivalent to it.

When equivalences are considered, the set of faults of type α-s-o which we can assume to have different fault effects reduces to

1.1. A-s-o ↔ B-s-o ↔ C-s-o.
1.2. E-s-o ↔ G-s-o.
1.3. F-s-o ↔ H-s-o.
1.4. D-s-o.

Let us now consider the effects of short-circuit faults.

The X-s-Z fault has the effect of isolating the input and output by short-circuiting the excitation current source so that the output voltage definitely becomes zero. On the other hand, the Y-s-Z fault also separates the input from the output, with the output terminal now connected to ground and not sensing what is happening in the circuit. From the output voltage standpoint, however, the two faults are equivalent. The X-s-Y fault only bridges resistor R_2 and has no equivalent effect. Accordingly, when the equivalence is introduced, the set of faults of type α-s-β reduces to

2.1. Y-s-Z ↔ X-s-Z.
2.2. X-s-Y.

The fault dictionary is shown in Table 7.11.2. "FF" indicates a "fault free" circuit.

Since the circuit is linear and resistive, a change in the value of the excitation current or a transition to the frequency or time domain will not in any way contribute to a change in the picture of equivalence, that is, of the fault effects. Therefore, Table 7.11.2 can be directly used for testing, where the test signal is $I = 1$ mA. Of course, any measured value at the output that is different from 1.143 V would indicate that there are faults in the circuit, but if it is different from those listed in Table 7.11.2, its cause is not a catastrophic fault. ■

Problem 7.11.1.2

In the circuit of Fig. 7.11.1 the resistance values are subject to large changes which may be considered a fault. For each resistor in the circuit, determine the analytical expression for the increment of the output voltage, and then, using the values of the circuit elements from Problem 7.11.1.1, calculate the values of $V(R_x \pm \Delta R_x)$. Here, x stands in place of 1, 2, and 3, and ΔR_x is the increment calculated as $\Delta R_x = 0.2 \cdot R_x$.

Solution to Problem 7.11.1.2

By analysis of the circuit of Fig. 7.11.1 the faulty circuit function is easily obtained as:

Table 7.11.2 Overview of catastrophic fault effects

Cipher	FF	1.1	1.2	1.3	1.4	2.1	2.1
Representative	–	A-s-o	E-s-o	F-s-o	D-s-o	Y-s-Z	X-s-Y
Effect (value of V [V])	1.143	0	8	2	Undefined	0	1.6

$$V(R_x + \Delta R_x) = V \cdot \frac{1 + p_x}{1 + q_x}. \tag{7.11.1.2.1}$$

In the above expression, it holds

$$p_x = \Delta R_x / R_x \text{ and } q_x = \Delta R_x/(R_1 + R_2 + R_3), \quad \text{for } x = 1 \text{ and } 3 \tag{7.11.1.2.2}$$

and

$$p_2 = 0 \text{ and } q_2 = \Delta R_2/(R_1 + R_2 + R_3). \tag{7.11.1.2.3}$$

The numerical values obtained after analysis are given in Table 7.11.3.

By studying Table 7.11.3, we conclude that if we want to test the circuit of Fig. 7.11.1 in order to check for presence of parametric (soft) faults as given above, we need to have an instrument whose accuracy is better than 5%. In that case, a single measurement will establish the presence of any of the parametric faults given here and also the catastrophic faults from the previous problem.

The reader is encouraged to reproduce Table 7.11.3 using other values of ΔR_x. ■

Problem 7.11.1.3

In the circuit of Fig. 7.11.4, determine the set of catastrophic faults related to the capacitor; find appropriate equivalences and possible test signals. Use DC and AC analysis. Next, determine the fault effects related to large increments of the capacitance and discuss possible test signals.

Table 7.11.3 Results of faulty circuit analysis

Cipher	Variable	Increment (Ω)	$V(R_x + \Delta R_x)$	ΔV (%)
3.1	R_1	400	1.333	−16.62
3.2	R_1	−400	0.941	17.67
3.3	R_2	800	1.081	5.424
3.4	R_2	−800	1.212	−6.036
3.5	R_3	1600	1.231	−7.699
3.6	R_3	−1600	1.032	9.711

Fig. 7.11.4 Example RC circuit

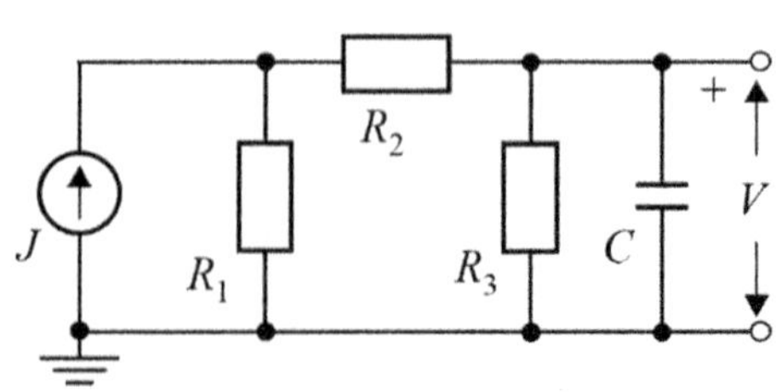

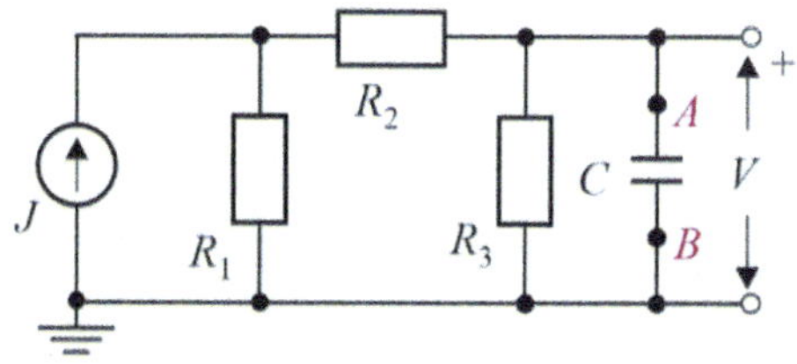

Fig. 7.11.5 Example RC circuit with nodes subject to potentially faulty behavior

Solution to Problem 7.11.1.3

According to Fig. 7.11.5 possible catastrophic capacitor-related faults are a branch break which can be expressed as *A*-s-o and *B*-s-o, as well as a short-circuited branch which can be expressed as *A*-s-*B*.

Faults *A*-s-o and *B*-s-o are equivalent, and the fault effect is the conversion of the circuit of Fig. 7.11.4 into the fault-free version of the circuit from Fig. 7.11.1. The fault *A*-s-*B* drives the output to ground and the fault effect is the same as the set of similar faults that were valid for the circuit of Fig. 7.11.1.

When generating the test, the *A*-s-*B* fault can be detected by any nonzero signal, but we will not know if it is actually present (if it is, it will certainly be detected). With an *A*-s-o or *B*-s-o fault, the situation is somewhat different.

Namely, if a DC signal is used, the response of the fault-free circuit and the faulty circuit will match, which means that we have to abandon the DC domain and move to the frequency or time domain. Here, the task is to search for the test in the frequency domain, so first of all, we determine the frequency response of the circuit. It reads

$$V(j\omega) = V_o \cdot \frac{1}{1 + j \cdot \omega \cdot \tau}, \tag{7.11.1.3.1}$$

where

$$V_o = J \cdot \frac{R_1 R_3}{R_1 + R_2 + R_3}, \tag{7.11.1.3.2}$$

and

$$\tau = \frac{(R_1 + R_2) \cdot R_3}{R_1 + R_2 + R_3} \cdot C. \tag{7.11.1.3.3}$$

For testing purposes, it is possible to measure the amplitude:

$$|V(j\omega)| = |V_o| \cdot \frac{1}{\sqrt{1 + \omega^2 \tau^2}} \tag{7.11.1.3.4}$$

or phase characteristic:

$$\varphi = -\text{arctg}(\omega \cdot \tau). \tag{7.11.1.3.5}$$

In both cases, if the frequency of the signal is low, i.e., if $\omega \ll \omega_g = 1/\tau$, there will be no noticeable difference between the DC value (which corresponds to the response of the faulty circuit) and the modulus, while the phase will be close to zero (which also corresponds to the response of the faulty circuit). This leads us to the conclusion that the angular frequency of the signal must be equal or higher than ω_g.

If, for example, $\omega = 2\omega_g$ was chosen, the responses of the fault-free circuit will have values

$$|V(j\omega)| = |V_o| \cdot \frac{1}{\sqrt{1+(2/\tau)^2\tau^2}} = |V_o|/\sqrt{5} = 0.447 \cdot |V_o| \qquad (7.11.1.3.6)$$

and

$$\varphi = -\text{arctg}\left(\frac{2}{\tau}\cdot\tau\right) = -\text{arctg}(2) = -1.107 \text{ rad} \qquad (7.11.1.3.7)$$

or $\varphi = 63.42°$.

Based on this, we conclude that any of the faults *A*-s-o or *B*-s-o can be easily detected by a sinusoid whose frequency is $2\omega_g$, whether we measure the amplitude or the phase of the output signal.

The effects of parametric faults $\pm\Delta C$ can be obtained by substituting the new capacitance value into expressions (7.11.1.3.3), (7.11.1.3.4), and (7.11.1.3.5). One gets:

$$\tau + \Delta\tau = \frac{(R_1+R_2)\cdot R_3}{R_1+R_2+R_3}\cdot(C+\Delta C). \qquad (7.11.1.3.8)$$

For the amplitude characteristic we now have

$$|V(j\omega)|_F = |V_o| \cdot \frac{1}{\sqrt{1+\omega^2(\tau+\Delta\tau)^2}} = |V_o| \cdot \frac{1}{\sqrt{1+(\omega\cdot\tau)^2\left(1+\frac{\Delta\tau}{\tau}\right)^2}}, \qquad (7.11.1.3.9a)$$

where the index "F" denotes faulty.

If a test signal is used for which $\omega \cdot \tau \gg 1$ is valid, this expression reduces itself into

$$|V(j\omega)|_F \approx |V_o| \cdot \frac{1}{\sqrt{1+\omega^2\tau^2}} \frac{1}{\sqrt{1+\frac{\Delta C}{C}\left(1+\frac{\Delta C}{C}\right)}}, \qquad (7.11.1.3.9b)$$

which means that the faulty circuit, regardless of frequency, will have $\sqrt{1+\frac{\Delta C}{C}\left(1+\frac{\Delta C}{C}\right)}$ times smaller amplitude than the fault-free.

Of course, when $\Delta C < 0$, the expression $\sqrt{1+\frac{\Delta C}{C}\left(1+\frac{\Delta C}{C}\right)}$ is less than unity, so the response of the faulty circuit will be larger than the response of the fault-free circuit.

For the phase characteristic of the faulty circuit, we get

$$\varphi_F = -\mathrm{arctg}[\omega \cdot (\tau + \Delta\tau)] = -\mathrm{arctg}\left[(\omega \cdot \tau) \cdot \left(1 + \frac{\Delta\tau}{\tau}\right)\right]. \quad (7.11.1.3.10)$$

When reasoning about the test signal, during this measurement, it should be considered that the largest (by absolute value) phase is $\pi/2$ and that the phase characteristic tends to this value slowly, that is, the phase changes slowly with increased frequency. This makes it difficult to recognize the difference between the response of a good and a faulty circuit.

As an illustration, we will take the case when $(\Delta C/C) = 0.2$. If a signal of frequency $2\omega_g$ is used, it comes that the amplitude reduction is

$$\frac{|V(j\omega)|_F}{|V(j\omega)|} = \frac{\sqrt{1+(\omega\cdot\tau)^2}}{\sqrt{1+(\omega\cdot\tau)^2\left(1+\frac{\Delta\tau}{\tau}\right)^2}} = \frac{\sqrt{1+(2)^2}}{\sqrt{1+(2)^2(1+0.2)^2}} = 0.86. \quad (7.11.1.3.11)$$

The value of the amplitude is decreased by 14%.

The change of the phase, for the same case, is

$$\Delta\varphi_F = -\mathrm{arctg}[\omega \cdot (\tau + \Delta\tau)] + \mathrm{arctg}(\omega \cdot \tau)]. \quad (7.11.1.3.12)$$

When numerical values are substituted, one gets $\Delta\varphi_F = -0.0691$ rad or $\Delta\varphi_F = -3.959°$. This increase expressed as a percentage is $[(100 \cdot \Delta\varphi_F)/\varphi] = 6\%$.

Since the amplitude and phase characteristics are nonlinear functions, the negative increment of the capacitance will be differently reflected in the increment of the response as compared to the positive one. Calculation of increments for negative ΔC and corresponding conclusions about the applicability of the test are left to the reader. ■

Problem 7.11.1.4

Figure 7.11.6 depicts the circuit schematic of a biquad (second-order transfer function) filter cell in which ideal operational amplifiers with infinite gain are used. The nominal values of the elements in the circuit are: $R_1 = 10\ \mathrm{k\Omega}$, $R_2 = 10\ \mathrm{k\Omega}$, $R_3 = 10\ \mathrm{k\Omega}$, $R_4 = 10\ \mathrm{k\Omega}$, $R_5 = 10\ \mathrm{k\Omega}$, $R_6 = 10\ \mathrm{k\Omega}$, $C_1 = 10$ pF, and $C_2 = 10$ pF.

Subject of variations in the circuit is the capacitance C_1, which varies by $\Delta C_1 = +20\%$ in relation to its nominal value.

Determine the frequency of the signal for which the amplitude characteristic has the highest value of sensitivity to changes of the frequency (that is where the amplitude characteristic reaches its maximum) and for that frequency determine the

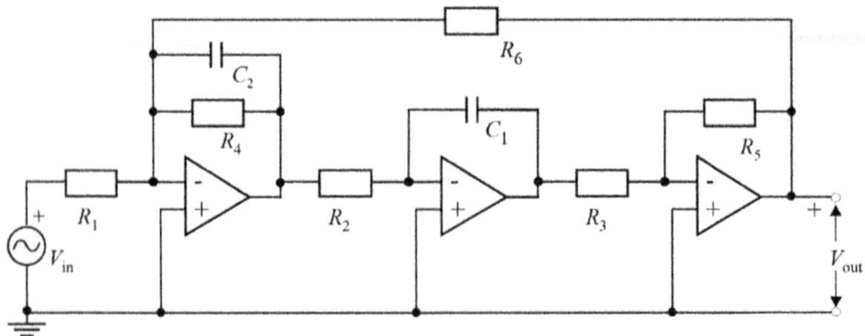

Fig. 7.11.6 Second-order active RC filter (Tow-Thomas biquad)

response of the fault-free and the faulty circuit using a faulted value $C_1 + \Delta C_1$. Based on this judge the test signal for this fault.

Solution to Problem 7.11.1.4

The circuit function (voltage gain), obtained after circuit analysis is:

$$T(s) = -\frac{R_5/(C_1C_2R_1R_2R_3)}{s^2 + s/(R_4C_2) + R_5/(C_1C_2R_2R_3R_6)} \tag{7.11.1.4.1a}$$

which can be expressed as

$$T(s) = \frac{T_0}{s^2 + 2\cdot\sigma\cdot\omega_0\cdot s + \omega_0^2}, \tag{7.11.1.4.1b}$$

where

$$T_0 = -R_5/(C_1C_2R_1R_2R_3), \tag{7.11.1.4.1c}$$

$$\omega_0^2 = -R_5/(C_1C_2R_2R_3R_6), \tag{7.11.1.4.1d}$$

and

$$\sigma = \frac{1}{2}\sqrt{\frac{C_1R_2R_3R_6}{C_2R_4^2R_5}}. \tag{7.11.1.4.1e}$$

For convenience, after combining (7.11.1.4.1c) and (7.11.1.4.1d), we will use

$$T_0 = -\frac{R_6}{R_1}\omega_0^2. \tag{7.11.1.4.1f}$$

When numerical values of the circuit elements are substituted one gets

$$T(s) = \frac{V_{\text{out}}}{V_{\text{in}}} = \frac{-1 \times 10^{14}}{s^2 + 1 \times 10^7 s + 1 \times 10^{14}}, \quad (7.11.1.4.2)$$

so that $T_0 = -10^{14}$, $\omega_0 = 10^7$, and $\sigma = 0.5$.

The amplitude characteristic is obtained from the formula:

$$|T(s)|^2 = T(s) \cdot T(-s)_{|s=j\omega} \quad (7.11.1.4.3a)$$

or

$$|T(s)|^2_{s=j\omega} = \frac{T_0^2}{[\omega_0^2 - \omega^2]^2 + 4\sigma^2\omega_0^2\omega^2} \quad (7.11.1.4.3b)$$

or

$$|T(s)|^2_{s=j\omega} = \frac{[R_5/(C_1C_2R_1R_2R_3)]^2}{\left[R_5/(C_1C_2R_2R_3R_6) - \omega^2\right]^2 + \omega^2/(R_4C_2)^2}. \quad (7.11.1.4.3c)$$

Differentiating with respect to ω and equating the derivative to zero, gives:

$$\omega_{\max} = \omega_0\sqrt{1 - 2 \cdot \sigma^2} = \sqrt{\frac{R_5}{C_1C_2R_2R_3R_6} - \frac{1}{2R_4^2C_2^2}} = 10^7 \cdot \sqrt{\frac{1}{2}} = 7.071 \times 10^6 \text{ rad/s} \quad (7.11.1.4.4)$$

or $f_{\max} = 1.1254$ MHz.

The value of the amplitude characteristic of the fault-free circuit at this frequency is

$$\sqrt{|T(j\omega_{\max})|^2} = \frac{|T_0|}{2\omega_0\sigma\sqrt{1 - 2\sigma^2}} = \frac{2}{\sqrt{3}} = 1.1547, \quad (7.11.1.4.5)$$

and the value of the amplitude characteristic when a fault $\Delta C_1 = 0.2 \cdot C_1$ is present is

$$\sqrt{|T_F(j\omega_{\max})|^2} = 1. \quad (7.11.1.4.6)$$

Based on this, we conclude that the responses of the fault-free and the faulty circuit at the frequency $\omega_{\max}$ differ by 13.37%, which means that the sinusoidal signal whose frequency is $f_{\max} = \omega_{\max}/(2\pi)$ certainly represents a test for this fault.

The reader is recommended to repeat this analysis but to calculate with the phase instead of the amplitude characteristic. At the same time, in order to obtain correct results, it is necessary to first decompose the circuit function into factors in the form:

$$T(s) = T_0\frac{1}{s - s_1} \cdot \frac{1}{s - s_2}, \quad (7.11.1.4.7)$$

where

$$s_{1,2} = \alpha \pm j \cdot \beta, \tag{7.11.1.4.8a}$$

$$\alpha = -\sigma \cdot \omega_0, \tag{7.11.1.4.8b}$$

and

$$\beta = \omega_0\sqrt{1 - \sigma^2}. \tag{7.11.1.4.8c}$$

The phase characteristic is now expressed as

$$\varphi(\omega) = \arg\{T_0\} - \mathrm{arctg}\left(\frac{\omega - \beta}{-\alpha}\right) - \mathrm{arctg}\left\{\frac{\omega + \beta}{-\alpha}\right\}. \tag{7.11.1.4.9}$$

In addition, since the phase characteristic is a monotonic function, when searching for a new frequency f_{max}, one needs to find the zero of the derivative of the group delay, which is defined as:

$$\tau_D = -\frac{d\varphi}{d\omega}, \tag{7.11.1.4.10}$$

so that the new ω_{max} is obtained as the solution of the following equation:

$$\frac{d\tau_D}{d\omega}_{|\omega=\omega_{max}} = 0. \tag{7.11.1.4.11}$$

In other words, the new f_{max} will be at the point where the second derivative of the phase is equal to zero. ■

Problem 7.11.1.5

For the circuit of Fig. 7.11.7 determine the increments of the parameters of the circuit function σ, ω_n, Q, as a consequence of changes in all elements of the circuit by + 10%. It is known that $R_1 = 10$ kΩ, $R_2 = 10$ kΩ, $R_3 = 10$ kΩ, $R_4 = 10$ kΩ, $R_5 = 10$ kΩ, $R_6 = 3$ kΩ, $R_7 = 7$ kΩ, $C_1 = 20$ pF, and $C_2 = 20$ pF. The quantity Q is defined as

$$Q = -\frac{\omega_0}{2 \cdot \alpha} = \frac{1}{2 \cdot \sigma} \tag{7.11.1.5.1}$$

and is referred to as the Q-factor of the circuit function.

Solution to Problem 7.11.1.5

The required numerical values of increments of the circuit function parameters are given in Table 7.11.4. It is recommended to the reader to generate the same table

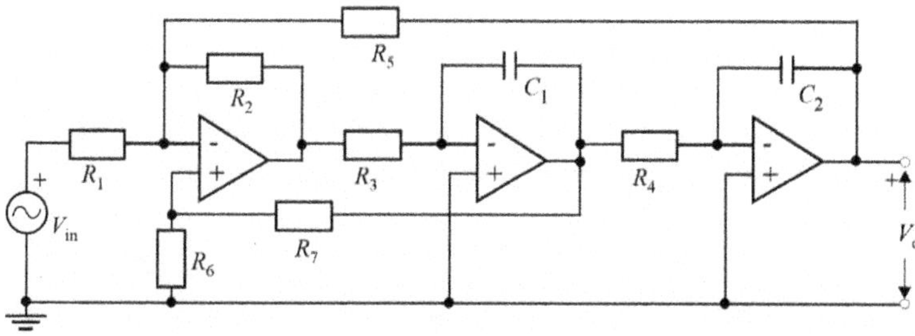

Fig. 7.11.7 Multiple feedback active RC filters

with negative values of increments of the circuit elements, i.e., with increments of −10%. In order to do this, the reader should first determine the transfer function of the circuit.

By analyzing this table, we conclude that not all circuit function parameters are equally sensitive to changes in individual circuit elements. If the parameters were the final indicator of testability, then one would choose the following:

1. For testing R_1 one should observe the increment of Q.
2. For testing R_2 one should observe the increment of ω_0.
3. For testing R_3 one should observe the increment of Q.
4. For testing R_4 one should observe the increment of σ.
5. For testing R_5 one should observe the increment of ω_0.
6. For testing R_6 one should observe the increment of σ.
7. For testing R_7 one should observe the increment of Q.
8. For testing C_1 one should observe the increment of Q.
   and
9. For testing R_2 one should observe the increment of σ. ■

Table 7.11.4 Increments of the transfer function parameters

$100 \cdot \Delta p/p$ increment of the parameter	ω_0	$\Delta\omega_0/\omega_0$	Q	$\Delta Q/Q$	σ	$\Delta\sigma/\sigma$
Nominal value	5000	0	1.1111	0	0.45	0
+10% of R_1	5000	0	1.1458	0.03123	0.4363	−0.03044
+10% of R_2	5244	0.0488	1.0925	−0.01674	0.45766	0.01702
+10% of R_3	4767.3	−0.04654	1.1653	0.04878	0.429	−0.04666
+10% of R_4	4767.3	−0.04654	1.0594	−0.04653	0.47196	0.0488
+10% of R_5	4767.3	−0.0488	1.0925	−0.01674	0.45766	0.01702
+10% of R_6	5000	0	1.0404	−0.06363	0.48	0.06666
+10% of R_7	5000	0	1.1889	0.07	0.42056	−0.06666
+10% of C_1	4767.3	−0.04654	1.1653	0.04878	0.429	−0.04666
+10% of C_2	4767.3	−0.04654	1.0594	−0.04653	0.4719	0.0488

Problem 7.11.1.6

In the circuit of Fig. 7.11.5 determine the fault effects related to large increments in capacitance C, $\Delta C = 0.2 \cdot C$, and discuss possible test signals for time domain measurements. As an excitation signal, use the step function of amplitude $I = 1$ mA. Use $C = 10$ pF.

Solution to Problem 7.11.1.6

In order to determine the response to the step function, we will apply the usual theory, namely the Laplace transform (convolution). The image of the circuit function in the s-domain is given by

$$V(s) = V_o \cdot \frac{1}{1 + s \cdot \tau}, \tag{7.11.1.6.1}$$

where

$$V_o = I \cdot \frac{R_1 R_3}{R_1 + R_2 + R_3} \tag{7.11.1.6.2}$$

and τ is given by (7.11.1.3.3). The excitation image in the s-domain is

$$J(s) = I/s. \tag{7.11.1.6.3}$$

After convolution for the response of the circuit without the initial values of the voltage on the capacitor, one gets:

$$v(t) = V_o \cdot \left(1 - e^{-t/\tau}\right). \tag{7.11.1.6.4}$$

When evaluating the properties of this response from the testing against changes in the capacitance value point of view, we conclude that the change in the capacitance, since it only changes the time constant, affects only the rate of change of the output voltage, and not the shape of the curve, that is, the final established value of it. Therefore, we can use delay time and rise time as measured information about this response. The delay time will be used here since it is only expressed by one measured quantity. The delay time is the time it takes for the response to reach half of its steady-state value so that t_d represents the solution to the following equation:

$$\frac{V_o}{2} = V_o \cdot \left(1 - e^{-t_d/\tau}\right), \tag{7.11.1.6.5a}$$

which is

$$t_d = -\tau \cdot \ln(1/2). \tag{7.11.1.6.5b}$$

For the fault-free circuits one gets $t_d = 3.43 \times 10^{-8} \cdot 0.693 = 23.8$ ns, while for the faulty circuit one get $\{t_d\}_F = t_D \cdot 1.2 = 28.53$ ns.

Given that the measured quantities for the fault-free and the faulty circuit differ by all of 20%, in this case, the proposed signal can be accepted as a test signal. ■

Problem 7.11.1.7

The circuit of Fig. 7.11.7 is excited by the step function. The response of the circuit is given by [with reference to (7.11.1.4.7) and (7.11.1.4.8)]

$$\frac{v_{out}}{v_{in}} \cdot \frac{\omega_0^2}{|T_0|} = 1 - \frac{1}{\sqrt{1-\sigma^2}} \cdot e^{-\alpha \cdot t} \cdot \cos(\beta \cdot t + \theta), \quad (7.11.1.7.1)$$

where

$$tg(\theta) = -\alpha/\beta, \quad (7.11.1.7.2)$$

and the rest of the quantities are defined in relation to the function of the circuit from Fig. 7.11.7 as well as for the function of the circuit of Fig. 7.11.6.

This function is calculated for the circuit of Fig. 7.11.7 and given in Fig. 7.11.8. The indicated quantities (parameters of the step response) can be approximately calculated as follows:

(a) The time of the first maximum:

$$t_{max} = \pi/\beta. \quad (7.11.1.7.3)$$

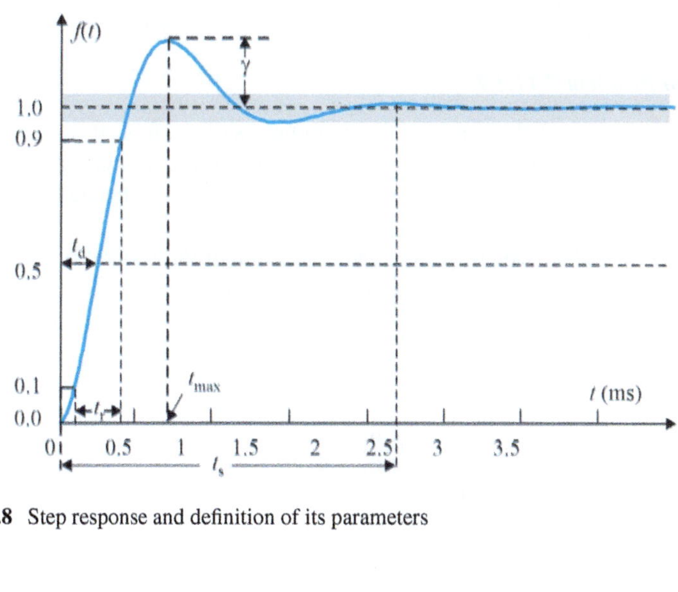

Fig. 7.11.8 Step response and definition of its parameters

Table 7.11.5 Step response parameters of the fault-free circuit

t_{max} (ms)	t_s (ms)	γ (%)	t_r (ms)
0.7036	1.4222	20.5364	0.385

(b) Overshoot:

$$\gamma(\%) = 100 \cdot \exp\left\{-\frac{\sigma \cdot \pi}{\sqrt{1-\sigma^2}}\right\}. \tag{7.11.1.7.4}$$

(c) Rise time:

$$t_r = \frac{0.8 + 2.5 \cdot \sigma}{\omega_0}. \tag{7.11.1.7.5}$$

(d) Settling time:

$$t_s = \frac{3.2}{\sigma \cdot \omega_0} \quad \text{for } 0 < \sigma < 0.69, \tag{7.11.1.7.6a}$$

and

$$t_s = \frac{4.5}{\sigma \cdot \omega_0} \quad \text{for } \sigma > 0.69. \tag{7.11.1.7.6b}$$

Using the given formulas, calculate, first, the nominal values of the moment of the first maximum, overshoot, rise time, and settling time.

Then, determine the increments of all these quantities if all elements in the circuit change by +10%.

Solution to Problem 7.11.1.7

Table 7.11.5 gives the values of response parameters of the fault-free circuit.

Table 7.11.6 contains the logarithmic sensitivities of the response parameters for given relative increments of the circuit elements when they are changed one at a time.

Based on this table, we can draw conclusions about the testability of individual faults and about the response parameter that should be observed when considering testability. Moreover, since the definition of settling time is relatively vague, only the following parameters will be discussed here: Δt_{max}, $\Delta\gamma$ and Δt_r.

We choose the following:

1. For testing R_1 one should observe the increment of γ.
2. For testing R_2 one should observe the increment of t_{max}, although $\Delta\gamma$ can serve too.
3. For testing R_3 one should observe the increment of γ.
4. For testing R_4 one should observe the increment of γ.

Table 7.11.6 Logarithmic sensitivities of the response parameters

$100 \cdot \Delta p/p$ increment of the parameter	$\frac{\Delta t_{max}/t_{max}}{\Delta p/p}$	$\frac{\Delta t_s/t_s}{\Delta p/p}$	$\frac{\Delta \gamma/\gamma}{\Delta p/p}$	$\frac{\Delta t_r/t_r}{\Delta p/p}$
+10% of R_1	−0.0753	0.3140	0.6157	−0.1779
+10% of R_2	−0.4235	0.6249	−0.3344	−0.3704
+10% of R_3	0.3688	1.0015	0.9532	0.2021
+10% of R_4	0.6238	0.0014	−0.9402	0.7864
+10% of R_5	0.5341	0.3126	−0.3296	0.5925
+10% of R_6	0.1796	−0.6250	−1.2705	0.3896
+10% of R_7	−0.1597	0.7000	1.3785	−0.3896
+10% of C_1	0.3688	1.0015	0.9532	0.2021
+10% of C_2	0.6238	0.0014	−0.9402	0.7864

5. For testing R_5 one should observe the increment of t_r,
   although Δt_{max} can serve too.
6. For testing R_6 one should observe the increment of γ.
7. For testing R_7 one should observe the increment of γ.
8. For testing C_1 one should observe the increment of γ.
   and
9. For testing R_2 one should observe the increment of γ.

In other words, the overshoot in the step response of the second-order filter is very sensitive to changes in parameters, so it is best to be observed in order to judge the correctness of the circuit.

It is recommended to the reader to generate the same table but with negative values of increments of circuit elements, i.e., with increments of −10%. To be able to do this, the reader will first, as before, determine the transfer function of the circuit. ■

Problem 7.11.1.8

In the case of the amplifier depicted in Fig. 7.11.9 the voltage gain is measured. Determine how the following faults are mapped onto the gain value.

1. Channel length T_1 increased for 10%.
2. Channel length T_1 decreased for 10%.
3. Channel length T_2 increased for 10%.
4. Channel length T_2 decreased for 10%.
5. Channel width T_1 increased for 10%.
6. Channel width T_1 decreased for 10%.
7. Channel width T_2 increased for 10%.
8. Channel width T_2 decreased for 10%.
9. Threshold voltage T_1 increased for 10%.
10. Threshold voltage T_1 decreased for 10%.
11. Threshold voltage T_2 increased for 10%.
12. Threshold voltage T_2 decreased for 10%.

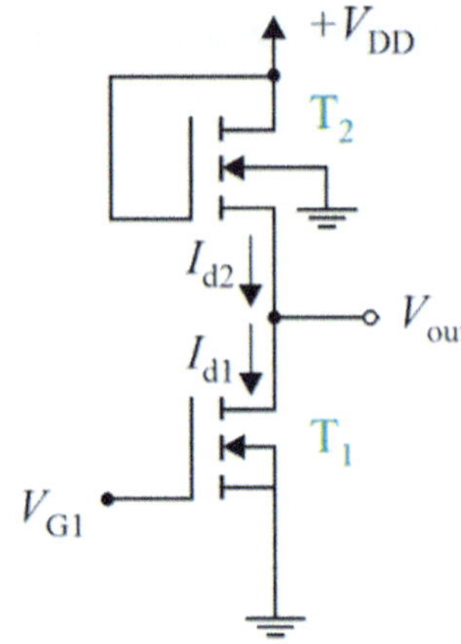

Fig. 7.11.9 NMOS inverter

The characteristic of the transistor is given by

$$I_D = A \cdot (V_{GS} - V_T)^2, \tag{7.11.1.8.1a}$$

where

$$A = \mu \cdot C'_{ox} W/(2 \cdot L), \tag{7.11.1.8.1b}$$

and μ is the mobility of the majority carriers (in this case electrons) in the channel, C'_{ox} is the capacitance of the oxide per unit area, W is the width, L is the length of the channel, and V_T is the threshold voltage.

Transistors with $(W/L)_1 = 4.5$, $(W/L)_2 = 0.5$, and $V_{T1} = V_{T2} = 1$ V are used. It is given that $V_{G1} = 3$ V and $V_{DD} = 12$ V.

Solution to Problem 7.11.1.8

The gate of transistor T_2 is connected to the drain, so it is valid

$$V_{DS2} = V_{GS2} = V_{DD} - V_{out}. \tag{7.11.1.8.2}$$

It represents the load and behaves as a nonlinear conductance whose characteristic is given by

$$I_{D2} = A_2 \cdot (V_{DS2} - V_T)^2. \tag{7.11.1.8.3}$$

The lower transistor (T_1) is an excitation or amplifying transistor.

The dynamic resistance of the load is

$$R_D = r_{D2} = \frac{\partial V_{DS2}}{\partial I_{D2}} = 1/[2A_2(V_{DS2} - V_{T2})]. \tag{7.11.1.8.4}$$

On the other hand, for the transconductance of the excitation transistor we have

$$g_m = \frac{\partial I_D}{\partial V_{GS}} = 2A_1(V_{GS1} - V_{T1}), \tag{7.11.1.8.5}$$

so, for gain we have

$$A = -g_m R_D = -\frac{A_1}{A_2} \frac{V_{GS1} - V_{T1}}{V_{DD} - V_{out} - V_{T2}} = -\frac{(W/L)_1}{(W/L)_2} \frac{V_{G1} - V_{T1}}{V_{DD} - V_{out} - V_{T2}}. \tag{7.11.1.8.6}$$

In order to be able to analyze the fault effects, we first need to determine the value of the DC output voltage V_{out}. We get it by equalizing the transistor currents

$$I_{D1} = I_{D2} \tag{7.11.1.8.7}$$

or

$$A_1(V_{GS1} - V_{T1})^2 = A_2(V_{DD} - V_{out} - V_{T2})^2 \tag{7.11.1.8.8}$$

or

$$V_{out} = V_{DD} - V_{T2} - \sqrt{\frac{W_1/L_1}{W_2/L_2}} \cdot (V_{G1} - V_{T1}). \tag{7.11.1.8.9}$$

By substituting the numerical values for the nominal value of the output voltage, $V_{out\_o} = 6$ V is obtained, while for the nominal gain one gets $A_o = -9$.

The values of these two quantities that arise when the transistor parameters are changed are given in Table 7.11.7.

When analyzing this table, while observing the gain increments, we conclude that only in three cases can we talk about the clear testability of the faults. This applies to positive increment L_1, positive increment W_2, and negative increment W_2. In the rest of the cases, either the increments are small or equal to zero, so there is no question of observability and testability.

The incidental result, the value of the output voltage, however, speaks of an alternative possibility of testing this nonlinear resistive circuit. Namely, if we measure the DC value of the output voltage, we can more easily draw conclusions about the correctness of the circuit than when we measure the gain. As can be seen from the table, the relative increase of the output voltage is less than 5% only in the last two rows of the table, which means that in all cases, except for the last two, by measuring the output voltage, the circuit can be tested for changes in transistor parameters.

Table 7.11.7 Faults and fault effects for the amplifier depicted in Fig. 7.11.9

Δp increment of the parameter	$(W/L)_1$	$(W/L)_2$	V_{T1} (V)	V_{T2} (V)	V_{out} (V)	$\frac{\Delta V_{out}}{V_{out}}$ %	$-A$	$\Delta A/A$ %
Nominal values	4.5000	0.5000	1	1	5.0000		3.00	
L_1 increased for 10%	4.0909	0.5000	1	1	5.2792	5.58	2.86	−4.67
L_1 decreased for 10%	5.0000	0.5000	1	1	4.6754	−6.49	3.16	5.33
L_2 increased for 10%	4.5000	0.4545	1	1	4.7068	−5.86	3.15	3.00
L_2 decreased for 10%	4.5000	0.5556	1	1	5.3079	6.16	2.85	−3.00
W_1 increased for 10%	4.9500	0.5000	1	1	4.7071	−5.86	3.15	3.00
W_1 decreased for 10%	4.0500	0.5000	1	1	5.3079	6.16	2.85	−3.00
W_2 increased for 10%	4.5000	0.5500	1	1	5.2972	6.06	2.87	−4.33
W_2 decreased for 10%	4.5000	0.4500	1	1	4.6754	−6.49	3.16	5.33
V_{T1} increased for 10%	4.5000	0.5000	1.1	1	5.3000	6.00	3.00	0.00
V_{T1} decreased for 10%	4.5000	0.5000	0.9	1	4.7000	−6.00	3.00	0.00
V_{T2} increased for 10%	4.5000	0.5000	1	1.1	4.9000	−2.00	3.00	0.00
V_{T2} decreased for 10%	4.5000	0.5000	1	0.9	5.1000	2.00	3.00	0.00

The reason that the variations of V_{T2} are not testable (at least not clearly and simply) lies in the fact that the current of the transistor T_2 is determined by the transistor T_1, which acts as a constant current source, so that the variation of V_{T2} is compensated by the variation of the output voltage, which can be seen from (7.11.1.8.8). The value of the output voltage itself is, however, much higher than the increment of V_{T2}, so its influence is neutralized. ∎

Problem 7.11.1.9

In the circuit of Fig. 7.11.10 the following faults are present:

1. Channel length T_2 increased for 10%.
2. Channel length T_2 decreased for 10%.
3. Channel length T_3 increased for 10%.
4. Channel length T_3 decreased for 10%.

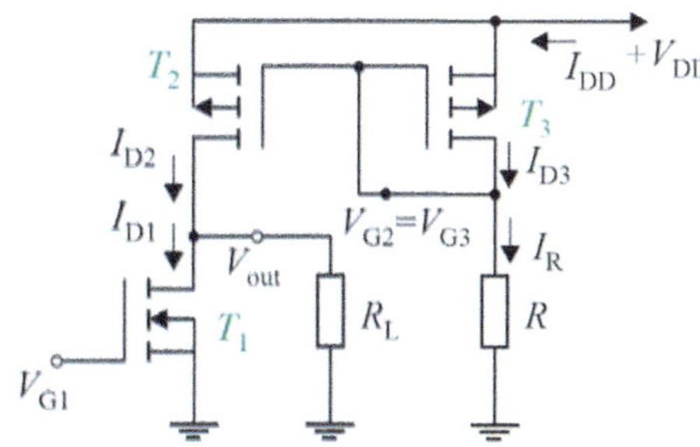

Fig. 7.11.10 CMOS amplifier using a current mirror as a load

5. Threshold voltage T_3 increased for 10%.
6. Threshold voltage T_3 decreased for 10%.
7. Threshold voltage T_2 increased for 10%.
8. Resistance R increased for 10%.
9. Resistance R decreased for 10%.

Check if the testability for these faults can be established by measuring the DC supply current (I_{DD}). The nominal values of the transistor parameters are $(W_1/L_1) = 3$, $(W_2/L_2) = 0.5$, $(W_3/L_3) = 0.5$, $V_{T1} = -V_{T2} = -V_{T3} = 1$ V, and $(\mu \cdot C'_{ox})/2 = 1$ mA/V^2. It is given that $V_{DD} = 12$ V and $R = 2$ kΩ.

Solution to Problem 7.11.1.9

To determine the supply current, we first determine the voltage $V_{GS3} = V_{GS2} = V_{G3} - V_{S3} = V_{G3} - V_{DD}$, between the gate and the source of the P-channel transistors. To that end, we formulate the equation

$$I_{D3} = I_R, \tag{7.11.1.9.1a}$$

or

$$(V_{DD} + V_{GS3})/R = A_3(V_{GS3} - V_{T3})^2. \tag{7.11.1.9.1b}$$

The solution of this equation is

$$V_{GS3} = V_{GS2} = V_{T3} - \frac{1}{2R \cdot A_3}\left[\sqrt{1 + 4R \cdot A_3(V_{T3} + V_{DD})} - 1\right]. \tag{7.11.1.9.2}$$

Based on this, the supply current can be expressed as

$$I_{DD} = (V_{DD} + V_{GS3})/R + A_2(V_{GS2} - V_{T2})^2. \tag{7.11.1.9.3}$$

The calculated numerical values needed for conclusion are given in Table 7.11.8.

Analyzing this table, we conclude that only three of the ten listed faults cannot be tested by measuring the DC component of the supply current since its relative

Table 7.11.8 IDDQ testing fault dictionary

Δp parameter increment	$(W/L)_2$	$(W/L)_3$	V_{T2} (V)	V_{T3} (V)	R (kΩ)	V_{GS3} (V)	I_{DD} (mA)	$\frac{\Delta I_{DD}}{I_{DD}}$ %
Nominal values	0.5000	0.5000	−1	−1	2	−3.8541	8.1459	0.0
L_2 increased for 10%	0.4545	0.5000	−1	−1	2	−3.8541	7.7756	8.08
L_2 decreased for 10%	5.5556	0.5000	−1	−1	2	−3.8541	8.5984	−1.65
L_3 increased for 10%	0.5000	0.4545	−1	−1	2	−3.9717	8.4296	−0.34
L_3 decreased for 10%	0.5000	0.5556	−1	−1	2	−3.7284	7.8579	7.10
V_{T3} increased for 10%	0.5000	0.5000	−1	−1.1	2	−3.9392	8.3498	−1.29
V_{T3} decreased for 10%	0.5000	0.5000	−1	−0.9	2	3.7690	7.9492	6.03
V_{T2} increased for 10%	0.5000	0.5000	−1.1	-1	2	−3.8541	7.8655	7.02
R increased for 10%	0.5000	0.5000	1	1	2.2	−3.7402	7.5088	11.23
R decreased for 10%	0.5000	0.5000	1	0.9	1.8	−3.9843	8.9062	−5.29

increment is smaller than 1.5%. They are the negative increment of L_2, the positive increment L_3, and the positive increment V_{T3}.

In addition, when considering this table, we notice that the nonlinearity of the components has a great influence on the sensitivity of the output value to parameter changes. That, in general, makes difficult the testing of circuit in which the nonlinear components are built-in.

For the sake of exercise, the reader is recommended to repeat this same task, but with twice the values of the increments of the specified parameters. ■

Problem 7.11.1.10 (as an exercise only)

The nominal values of the parameters in the circuit of Fig. 7.11.10 are: $(W_1/L_1) = 3$, $(W_2/L_2) = 0.5$, $(W_3/L_3) = 0.5$, $V_{T1} = -V_{T2} = -V_{T3} = 1$ V, and $(\mu \cdot C'_{ox})/2 = 1$ mA/V^2. It is known that $V_{G1} = 2$ V, $V_{DD} = 12$ V, $R_L = 5$ kΩ, and $R = 2$ kΩ.

Check whether testability can be established for the faults listed in Task 7.11.1.9 by measuring the DC output voltage V_{out}. ■

Problem 7.11.1.11

In the circuit of Fig. 7.11.9 which is here, in Fig. 7.11.11, redrawn to show the faults Out-s-Gr and Out-s-V_{DD}, determine the value of the output voltage and supply current of the fault-free and the faulty circuit and discuss the testability, i.e., the

Fig. 7.11.11 CMOS inverter

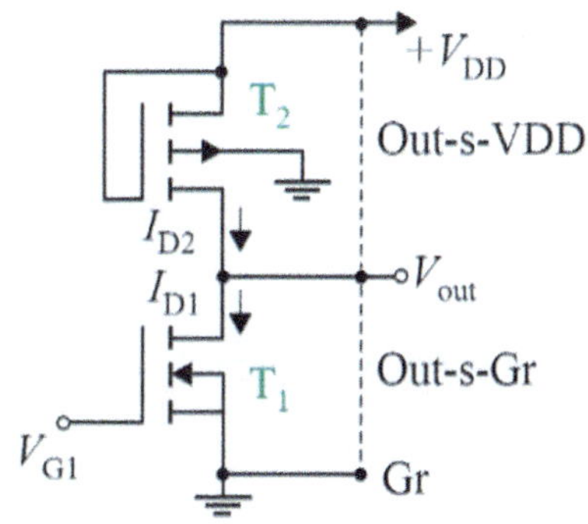

choice of the quantity that should be measured in order to recognize the presence of the mentioned faults (the faults to become detectable and observable). It is known that $(W_1/L_1) = 4$, $(W_2/L_2) = 1$, $V_{T1} = -V_{T2} = 0.5$ V, $(\mu \cdot C'_{ox})/2 = 1$ mA/V^2, $V_{G1} = 1.5$ V, and $V_{DD} = 3$ V.

Solution to Problem 7.11.1.11

For the output voltage in the fault-free circuit, as before, we have

$$V_{out} = V_{DD} - V_{T2} - \sqrt{\frac{W_1/L_1}{W_2/L_2}} \cdot (V_{G1} - V_{T1}). \quad (7.11.1.11.1)$$

For the supply current in the fault-free circuit it is valid:

$$I_{DD} = I_{D1} = I_{D2}. \quad (7.11.1.11.2)$$

When the Out-s-V_{DD} fault is present in the circuit, the values of these two quantities change and become

$$V_{out|Out\text{-}s\text{-}V_{DD}} = V_{DD} \quad (7.11.1.11.3)$$

and

$$I_{DD|Out\text{-}s\text{-}V_{DD}} = I_{D1} = A_1(V_{GS1} - V_{T1})^2. \quad (7.11.1.11.4)$$

Finally, in the presence of the fault Out-s-Gg is formed

$$V_{out|Out\text{-}s\text{-}Gr} = 0 \quad (7.11.1.11.5)$$

and

$$I_{DD|Out\text{-}s\text{-}Gr} = I_{D2} = A_2 \cdot (V_{GS1} - V_{T2})^2 = A_2 \cdot (V_{DD} - V_{T2})^2. \quad (7.11.1.11.6)$$

Table 7.11.9 Fault effects in the CMOS inverter

Type of fault	V_{out} (V)	I_{DD} (mA)
Nominal values	1.5	4
Out-s-V_{DD}	3	4
Out-s-Gg	0	12.25

The values of these two quantities that arise in the presence of the listed catastrophic faults are given in Table 7.11.9.

Analyzing this table, we observe that short-circuit faults are mapped into increments of the output voltage being equal to its nominal value (positive or absolute negative), so that $\Delta V_{out}/V_{out}$ in both cases is equal to 1 or 100% in absolute value.

When it comes to supply current, however, the situation is not like that. While in the case of Out-s-V_{DD} the value of the supply current does not change (at least not noticeably if one considers its increase due to channel shortening, which is neglected here), in the case of Out-s-Gr its value changes dramatically. Namely, in the first case, due to a short circuit between the power supply and the output terminal, the transistor T_1 acts as a constant current source that does not depend on the voltage on it. In the second case, the short circuit transforms T_2 into a nonlinear resistor with a parabolic characteristic being found between the power supply and the ground (to which the output is connected now), so that the current increases sharply due to the increase of the transistor voltage.

The conclusion would be that it is enough to observe the output voltage in order to test the circuit for the presence of these two faults. However, we can detect one of them by observing the supply current, too. ■

Problem 7.11.1.12 (as exercise only)

In the circuit of Fig. 7.11.10 which is here, in Fig. 7.11.12, redrawn to show the faults V_{G3}-s-G and V_{G3}-s-Out, determine the value of the output voltage and supply current of the fault-free and the faulty circuit and discuss the testability, i.e., the choice of the quantity that should be measured in order to recognize the individual presence of the mentioned faults (the faults to become detectable and observable). Data for the calculation should be taken from Problem 7.11.1.10. Assume that the breakdown voltage between the gate and the channel is larger than the supply voltage. ■

Problem 7.11.1.13

Figure 7.11.13 shows again the circuit from Fig. 7.11.9 with the difference that some current signal values are indicated.

The characteristic of the transistor is given by

$$i_D = A \cdot (v_{GS} - V_T)^2, \qquad (7.11.1.13.1a)$$

where

$$A = \mu \cdot C'_{ox} W/(2 \cdot L), \qquad (7.11.1.13.1b)$$

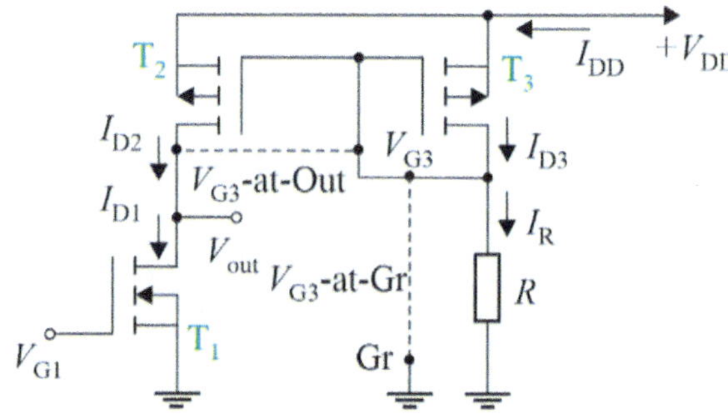

Fig. 7.11.12 CMOS amplifier with faults

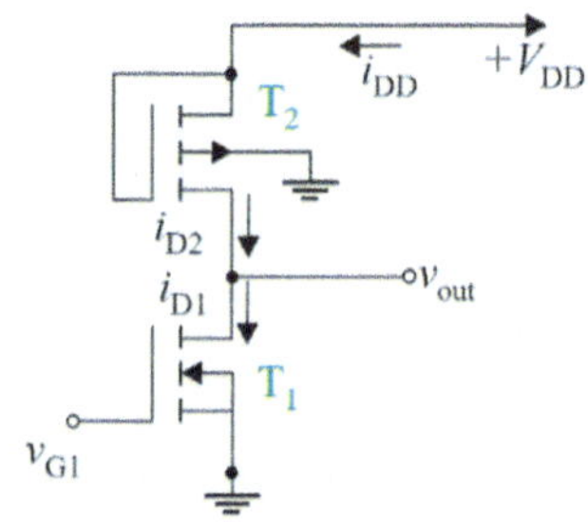

Fig. 7.11.13 CMOS inverter with instantaneous values shown

and μ is the mobility of the majority carriers (in this case electrons) in the channel, C'_{ox} is the capacitance of the oxide per unit area, W is the width, L is the length of the channel, and V_T is the threshold voltage.

The used transistors have $(W/L)_1 = 4.5$, $(W/L)_2 = 0.5$, $\mu \cdot C'_{ox}/2 = 1$ mA/V^2, and $V_{T1} = V_{T2} = 1$ V. It is also known that $v_{G1} = [3 + 1 \cdot \sin(2 \cdot \pi \cdot 1000 \cdot t)]$ V and $V_{DD} = 12$ V.

Based on the value of the expressions

$$R_{i_{DD}i_{DD}}(\tau) = R_i = \frac{1}{T}\int_0^T i_{DD}(t)i_{DD}(t+\tau)\cdot dt, \qquad (7.11.1.13.1c)$$

$$R^L_{i_{DD}i^L_{DD}}(\tau) = R^L_i = \frac{1}{T}\int_0^T i_{DD}(t)i^L_{DD}(t+\tau)\cdot dt, \qquad (7.11.1.13.1d)$$

$$R_{v_{out}v_{out}}(\tau) = R_v = \frac{1}{T}\int_0^T v_{out}(t)v_{out}(t+\tau)\cdot dt, \qquad (7.11.1.13.2a)$$

$$R^L_{v_{out}v^L_{out}}(\tau) = R^L_v = \frac{1}{T}\int_0^T v_{out}(t)v^L_{out}(t+\tau)\cdot dt, \qquad (7.11.1.13.2b)$$

and

$$R^{\mathrm{L}}_{v^{\mathrm{L}}_{\mathrm{out}} i^{\mathrm{L}}_{\mathrm{DD}}}(\tau) = R^{\mathrm{L}}_{\mathrm{v,i}} = \frac{1}{T}\int_0^T v^{\mathrm{L}}_{\mathrm{out}}(t) i^{\mathrm{L}}_{\mathrm{DD}}(t+\tau)\cdot \mathrm{d}t, \tag{7.11.1.13.3}$$

where the symbol L stands for a faulty circuit, in the presence of one of the following faults:

1. W_1 increased for 20%.
2. W_1 decreased for 20%.
3. W_2 increased for 20%.
   and
4. W_2 decreased for 20%,

Determine the relative increments of the correlation coefficients of the fault-free and the faulty circuit.

Use the following:

$$v_{\mathrm{out}} = V_{\mathrm{DD}} - V_{\mathrm{T2}} - \sqrt{\frac{W_1/L_1}{W_2/L_2}}\cdot (v_{\mathrm{GS1}} - V_{\mathrm{T1}}) \tag{7.11.1.13.4}$$

and

$$i_{\mathrm{DD}} = i_{\mathrm{D1}}. \tag{7.11.1.13.5}$$

For each fault, individually estimate the value of τ that leads to the largest difference between the correlation coefficients of the fault-free and the faulty circuit.

For each fault, determine which of the expressions (7.11.1.13.1c), (7.11.1.13.2b), and (7.11.1.13.3) is the most suitable for recognizing the fault effect.

Solution to Problem 7.11.1.13

For the instantaneous value of the supply current we have

$$i_{\mathrm{DD}} = A\cdot (v_{\mathrm{GS1}} - V_{\mathrm{T}})^2 = \frac{W_1}{L_1}\cdot [2+\sin(2000\cdot\pi\cdot t)]^2 \times 10^{-3}\ \mathrm{A}. \tag{7.11.1.13.6a}$$

For the fault-free circuit it is

$$i_{\mathrm{DD}} = 4.5\cdot [2+\sin(2000\cdot\pi\cdot t)]^2 \times 10^{-3}\ \mathrm{A}, \tag{7.11.1.13.6b}$$

while for the faulty circuit it becomes

$$i^{\mathrm{L}}_{\mathrm{DD}} = 4.5\cdot (1+p_x)\cdot [2+\sin(2000\cdot\pi\cdot t)]^2 \times 10^{-3} = (1+p_x)\cdot i_{\mathrm{DD}}. \tag{7.11.1.13.6c}$$

where $x = 1, 2, 3$, and 4, with $p_{1,2} = \pm 0.2$ and $p_{3,4} = 0$.

Accordingly, the autocorrelation coefficient is obtained as

$$R_{\mathrm{i}} = \alpha \int_{0}^{T} [2 + \sin(2000 \cdot \pi \cdot t)]^2 \{2 + \sin[2000 \cdot \pi \cdot (t + \tau)]\}^2 \cdot \mathrm{d}t, \quad (7.11.1.13.7\mathrm{a})$$

where $\alpha = 20.25 \times 10^{-6}/T = 20.25 \times 10^{-3}$ s, with $T = 1$ ms.

The correlation coefficient of the currents in the fault-free and the faulty circuit is

$$R_{\mathrm{i}}^{\mathrm{L}} = (1 + p_x) \cdot R_{\mathrm{i}}. \quad (7.11.1.13.7\mathrm{b})$$

Based on that, for the relative increment of the correlation coefficient, we have

$$\Delta_{\mathrm{i}} = 1 - R_{\mathrm{i}}^{\mathrm{L}}/R_{\mathrm{i}} = -p_x. \quad (7.11.1.13.8)$$

We conclude that when it comes to the supply current, in this simple circuit (where it is determined only by the current of T_1 which is acting as a constant current source), the correlation coefficient of the faulty circuit is $(1 + p_x)$ times higher than that of the fault-free circuit. Since, obviously, $R_{\mathrm{i}} \neq 0$, the relative difference between $R_{\mathrm{i}}^{\mathrm{L}}$ and R_{i} does not depend on the value of τ and amounts to $\Delta_{\mathrm{i}} = -p_x$.

Bearing in mind that $p_3 = p_4 = 0$, only the faults "1" and "2" will be able to be tested in this way.

The output voltage can be written:

$$v_{\mathrm{out}}^{\mathrm{L}} = 11 - 3 \cdot \sqrt{\frac{1 + p_x}{1 + q_x}} \cdot [2 + \sin(2000 \cdot \pi \cdot t)], \quad (7.11.1.13.9)$$

where $p_0 = q_0 = 0$, for the fault-free circuit, $p_{1,2} = \pm 0.2$ and $q_{1,2} = 0$, for the first two faults, and $p_{3,4} = 0$ and $q_{3,4} = \pm 0.2$, for the last two faults.

Accordingly, for the autocorrelation coefficient of the output voltage we have

$$R_{\mathrm{v}} = \frac{1}{T} \int_{0}^{T} \{11 - 3 \cdot [2 + \sin(\omega t)]\}(11 - 3 \cdot \{2 + \sin[\omega(t + \tau)]\}) \cdot \mathrm{d}t$$

$$(7.11.1.13.10\mathrm{a})$$

and

$$R_{\mathrm{v}}^{\mathrm{L}} = \frac{1}{T} \int_{0}^{T} \left\{ 11 - 3\sqrt{\frac{1 + p_x}{1 + q_x}} [2 + \sin(\omega t)] \right\}$$

$$\cdot\left(11-3\sqrt{\frac{1+p_x}{1+q_x}}\{2+\sin[\omega(t+\tau)]\}\right)\cdot \mathrm{d}t \tag{7.11.1.13.10b}$$

where $\omega = 2000\cdot\pi$ and $T = 1$ ms.

After getting rid of the brackets for the autocorrelation coefficient of the output voltage, we get

$$\begin{aligned}
R_\mathrm{v} &= \frac{1}{T}\int_0^T \{5-3\cdot\sin(\omega t)\}\{5-3\cdot\sin[\omega(t+\tau)]\}\cdot \mathrm{d}t\\
&= \frac{1}{T}\int_0^T \{25-15\cdot\sin(\omega t)-15\cdot\sin[\omega(t+\tau)]+9\cdot\sin(\omega t)\cdot\sin[\omega(t+\tau)]\}\cdot \mathrm{d}t\\
&= \frac{1}{T}\int_0^T 25\cdot \mathrm{d}t+\frac{9}{T}\int_0^T \sin(\omega t)\cdot\sin[\omega(t+\tau)]\cdot \mathrm{d}t\\
&= 25+\frac{9}{2\cdot T}\int_0^T [\cos(\tau)-\cos(2\omega t+\tau)]\cdot \mathrm{d}t = 25+4.5\cdot\cos(\tau).
\end{aligned} \tag{7.11.1.13.11a}$$

In a similar way we find

$$\begin{aligned}
R_\mathrm{v}^\mathrm{L} &= \frac{1}{T}\int_0^T \{A+B\cdot\sin(\omega t)\}\cdot\{A+B\cdot\sin[\omega(t+\tau)]\}\cdot \mathrm{d}t\\
&= \frac{1}{T}\int_0^T \left\{A^2+\frac{B^2}{2}\cdot[\cos(\tau)-\cos(2\omega t+\tau)]\right\}\cdot \mathrm{d}t\\
&= A^2+\frac{B^2}{2}\cdot\cos(\tau)
\end{aligned} \tag{7.11.1.13.11b}$$

where $B = -3\cdot\sqrt{(1+p_x)/(1+q_x)}$ and $A = 11+2\cdot B$.

The relative difference between R_v and R_v^L is obtained to be:

$$\Delta_\mathrm{v} = 1-R_\mathrm{v}^\mathrm{L}/R_\mathrm{v} = 1-\frac{A^2+\frac{B^2}{2}\cdot\cos(\tau)}{25+4.5\cdot\cos(\tau)}. \tag{7.11.1.13.12}$$

Since the value of τ that leads to the maximum deviation of the correlation coefficient of the faulty circuit from the autocorrelation coefficient of the fault-free one is of interest, we differentiate the expression (7.11.1.13.12) with respect to $\cos(\tau)$ and equate the result to zero. This leads to the solution: $\cos(\tau)_{1,2} = \pm 1$, which is a consequence of the monotonicity of the function we are observing. That is why we assume that the largest and smallest values of the function are at the ends of the variable interval (areas of definition), so we choose $\cos(\tau)_{1,2} = \pm 1$, i.e., $\tau_1 = 0$ and $\tau_2 = \pi$. Which of these two solutions will be chosen will depend on the relationship between the constants A and B in (7.11.1.13.12). Their values for all four faults together with the values of Δ_v and the selected value of τ are given in Table 7.11.10.

Analyzing Table 7.11.10, we easily conclude that in all four cases, the right choice is represented by $\tau = \pi$. If such a choice is made in all four cases, we will have a deviation of the correlation coefficient of the faulty circuit from the autocorrelation coefficient of the fault-free circuit larger than 30%, which is very favorable from the measurement point of view.

It remains to process Expression (7.11.1.13.3) which refers to the correlation coefficient $R^{L}_{v,i}$. By substituting the expression for the current (7.11.1.13.6c) and the voltage (7.11.1.13.9) in the faulty circuit, we get:

$$\begin{aligned} R^{L}_{v,i} &= \frac{1}{T}\int_0^T v^{L}_{out}(t)\, i^{L}_{DD}(t+\tau)\cdot dt \\ &= \frac{C}{T}\int_0^T \{A + B\cdot\sin(\omega t)\}\left\{1 + \sin(\omega t+\tau) + \frac{1}{4}\cdot\sin^2(\omega t+\tau)\right\}\cdot dt, \end{aligned} \tag{7.11.1.13.13}$$

Table 7.11.10 Choice of τ for maximum detectability

Fault	τ	A	A^2	B	B^2	Δ_v	Choice
1. $p = 0.2$; $q = 0$	0	4.428	19.61	−3.286	10.80	0.152	
1. $p = 0.2$; $q = 0$	π	4.428	19.61	−3.286	10.80	0.307	*
2. $p = -0.2$; $q = 0$	0	5.634	31.74	−2.683	7.199	−0.198	
2. $p = -0.2$; $q = 0$	π	5.634	31.74	−2.683	7.199	−0.373	*
3. $p = 0$; $q = 0.2$	0	5.522	30.49	−2.739	7.502	−0.161	
3. $p = 0$; $q = 0.2$	π	5.522	30.49	−2.739	7.502	−0.304	*
4. $p = 0$; $q = -0.2$	0	4.292	18.42	−3.354	11.25	0.185	
4. $p = 0$; $q = -0.2$	π	4.292	18.42	−3.354	11.25	0.376	*

where A and B are as before, and $C = 18 \times 10^{-3} \cdot (1 + p)$. This expression can be simplified as follows:

$$R^{L}_{v,i} = \frac{C}{T}(I_1 + I_2 + I_3 + I_4 + I_5 + I_6) = C \cdot \left[\frac{9A}{8} + \frac{B}{2}\cos(\tau)\right], \quad (7.11.1.13.14a)$$

where

$$I_1 = \int_0^T A \cdot dt = A \cdot T, \quad (7.11.1.13.14b)$$

$$I_2 = \int_0^T B \cdot \sin(\omega \cdot t)] \cdot dt = 0, \quad (7.11.1.13.14c)$$

$$I_3 = \int_0^T A \cdot \sin(\omega t + \tau) \cdot dt = 0, \quad (7.11.1.13.14d)$$

$$I_4 = \int_0^T B \cdot \sin(\omega t) \cdot \sin(\omega t + \tau) \cdot dt$$
$$= \frac{B}{2}\int_0^T [\cos(\tau) - \cos(2\omega t + \tau)] \cdot dt = \frac{B \cdot T}{2}\cos(\tau), \quad (7.11.1.13.14e)$$

$$I_5 = \frac{A}{4}\int_0^T \sin^2(\omega t + \tau) \cdot dt = \frac{A}{8}\int_0^T [1 - \cos(2\omega t + 2\tau)] \cdot dt = \frac{A \cdot T}{8}, \quad (7.11.1.13.14f)$$

and

$$I_6 = \frac{B}{4}\int_0^T \sin(\omega t) \cdot \sin^2(\omega t + \tau) \cdot dt = \frac{B}{4}(J_1 + J_2 + J_3 + J_4) = 0, \quad (7.11.1.13.14g)$$

since it is

$$J_1 = \frac{1}{2}\int_0^T \sin(\omega \cdot t) \cdot dt = 0, \quad (7.11.1.13.14h)$$

$$J_2 = -\frac{\cos(2\tau)}{2}\int_0^T \cos^2(\omega t)\cdot\sin(\omega t)\cdot dt = \frac{\cos(2\tau)}{6\omega}\cos^3(\omega t)\Big|_0^T = 0, \tag{7.11.1.13.14i}$$

$$J_3 = \frac{\cos(2\tau)}{2}\int_0^T \sin^2(\omega t)\cdot dt = -\frac{\cos(2\tau)}{2\omega}\cos(\omega t)\Big|_0^T + \frac{\cos(2\tau)}{6\omega}\cos^3(\omega t)\Big|_0^T = 0, \tag{7.11.1.13.14j}$$

and

$$J_4 = \sin(2\tau)\int_0^T \sin^2(\omega t)\cdot\cos(\omega t)\cdot dt = \frac{\sin(2\tau)}{3\omega}\sin^3(\omega t)\Big|_0^T = 0. \tag{7.11.1.13.14k}$$

After substitution of A, B, and C in (7.11.1.13.14a) one gets:

$$R_{\mathrm{v,i}}^{\mathrm{L}} = 27\times 10^{-3}\left(\frac{33}{4} - \cos(\tau)\cdot\sqrt{\frac{1+p_x}{1+q_x}}\right)(1+p_x). \tag{7.11.1.13.15}$$

It is obvious that the maximum value of this expression will occur when $\tau = \pi$ so that it will hold:

$$R_{\mathrm{v,i}}^{\mathrm{L}} = 27\times 10^{-3}\left(\frac{33}{4} + \sqrt{\frac{1+p_x}{1+q_x}}\right)(1+p_x). \tag{7.11.1.13.16}$$

For the fault-free circuit one gets

$$R_{\mathrm{v,i}} = R_{\mathrm{v,i}}^{\mathrm{L}}\Big|_{\substack{p=0\\ q=0}} = \frac{999}{4}\times 10^{-3} \approx 0.25, \tag{7.11.1.13.17}$$

so that for the relative increment one gets

$$\Delta_{\mathrm{v,i}} = 1 - R_{\mathrm{v,i}}^{\mathrm{L}}/R_{\mathrm{v,i}} = 1 - 0.892\cdot(1+p_x)\cdot\left[1 + \frac{4}{33}\sqrt{\frac{1+p_x}{1+q_x}}\right] \approx 1 - 0.892\cdot(1+p_x). \tag{7.11.1.13.18}$$

If $p_1 = 0.2$ one gets $\Delta_{\mathrm{v,i}} = -0.0704$, and for $p_2 = -0.2$ one gets $\Delta_{\mathrm{v,i}} = 0.2864$.

Summarizing the results, we can easily conclude that among the three quantities we calculated (R_i, R_v, and $R_{v,i}$), R_v with $\tau = \pi$ should be chosen as the most suitable for testing the mentioned faults.

7.11.2 Designing for Testability

Problem 7.11.2.1

One of the design concepts for the testability of analog circuits is the oscillation method. It consists in creating the conditions in advance so that parts of the circuit can be isolated from the rest of the circuit with an external control signal and, at the same time, feedback is introduced to the isolated part of the system so that it behaves as an oscillator. This principle is illustrated in a simplified form in Fig. 7.11.14. DUT is the part of the circuit under test (DUT = device under test). In response to the operating mode signal's control, the switch transitions to a state aligning the output signal with the input, simultaneously disengaging the input from the excitation. In the case of a fault-free circuit, the oscillator produced in this way will oscillate at a known frequency. In the presence of a fault, three situations can arise: that the oscillation frequency changes, that the circuit does not oscillate, and that it oscillates at the nominal frequency. In the first two cases, it will be possible to recognize the fault effect(s), and in the third, the oscillation method is inapplicable.

For the (second-order) bandpass filter circuit of Fig. 7.11.15 it is known that $R_1 = R_2 = R_3 = 20\ \text{k}\Omega$ and $C_1 = C_2 = 20$ nF. First, determine the value k needed to maintain the oscillations that occur when the circuit is transformed into an oscillator and determine the oscillation frequency ω_0. Then determine the oscillation frequency (if oscillations are possible) if there are stack-at-short and stack-at-open faults at each resistive and capacitive branch individually except for the branches in the local feedback loop [branches R and $(k - 1)R$].

Discuss the obtained results from the point of view of diagnostics.

Compile a mapping table of element value increments into the change of oscillation frequency for increments of ±20%. Analyze the usability of the table from the point of view of testing and diagnostics.

Solution to Problem 7.11.2.1

The oscillator created in this way is shown in Fig. 7.11.16 where the node voltages required for the formulation of the circuit equations are marked. In order to determine

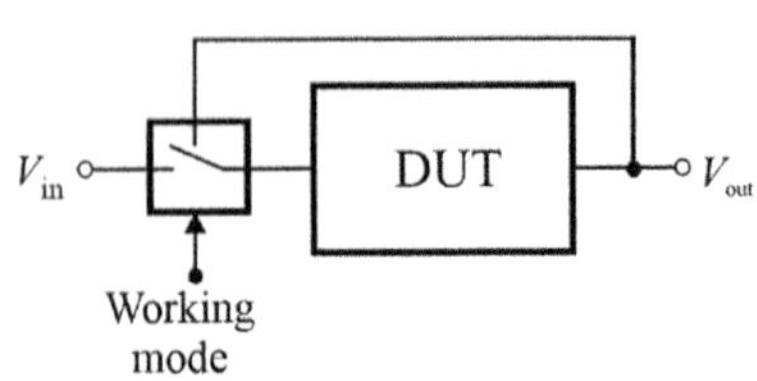

Fig. 7.11.14 Testing setup for oscillation-based testing

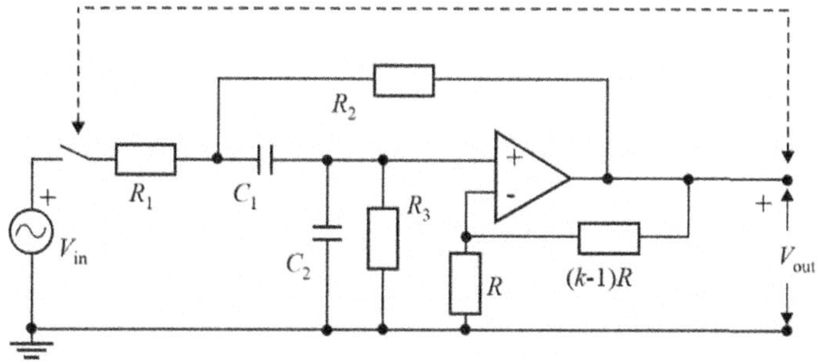

Fig. 7.11.15 Catastrophic faults and fault effects

the oscillation frequency and the oscillation condition, we first formulate the nodal equations. For the sake of simplicity, the abbreviation (which is easy to prove) $A = V_{out}/V_{in} = k$ will be introduced at the very beginning.

The nodal equations are:

$$\frac{1}{R_1}(V - A \cdot V_{in}) + \frac{1}{R_2}(V - A \cdot V_{in}) + sC_1(V - A \cdot V_{in}) = 0$$
$$-sC_1(V - A \cdot V_{in}) + \frac{1 + sC_2R_3}{R_3} \cdot V_{in} = 0. \qquad (7.11.2.1.1)$$

The system's determinant is

$$\Delta = \begin{vmatrix} \frac{1}{R_1} + \frac{1}{R_2} + sC_1 & -A \cdot \left(\frac{1}{R_1} + \frac{1}{R_2}\right) - sC_1 \\ -sC_1 & sC_1 + \frac{1+sC_2R_3}{R_3} \end{vmatrix} \qquad (7.11.2.1.2a)$$

or, in a developed form:

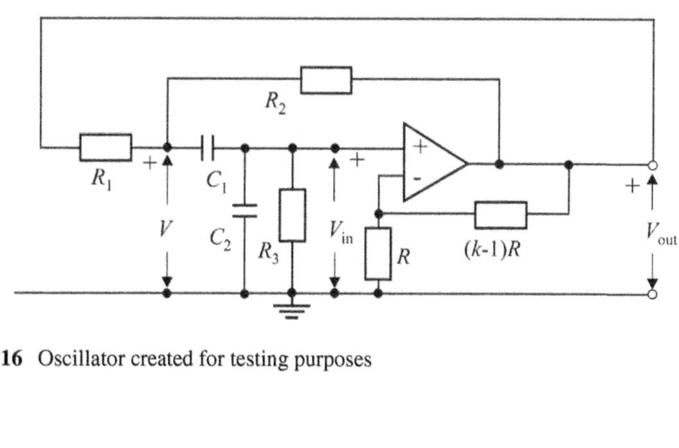

Fig. 7.11.16 Oscillator created for testing purposes

$$\Delta = \left(\frac{1}{R_1} + \frac{1}{R_2}\right) \cdot \frac{1}{R_3} + s \cdot \left[\left(\frac{1}{R_1} + \frac{1}{R_2}\right) \cdot (C_1(1 - A) + C_2) + \frac{C_1}{R_3}\right] + s^2 C_1 C_2 = 0. \tag{7.11.2.1.2b}$$

By equating to zero the real part of the determinant on the axis of real frequencies,

$$\text{Re}\{\Delta\}_{|s=j\omega} = 0, \tag{7.11.2.1.3}$$

one gets the oscillation frequency as

$$\omega_0 = \sqrt{\frac{R_1 + R_2}{C_1 C_2 R_1 R_2 R_3}}. \tag{7.11.2.1.4}$$

The necessary value of the gain is obtained by equating with zero the imaginary part of the determinant on the axis of real frequencies:

$$A_{\text{min}} = 1 + \frac{C_2}{C_1} + \frac{R_1 R_2}{(R_1 + R_2) R_3}. \tag{7.11.2.1.5}$$

By analyzing these expressions, we conclude that a very small number of catastrophic faults will create conditions for oscillations to occur in the faulty circuit as well. An overview of these faults is given in Table 7.11.11 together with the oscillation frequencies that occur when these faults are present.

All other catastrophic faults will cause the oscillation to stop which is just equally well from a testing point of view, since it can be considered as if a large frequency change has occurred: it was ω_0 and it has dropped to zero.

As can be seen from Table 7.11.11, only one of the obtained data can be used for diagnostic purposes. Namely, by measuring the frequency, if it turns out to be equal to 984.37 Hz, we can claim that the fault "C_2: open-circuited" is present. In all other cases, there are multiple faults that have the same effect, so they are unrecognizable from a diagnostic point of view.

Table 7.11.12 contains numerical values of the oscillation frequency in the oscillator if parametric faults were present in the circuit.

By analyzing Table 7.11.12 we conclude that the oscillation method can also be used for testing parametric faults. The frequency increment values range between 4.26 and 11.8%.

Table 7.11.11 Faults that lead to oscillations

No.	Type of fault	Oscillation frequency (Hz)
1	Fault-free	562.7
2	R_1: short-circuited	1447.6
3	R_2: short-circuited	1447.6
4	C_2: open-circuited	984.37

Table 7.11.12 Oscillation frequency and its variation for different parametric faults

No.	Type of fault	f_0 (Hz)	$\Delta = (\partial f_0/f_0) \cdot 100$
1	Fault-free	562.7	0
2	$1.2 \cdot R_1$	538.74	+4.26
3	$0.8 \cdot R_1$	596.83	−6.07
4	$1.2 \cdot R_2$	538.74	+4.26
5	$0.8 \cdot R_2$	596.83	−6.07
6	$1.2 \cdot R_3$	513.67	+8.71
7	$0.8 \cdot R_3$	629.12	−11.8
8	$1.2 \cdot C_1$	513.67	+8.71
9	$0.8 \cdot C_1$	629.12	−11.8
10	$1.2 \cdot C_2$	513.67	+8.71
11	$0.8 \cdot C_2$	629.12	−11.8

When comparing the application of this and some other method for testing faults in a given filter, it must not be forgotten that no external excitation signal is connected here, but only one control signal transforms the filter into an oscillator. In addition, only one measurement is performed: the frequency is measured.

Table 7.11.12 shows at the same time that, from the point of view of diagnostics, the oscillation method cannot provide reliable information based on which one may conclude which circuit element is faulty. Namely, several parametric faults have the same fault effect, that is, several increments in circuit element values lead to the same increments in the frequency of oscillations, which hides the component that is cause of the change of the frequency. ■

Problem 7.11.2.2

For the notch filter circuit of Fig. 7.11.17 it is known that $R_1 = R_2 = R_3 = 20\ \mathrm{k\Omega}$, $C_1 = C_2 = C_3$ 20 nF, and $k = 2$. First, determine the oscillation frequency ω_0. Then determine the oscillation frequency (if oscillations are possible) in presence of short-circuit and open-circuit faults for each resistive and capacitive branch individually, except for branches R and $(k-1)R$.

Discuss the obtained results from the point of view of diagnostics.

Compile a table mapping element value increments into changes of the oscillation frequency for soft faults (increments) of ±20%. Analyze the usability of the table from the point of view of testing and diagnostics.

Solution to Problem 7.11.2.2

The oscillator created in this way is shown in Fig. 7.11.18, where the node voltages required for the formulation of the circuit equations are marked. In order to determine the oscillation frequency, we first formulate the nodal equations. For the sake

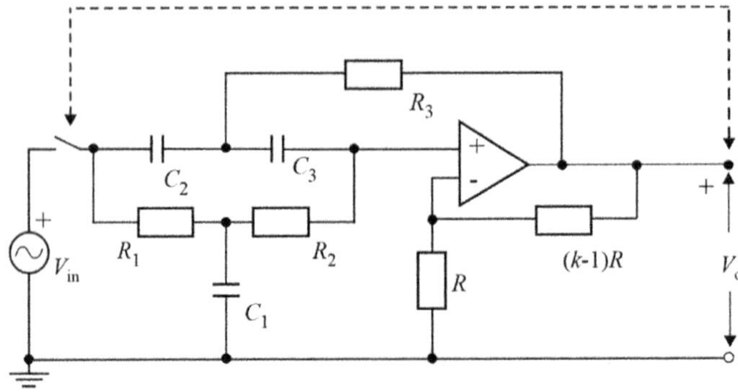

Fig. 7.11.17 Notch filter biquad

of simplicity, the abbreviation $A = V_{\text{out}}/V_{\text{in}} = k$ (which is easy to prove) will be introduced at the very beginning.

The nodal equations are

$$sC_2(V_1 - A \cdot V_{\text{in}}) + sC_3(V_1 - V_{\text{in}}) + \frac{1}{R_3}(V_1 - A \cdot V_{\text{in}}) = 0.$$
$$\frac{1}{R_1}(V_2 - A \cdot V_{\text{in}}) + \frac{1}{R_2}(V_2 - V_{\text{in}}) + sC_1V_2 = 0$$
$$sC_3(V_{\text{in}} - V_1) + \frac{1}{R_2}(V_{\text{in}} - V_2) = 0. \qquad (7.11.2.2.1)$$

The corresponding determinant is

$$\Delta = \begin{vmatrix} \frac{1}{R_3} + s(C_2 + C_3) & 0 & -\frac{A}{R_1} - s(A \cdot C_2 + C_3) \\ 0 & \frac{1}{R_1} + \frac{1}{R_2} + sC_1 & -\frac{A}{R_1} - \frac{1}{R_2} \\ -sC_3 & -\frac{1}{R_2} & \frac{1}{R_2} + sC_3 \end{vmatrix}. \qquad (7.11.2.2.2)$$

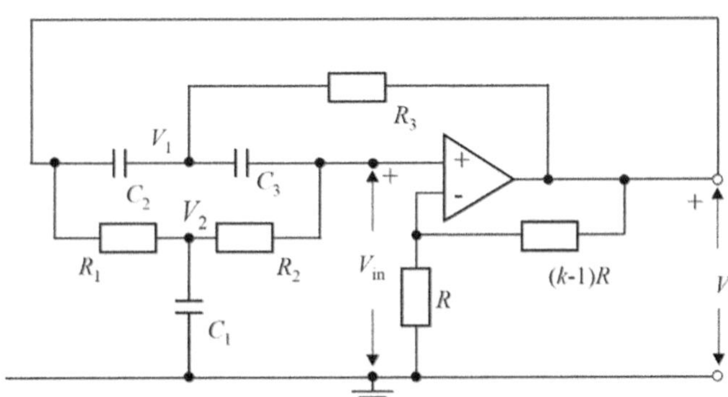

Fig. 7.11.18 Oscillator obtained from a notch filtering cell

Equating the real part of the determinant to zero gives

$$\omega_0^2 = \frac{C_3(R_1 + R_2) - \frac{1}{A-1}[C_1R_1 + (C_2 + C_3)R_3]}{C_1C_2C_3R_1R_2R_3}, \qquad (7.11.2.2.3a)$$

and equating the imaginary part to zero gives

$$\omega_0^2 = \frac{A-1}{(A-1)C_3R_3(C_1R_1 + C_2R_1 + C_2R_2) - C_1R_1R_2(C_2 + C_3)}. \qquad (7.11.2.2.3b)$$

Since the right-hand side of (7.11.2.2.3a) is negative for the given numerical values of the circuit parameters, it remains to calculate the oscillation frequency from (7.11.2.2.3b). Its nominal value is given in the first row of Table 7.11.13.

Care should be taken when considering the effects of catastrophic faults for this circuit as they change the configuration of the feedback circuit which has two feedback loops so that the overall feedback can become negative. That is why, in the following text, the expression (7.11.2.2.3a) will also be monitored, so as not to mistakenly conclude that the oscillations have disappeared.

Table 7.11.13 provides an overview of oscillation frequencies when certain catastrophic faults are present.

By analyzing Table 7.11.13, we conclude that all catastrophic faults except "R_1: short-circuit" and "R_2: open-circuit" can be tested by the oscillation method.

When it comes to diagnostics, we can easily conclude that two faults can be diagnosed: "C_1: open-circuit" and "C_3: short-circuit." These two faults will be recognized by the specific value of the oscillation frequency.

Table 7.11.13 Fault effects for catastrophic faults

No.	Fault type	Oscillation frequency (Hz)	Comment
1	Fault-free	397.89	
2	R_1: short-circuit	397.89	
3	R_2: short-circuit	281.35	
4	R_3: short-circuit	No oscillations	Valid for both expressions
5	R_1: open-circuit	No oscillations	Valid for both expressions (if $C_3 > C_1$ one would use (7.11.2.2.3a))
6	R_2: open-circuit	397.89	Obtained from (7.11.2.2.3a)
7	R_3: open-circuit	No oscillations	Valid for both expressions
8	C_1: open-circuit	281.35	
9	C_2: open-circuit	No oscillations	Valid for both expressions
10	C_3: open-circuit	No oscillations	Valid for both expressions
11	C_1: short-circuit	No oscillations	Valid for both expressions
12	C_2: short-circuit	No oscillations	Valid for both expressions
13	C_3: short-circuit	562.57	Obtained from (7.11.2.2.3a)

Table 7.11.14 Fault effects of parametric faults

No.	Type	f_0 (Hz)	$\Delta = (\partial f_0/f_0) \cdot 100$
1	Fault-free	397.89	0
2	$1.2 \cdot R_1$	397.89	0
3	$0.8 \cdot R_1$	397.89	0
4	$1.2 \cdot R_2$	444.85	−11.8
5	$0.8 \cdot R_2$	363.22	+8.71
6	$1.2 \cdot R_3$	314.56	17.2
7	$0.8 \cdot R_3$	629.12	−81.78
8	$1.2 \cdot C_1$	444.85	−11.8
9	$0.8 \cdot C_1$	363.22	+8.71
10	$1.2 \cdot C_2$	363.22	+8.71
11	$0.8 \cdot C_2$	444.85	−11.8
12	$1.2 \cdot C_3$	336.28	11.48
13	$0.8 \cdot C_3$	513.67	−35.22

To test for parametric faults, we generate a fault dictionary as the one depicted in Table 7.11.14.

By analyzing Table 7.11.14, we can easily conclude that the oscillation method can be successfully used to test parametric faults in this notch filter. In some cases, such as $\Delta R_3 = -0.2 \cdot R_3$ and $\Delta C_3 = -0.2 \cdot C_3$, the values of the increments in the oscillation frequency are so large that the fault is easily recognized during measurement. However, there is one exception. As can be seen from the second and third rows of Table 7.11.14, the parametric faults of R_1 cannot be tested by this method.

When it comes to diagnostics, the situation is less favorable. Namely, we have clearly recognizable oscillation frequency values, that is, oscillation frequency values that we can unequivocally associate with only one fault, only in the mentioned cases $\Delta R_3 = -0.2 \cdot R_3$ and $\Delta C_3 = -0.2 \cdot C_3$. If these frequencies are obtained by measurement, we can easily diagnose them. In all other cases of parametric faults, diagnosis is impossible or insufficiently reliable. ■

Problem 7.11.2.3

The phase-shift oscillator circuit from Fig. 7.11.19 is used to test the transistor, i.e., the size of its transconductance. How can it be established, by measuring the oscillation frequency, that there are catastrophic faults related to the transistor terminals, and how parametric faults in the RC feedback circuit can be detected? It is given that $R = 10$ kΩ, $C = 6.5$ nF, $R_D = 10$ kΩ, $R_S = 3$ kΩ, $R_{G1} = 100$ MΩ, $R_{G2} = 45.5$ MΩ, and $C_s \to \infty$.

Solution to Problem 7.11.2.3

From LNAE_Book 2, it is known that the oscillation frequency of the phase-shift oscillator is given by

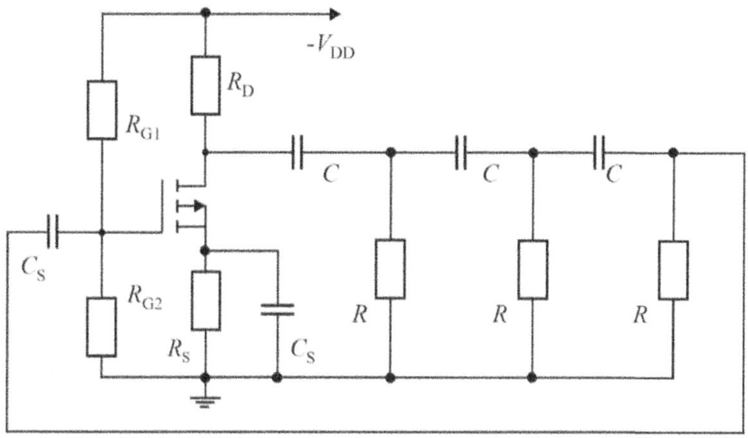

Fig. 7.11.19 Phase-shift oscillator using PMOS transistor

$$f_o = \frac{1}{2 \cdot \pi \sqrt{6} RC}, \tag{7.11.2.3.1a}$$

and the required amplifier gain value is $A = -g_m \cdot R_D = -27$.

That, however, was developed under the condition $R \gg R_D$ which is here not satisfied. Instead, for this case, one may use

$$f_o = \frac{1}{2 \cdot \pi \sqrt{12} RC}, \tag{7.11.2.3.1b}$$

For a fault-free circuit it would be $f_o \approx 706.8$ Hz, and the required transconductance of the transistor is $g_{m\_min} = 5.4$ mA/V.

Since the transistor does not affect the frequency of oscillation while only the frequency is measured, catastrophic faults related to the transistor will be detected only if the oscillations disappear. Possible catastrophic faults that can be identified in a discrete PMOS transistor are illustrated in Fig. 7.11.20a. They are: *G*-s-*D*, *G*-s-*S*, *D*-s-*S*, *G*-s-o, *S*-s-o, and *D*-s-o. Any of these faults, if present in the circuit, will cause the oscillations to cease. Therefore, the conclusion about the presence of one of them is easily established by the fact that there are no oscillations.

A breakdown between the gate and the channel is a very specific fault. One of the consequences of such a fault is the reduction of the transconductance of the transistor. This leads to a reduction in the gain of the basic amplifier and can lead to the cessation of oscillations. Another consequence is a dramatic decrease in the input resistance of the transistor. Namely, the breakdown results in the creation of a conductive path between the gate and the channel of the transistor, which is illustrated in Fig. 7.11.20b. If the value of the resulting resistance, R_{GS}, is of the order of resistance R or less, there will be a change in the frequency of oscillations. In this case, since the feedback circuit should provide a phase shift of 180°, the oscillations may even stop.

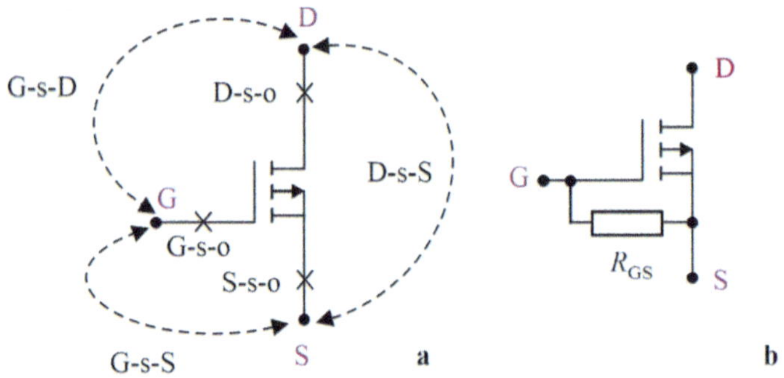

Fig. 7.11.20 **a** Catastrophic faults related to the PMOS transistor and **b** modeling of the gate breakdown

The situation is similar with the resistors in the feedback circuit. If the oscillation frequency deviates from the above, it means that some of the resistance values were incremented.

Changes in capacitance result in a change in frequency as long as the feedback circuit can provide an 180° phase shift.

Finally, any open or short circuit in the phase-shift circuit will cause the oscillations to stop.

The reader is recommended to perform a quantitative analysis of the effect of parametric faults on the version of this circuit shown in Fig. 7.11.21.

The numerical values of the circuit parameters are $R_1 = R_2 = R_3 = 10\ \text{k}\Omega$, $C_1 = C_2 = C_3 = 6.5$ nF, $R_D = 10\ \text{k}\Omega$, $R_S = 3\ \text{k}\Omega$, $R_{G1} = 100\ \text{M}\Omega$, $R_{G2} = 45.5\ \text{M}\Omega$, and $V_{DD} = 20$ V. The transistor parameters are $V_T = -1$ V i $A = 5\ \text{mA/V}^2$. $C_s \to \infty$. Change the values of the parameters of the phase-shift circuit by ±20%. In

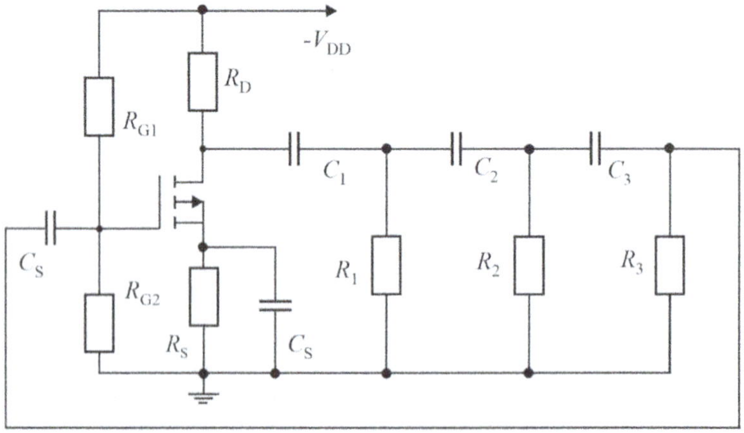

Fig. 7.11.21 Version of the phase-shift oscillator

particular, study the influence of parametric faults related to R_D and R_S that also change by ±20%. ■

7.11.3 Diagnosis

Problem 7.11.3.1

Using the data and results from Problems 7.11.1.1 and 7.11.1.2, compile a fault dictionary for the circuit shown in Fig. 7.11.1 considering both catastrophic and parametric faults and based on it discuss their diagnostics if only the output voltage V is measured.

Solution to Problem 7.11.3.1

To make it easier to follow the presentations, Figs. 7.11.22, 7.11.23 and 7.11.24 are shown here anew.

When looking at the catastrophic faults shown in Figs. 7.11.23 and 7.11.24, the part of the fault dictionary related to catastrophic faults is given in Table 7.11.2.

Here, one should be reminded that there are the following designations and equivalences of faults (Table 7.11.15):

1.1. *A*-s-o ↔ *B*-s-o ↔ *C*-s-o
1.2. *E*-s-o ↔ *G*-s-o
1.3. *F*-s-o ↔ *H*-s-o
1.4. *D*-s-o
and
2.1. *Y*-s-*Z* ↔ *X*-s-*Z*

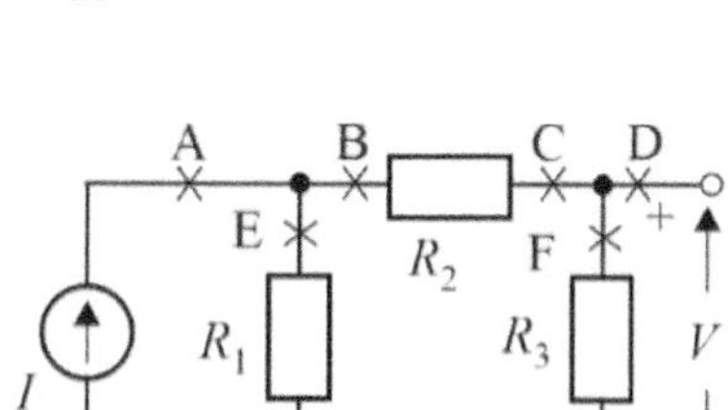

Fig. 7.11.22 Example resistive circuit

Fig. 7.11.23 Potential locations of stack-at-open faults

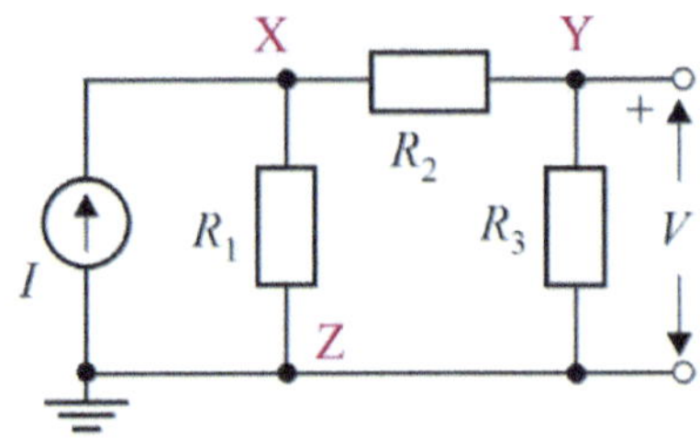

Fig. 7.11.24 Nodes potentially subject of stack-at-short faults

Table 7.11.15 Catastrophic faults and fault effects

Cipher	FF	1.1	1.2	1.3	1.4	2.1	2.1
Representative	–	A-s-o	E-s-o	F-s-o	D-s-o	Y-s-Z	X-s-Y
Effect (value of V [V])	1.143	0	8	2	Undefined	0	1.6

2.2. X-s-Y.

Considering these results, we easily conclude that among the catastrophic faults, if only they were possible in the circuit, we can diagnose only D-s-o and X-s-Y. The faults marked with 1.1, 1.2, 1.3, and 2.1 represent parts of the ambiguity groups, and the fault effect of D-s-o is such that it cannot be identified by measurement.

On the other hand, the dictionary of parametric faults can be taken in its entirety as given in Table 7.11.16 (repeated here).

Having all this in mind, a fault dictionary that contains only representatives of uncertainty groups including both catastrophic and parametric faults would be like the one shown in Table 7.11.17. Only minimal information related to faults is shown here, bearing in mind that the increments of all resistances are the same in absolute value, so there is no need to repeat their numerical values. R_{i+} indicates that R_i increases by 20%, and R_{i-} indicates that it decreases by the same amount.

By analyzing this table, we conclude that one catastrophic fault (X-s-Y) and all parametric ones will be diagnosed. In addition, the test can determine the presence of faults from one of the four ambiguity groups. ■

Table 7.11.16 Results of faulty circuit analysis (parametric faults)

Cipher	Variable	Increment (Ω)	$V(R_x + \Delta R_x)$
R_1	400	1.333	16.62
R_1	−400	0.941	17.67
R_2	800	1.081	5.424
R_2	−800	1.212	−6.036
R_3	1600	1.231	−7.699
R_3	−1600	1.032	9.711

Table 7.11.17 Complete fault dictionary

Cipher	Nature	V (V)
FF		1.143
1.1	*A*-s-o	0
1.2	*E*-s-o	8
1.3	*F*-s-o	2
2.1	*Y*-s-*Z*	0
2.1	*X*-s-*Y*	1.6
3.1	R_{1+}	1.333
3.2	R_{1-}	0.941
3.3	R_{2+}	1.081
3.4	R_{2-}	1.212
3.5	R_{3+}	1.231
3.6	R_{3-}	1.032

Problem 7.11.3.2

The nominal values of the elements of the circuit depicted in Fig. 7.11.25 are: $V_{DD} = 5$ V, $R_1 = 1$ kΩ, $R_2 = 2$ kΩ, $R_3 = 3$ kΩ, $R_4 = 4$ kΩ, $R_5 = 5$ kΩ, $R_6 = 6$ kΩ, $R_7 = 7$ kΩ, $R_8 = 8$ kΩ, $R_9 = 9$ kΩ, and $R_{10} = 10$ kΩ. Create a fault dictionary containing catastrophic faults. They are represented in the figure by a slash over the wire (representing open circuit latter denoted as OC) and by a dashed line connecting the terminals of the resistor (representing short circuit latter denoted as SC). Assume presence of a single fault in the circuit. If so, the overall number of faults will be $2 \times 10 + 2 = 22$. The last two are representing the catastrophic faults of V_{DD} and ground disconnected and should not be included in the fault dictionary since their fault effects are obvious.

The first version of this problem and its first solution were reported in [1].

Using the fault dictionary create an ANN that will perform diagnosis. The structure and the block diagram of the ANN are depicted in Fig. 7.11.26. Create the ANN by finding the number of hidden neurons (n), the weights, and the thresholds.

Verify the result by running the obtained ANN with erroneous "measured" input data the error being uniformly distributed statistical quantity with maximum deviation of 5%.

Solution to Problem 7.11.3.2

The fault dictionary obtained by repetitive simulation of the original circuit with faults inserted one at a time is depicted in Table 7.11.18.

A diagnostic task related to a simple resistive electrical circuit containing 10 resistors connected as shown in Fig. 7.11.25 is demonstrated. The element values are as follows: $V_{DD} = 5$ V, $R_1 = 1$ kΩ, $R_2 = 2$ kΩ, $R_3 = 3$ kΩ, $R_4 = 4$ kΩ, $R_5 = 5$ kΩ, $R_6 = 6$ kΩ, $R_7 = 7$ kΩ, $R_8 = 8$ kΩ, $R_9 = 9$ kΩ, and $R_{10} = 10$ kΩ. Catastrophic faults are modeled and diagnosed: short-circuited branches and open-circuited branches. To

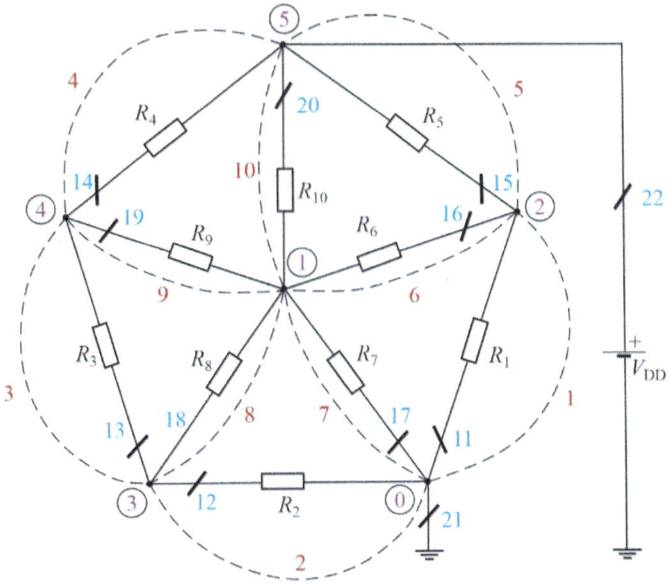

Fig. 7.11.25 Circuit to be diagnosed using ANNs

make the problem more difficult all possible faults of the above types were included in the fault dictionary, so the number of faults generated in this way is twice the number of elements in the circuit. In addition, the circuit is characterized by the fact that all output signals are affected by all faults making the diagnostic task very difficult.

The fault dictionary was generated by circuit analysis, so a code is assigned to every fault. It is supposed to be generated by the diagnostic system when the associated fault occurs in the original circuit. Having in mind the circuitry from Fig. 7.11.25, the diagnostic ANN has to have four inputs where the independent node voltages are to be used as input signals. The number of faults diagnosed imposes the number of outputs. Here we use conversion of the diagnostic signal into binary notation, so every output of the ANN represents a bit in the output word. Accordingly, we need five bits (outputs) to represent 20 possible faulty and the fault-free response.

The neural network used was a feed forward one. In the hidden layer its neurons have sigmoidal activation function:

$$\varphi(v) = \frac{1}{1 + \exp(-a \cdot v)}, \tag{7.11.3.2.1a}$$

where v is the input signal. In the output layer linear activation function was used (as explained in Chap. 7.12). Table 7.11.19 represents the responses of the ANN after training. Twelve neurons in the hidden layer were used ($n = 12$). To facilitate

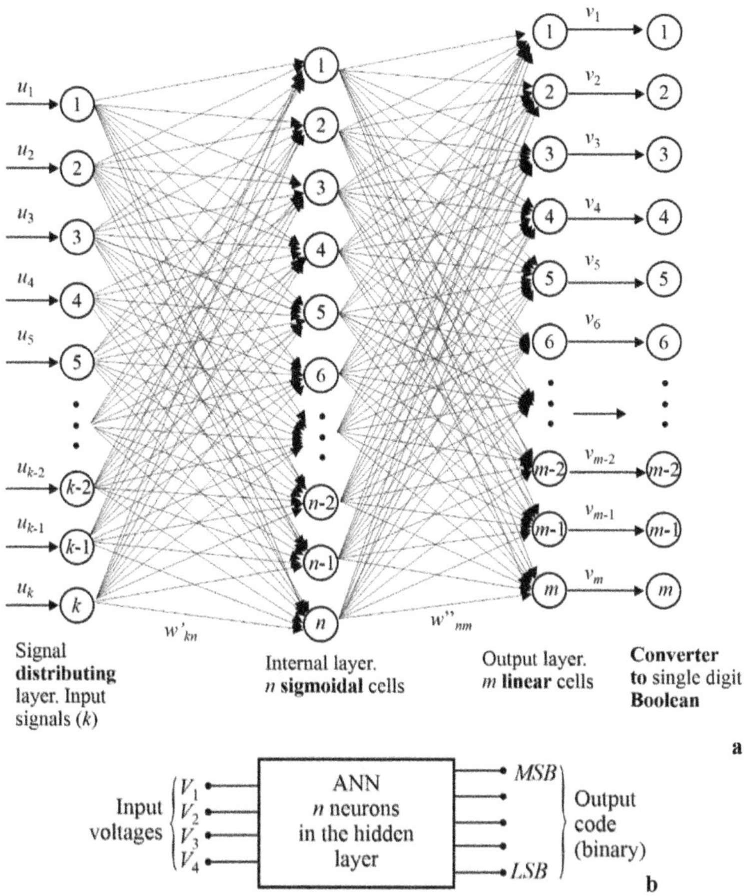

Fig. 7.11.26 Artificial neural network used for the solution of this problem. **a** General internal structure and **b** particular implementation (block schematic): $k = 4$, and $m = 5$

the evaluation of the quality of the responses the expected binary value is given in a separate column named "Code".

Excellent agreement between the obtained and expected responses is demonstrated in Table 7.11.19. Namely, we consider that every response smaller than 0.5 will be approximated by 0, and every response larger than 0.5 will approximated by 1. By inspection of this table we find that there are no responses with a value near to 0.5 which justifies the claim of excellent approximation (Table 7.11.20).

Tables 7.11.21 and 7.11.22 contains the weights and the thresholds of the ANN obtained after training.

To verify the results, the generalization property of the ANN was checked in the following way. In place of the circuit responses given in Table 7.11.18, "measured" values were brought to the input of the trained ANN. They were calculated according to the following formula:

Table 7.11.18 Fault dictionary for the circuit of Fig. 7.11.25

No.	Cipher	V_1	V_2	V_3	V_4	Code
0	FF	1.6839669	0.93706906	1.1277336	2.6107466	00000
1	SC_R_1	1.418367	0	1.0684034	2.5397723	00001
2	SC_R_2	1.6839669	0.93706906	1.1277336	2.6107466	00010
3	SC_R_3	1.6197006	0.9292317	1.6554211	1.6554211	00011
4	SC_R_4	2.2904527	1.0110307	2.037885	5	00100
5	SC_R_5	2.8355507	5	1.3849768	2.9184768	00101
6	SC_R_6	1.1311712	1.1311712	1.004249	2.4630268	00110
7	SC_R_7	0	0.73170725	0.75156569	2.1607513	00111
8	SC_R_8	1.3552735	0.89698452	1.3552735	2.6673748	01000
9	SC_R_9	2.131264	0.9916175	1.0193002	2.131264	01001
10	SC_R_{10}	5	1.3414633	1.8684758	3.4968681	01010
11	OC_R_1	2.5154545	3.8706609	1.3134731	2.8329396	01011
12	OC_R_2	3.3966578	4.2712078	4.0415773	4.2834222	01100
13	OC_R_3	1.7780038	0.94853699	0.3556007	4.0086162	01101
14	OC_R_4	1.1597996	0.87314624	0.34111753	545.78805	01110
15	OC_R_5	1.4782217	0.21117453	1.0817739	2.5557668	01111
16	OC_R_6	1.9794024	0.83333325	1.1937287	2.6896939	10000
17	OC_R_7	2.2502215	1.0061245	1.2542248	2.7620633	10001
18	OC_R_8	1.8178177	0.95339233	1.0350746	2.5876865	10010
19	OC_R_9	1.4948478	0.91400576	1.1735798	2.8134739	10011
20	OC_R_{10}	0.97495151	0.8506037	0.96935236	2.4212812	10100

$$\widehat{V} = V \cdot (1 + \delta), \tag{7.11.3.2.2}$$

where δ is a random number with uniform distribution in the interval $(-0.05, 0.05)$. Twenty groups of four random numbers were created and every member of the group was associated to a member of the corresponding row in Table 7.11.18.

To summarize, an ANN was successfully synthesized (with $k = 4$, $n = 12$, and $m = 5$) and trained to distinguish among the set of fault-free and faulty signals. Its robustness was tested by "measured" signals to which noise was added in a form of random numbers with amplitude of $\pm 5\%$ of the noise-free measured signal. As can be seen from Table 7.11.23 no miss can be observed. The most distant output values were for $\widehat{\mathrm{Out}}_1$ at FF, No. 16, and No. 20. These were, of course, easily classified.

Problem 7.11.3.3

Figure 7.11.27 depicts a two-stage amplifier. The first stage is a CS low noise amplifier using GaAs MESFET (T_1) while the second is a Class AB large signal amplifier using a GaN HEMT (T_2). The amplifier is a selective one behaving as a bandpass filter with a central frequency at $f_0 = 100$ MHz.

Table 7.11.19 ANN responses after training with noiseless signals

No.	Cipher	Out_1	Out_2	Out_3	Out_4	Out_5	Code
0	FF	−0.03452	0.064751	0.065323	0.010136	−0.025383	00000
1	SC_R_1	−0.00134	−0.027034	0.030538	−0.027342	1.01085	00001
2	SC_R_2	0.013668	0.023298	0.025062	1.00753	−0.017614	00010
3	SC_R_3	0.001038	0.013097	0.018179	1.01415	0.993868	00011
4	SC_R_4	−0.000933	0.015128	1.01269	−0.001061	−0.002895	00100
5	SC_R_5	0.005633	−0.039600	0.965766	0.008531	1.03518	00101
6	SC_R_6	0.016796	−0.042575	0.970832	0.999743	0.035932	00110
7	SC_R_7	0.000978	0.0059	0.990013	0.998441	1.00052	00111
8	SC_R_8	0.007604	0.946343	−0.023795	−0.007944	−0.005149	01000
9	SC_R_9	0.017858	1.03558	0.016499	−0.004687	1.00393	01001
10	SC_R_{10}	−0.002764	0.976286	0.047336	0.951114	0.026378	01010
11	OC_R_1	−0.013004	1.00117	−0.006362	1.01032	0.983848	01011
12	OC_R_2	−0.003832	1.01145	1.01578	0.008870	0.002211	01100
13	OC_R_3	−0.000564	1.01205	1.00523	0.005856	0.998328	01101
14	OC_R_4	−0.016972	1.00455	0.991052	1.01441	0.009889	01110
15	OC_R_5	0.000749	0.984323	1.0113	1.00015	0.997952	01111
16	OC_R_6	1.03963	−0.095547	−0.074141	−0.018981	0.067648	10000
17	OC_R_7	0.974028	0.019003	0.032691	0.022437	0.977526	10001
18	OC_R_8	0.994527	0.017434	0.028039	0.991792	−0.051093	10010
19	OC_R_9	1.00878	0.028469	0.015837	1.00453	0.9988	10011
20	OC_R_{10}	0.979762	0.049192	1.04671	0.016228	−0.025500	10100

Here is the list of the circuit element values: $V_{DD} = 12$ V, $V_{GG1} = 0.7$ V, $V_{GG2} = 2.1$ V, $R_{source} = 50\ \Omega$, $C_{source} = 0.1$ pF, $C_{S1} = 1$ pF, $L_{S11} = 0.1\ \mu$H, $L_g = 1$ nH, $L_{D1} = 57$ nH, $L_{Sa} = 25$ nH, $C_{CS2} = 100$ pF, $L_{con} = 0.05$ nH, $L_{S22} = 1$ mH, $L_{D2} = 10\ \mu$H, $C_{S3} = 100$ pF, $R_D = 50\ \Omega$, $L_D = 4.85$ nH, and $C_D = 450$ pF.

Use the following model parameters for the LTSpice model of the MESFET:

```
.model nmesfet nmf
+ beta = 80m vto = −1
+ cgd = 50f cgs = 300f
+ rd = 0.62 rs = 1.38
```

and the model for EPC8009 for the GaN HEMT.

The biasing used brings the quiescent operating points of T_1 and T_2 into positions as depicted in Fig. 7.11.28a, b, respectively.

Table 7.11.20 Perturbed (noisy) input signals used to validate the generalization

No.	Cipher	$\widehat{V}_1$	$\widehat{V}_2$	$\widehat{V}_3$	$\widehat{V}_4$	Code
0	FF	1.697944	0.903241	1.162017	2.66009	00000
1	SC_R_1	1.464464	0	1.0871	2.612664	00001
2	SC_R_2	1.331423	0.856079	0	1.952547	00010
3	SC_R_3	1.55637	0.943449	1.615857	1.732729	00011
4	SC_R_4	2.278084	0.974937	1.939251	4.839	00100
5	SC_R_5	2.849728	5.061	1.373481	2.905636	00101
6	SC_R_6	1.097802	1.157754	1.050043	2.473125	00110
7	SC_R_7	0	0.713049	0.725336	2.159023	00111
8	SC_R_8	1.406774	0.889001	1.412059	2.655905	01000
9	SC_R_9	2.070097	1.040703	0.99341	2.184546	01001
10	SC_R_{10}	4.7855	1.34428	1.786263	3.56226	01010
11	OC_R_1	2.606765	3.812988	1.359445	2.940591	01011
12	OC_R_2	3.462213	4.132394	4.10503	4.143783	01100
13	OC_R_3	1.773203	0.989514	0.369149	4.003405	01101
14	OC_R_4	1.109116	0.889474	0.332248	0.538966	01110
15	OC_R_5	1.472604	0.205726	1.099299	2.558578	01111
16	OC_R_6	1.891517	0.8715	1.210799	2.685121	10000
17	OC_R_7	2.171239	1.050595	1.277303	2.849621	10001
18	OC_R_8	1.860173	0.9512	0.992637	2.628572	10010
19	OC_R_9	1.520858	0.94362	1.157384	2.908288	10011
20	OC_R_{10}	0.988113	0.877908	0.964409	2.493193	10100

The amplifier was synthesized based on the methods described in [2, 3].
The following set of faults should be considered:

- Soft faults related to all passive elements and batteries excluding the elements of the excitation circuit and the power supply battery.
- Short-circuited all passive elements and all combinations of transistor terminals (excluding the excitation circuit and the power supply battery).
- Open-circuited all passive elements and all transistor terminals (excluding the excitation circuit and the power supply battery).

For the schematic of Fig. 7.11.27 there will be 72 such faults.

Create a fault dictionary based on simulation of every faulty circuit and performed reduction using proper representation of the ambiguity sets. For every fault create a signature containing the following quantities: output voltage amplitude (Amp), output voltage phase (Phase), output voltage THD (THD_v), supply current DC component (I_{DDQ}), and supply current THD (THD_i).

Code the faults randomly with uniform distribution and represent the codes as binary numbers.

Table 7.11.21 Weights of the ANN

Weights between the input and the hidden layers		Weights between hidden layer and the output layer	
$w(1, 1)(2, 1) = -2.76418$ $w(1, 2)(2, 1) = 18.648$	$w(1, 3)(2, 1) = 12.4319$ $w(1, 4)(2, 1) = -59.2299$	$w(2, 1)(3, 1) = -12.5775$ $w(2, 2)(3, 1) = 650.041$	$w(2, 3)(3, 1) = -133.761$ $w(2, 4)(3, 1) = -109.583$
$w(1, 1)(2, 2) = -46.2319$ $w(1, 2)(2, 2) = 245.899$	$w(1, 3)(2, 2) = 12.0754$ $w(1, 4)(2, 2) = -62.0151$	$w(2, 5)(3, 1) = 111.678$ $w(2, 6)(3, 1) = 4.75673$	$w(2, 7)(3, 1) = 351.461$ $w(2, 8)(3, 1) = -6.47534$
$w(1, 1)(2, 3) = -9.1737$ $w(1, 2)(2, 3) = 0.616269$	$w(1, 3)(2, 3) = 10.0854$ $w(1, 4)(2, 3) = 1.28262$	$w(2, 9)(3, 1) = 67.1201$ $w(2, 10)(3, 1) = -403.131$	$w(2, 11)(3, 1) = 122.246$ $w(2, 12)(3, 1) = -652.777$
$w(1, 1)(2, 4) = 11.9616$ $w(1, 2)(2, 4) = 30.8217$	$w(1, 3)(2, 4) = -26.8202$ $w(1, 4)(2, 4) = -17.3511$	$w(2, 1)(3, 2) = 54.154$ $w(2, 2)(3, 2) = 1083.65$	$w(2, 3)(3, 2) = -102.602$ $w(2, 4)(3, 2) = -149.017$
$w(1, 1)(2, 5) = 11.2978$ $w(1, 2)(2, 5) = 29.53$	$w(1, 3)(2, 5) = -26.0049$ $w(1, 4)(2, 5) = -17.5981$	$w(2, 5)(3, 2) = 146.783$ $w(2, 6)(3, 2) = -53.0022$	$w(2, 7)(3, 2) = 2838.12$ $w(2, 8)(3, 2) = -155.895$
$w(1, 1)(2, 6) = 279.601$ $w(1, 2)(2, 6) = 13.7036$	$w(1, 3)(2, 6) = 697.841$ $w(1, 4)(2, 6) = -1641.72$	$w(2, 9)(3, 2) = 1033.13$ $w(2, 10)(3, 2) = -527.106$	$w(2, 11)(3, 2) = 87.1595$ $w(2, 12)(3, 2) = -1089.43$
$w(1, 1)(2, 7) = -196.943$ $w(1, 2)(2, 7) = 95.9205$	$w(1, 3)(2, 7) = -514.43$ $w(1, 4)(2, 7) = -970.926$	$w(2, 1)(3, 3) = -57.0926$ $w(2, 2)(3, 3) = -1381.51$	$w(2, 3)(3, 3) = 96.1202$ $w(2, 4)(3, 3) = 225.987$
$w(1, 1)(2, 8) = -70.4899$ $w(1, 2)(2, 8) = 124.528$	$w(1, 3)(2, 8) = 67.5068$ $w(1, 4)(2, 8) = -6.64319$	$w(2, 5)(3, 3) = -225.961$ $w(2, 6)(3, 3) = 58.5109$	$w(2, 7)(3, 3) = -1566.96$ $w(2, 8)(3, 3) = 197.124$
$w(1, 1)(2, 9) = -126.553$ $w(1, 2)(2, 9) = 347.951$	$w(1, 3)(2, 9) = -177.951$ $w(1, 4)(2, 9) = 150.782$	$w(2, 9)(3, 3) = -1289.73$ $w(2, 10)(3, 3) = 488.826$	$w(2, 11)(3, 3) = -79.8134$ $w(2, 12)(3, 3) = 1385.79$
$w(1, 1)(2, 10) = -480.095$ $w(1, 2)(2, 10) = 450.228$	$w(1, 3)(2, 10) = -604.569$ $w(1, 4)(2, 10) = -104.915$	$w(2, 1)(3, 4) = 40.5465$ $w(2, 2)(3, 4) = 1528.44$	$w(2, 3)(3, 4) = -140.515$ $w(2, 4)(3, 4) = -306.895$
$w(1, 1)(2, 11) = -10.4852$ $w(1, 2)(2, 11) = 1.07533$	$w(1, 3)(2, 11) = 10.4431$ $w(1, 4)(2, 11) = 1.03003$	$w(2, 5)(3, 4) = 314.31$ $w(2, 6)(3, 4) = -48.9528$	$w(2, 7)(3, 4) = 2736.01$ $w(2, 8)(3, 4) = -244.722$

(continued)

Table 7.11.21 (continued)

Weights between the input and the hidden layers		Weights between hidden layer and the output layer	
w(1, 1)(2, 12) = 66.0218 w(1, 2)(2, 12) = 205.877	w(1, 3)(2, 12) = −194.7 w(1, 4)(2, 12) = −103.535	w(2, 9)(3, 4) = 1611.98 w(2, 10)(3, 4) = −572.845	w(2, 11)(3, 4) = 127.907 w(2, 12)(3, 4) = −1532.4
		w(2, 1)(3, 5) = 30.5726 w(2, 2)(3, 5) = −674.353	w(2, 3)(3, 5) = −67.8161 w(2, 4)(3, 5) = 59.6165
		w(2, 5)(3, 5) = −64.3659 w(2, 6)(3, 5) = −27.8249	w(2, 7)(3, 5) = −1504.69 w(2, 8)(3, 5) = 121.274
		w(2, 9)(3, 5) = −791.488 w(2, 10)(3, 5) = −127.589	w(2, 11)(3, 5) = 60.5692 w(2, 12)(3, 5) = 667.501

Table 7.11.22 Thresholds of the ANN

Thresholds of the hidden neurons	Thresholds of the output neurons
theta(2, 1) = 21.2018	theta(3, 1) = −45.2249
theta(2, 2) = 21.9813	theta(3, 2) = −863.104
theta(2, 3) = −0.801217	theta(3, 3) = 1080.67
theta(2, 4) = 4.79756	theta(3, 4) = −1354.47
theta(2, 5) = 5.20862	theta(3, 5) = 683.576
theta(2, 6) = 469.329	
theta(2, 7) = 65.5243	
theta(2, 8) = 3.31506	
theta(2, 9) = 17.3652	
theta(2, 10) = 79.4694	
theta(2, 11) = −0.649413	
theta(2, 12) = 65.7562	

Create an ANN with 5 inputs and 7 outputs and a single hidden layer. It should be trained to accept the signatures for every fault as input and to learn the binary code so that every binary figure is associated to a single output terminal.

Train the ANN for various numbers of hidden neurons until best performance is reached.

Insert noise in the signatures and check the generalization property of the ANN so obtained.

Table 7.11.23 ANN responses to noisy signals

No.	Cipher	Out_1	Out_2	Out_3	Out_4	Out_5	Code
0	FF	0.477204	0.226303	0.246503	−0.007106	−0.10325	00000
1	SC_R_1	0.027803	0.018728	0.017649	0.002149	0.990238	00001
2	SC_R_2	0.109252	0.059412	0.054502	1.00085	−0.03926	00010
3	SC_R_3	−0.040758	−0.018866	−0.008910	1.00488	1.00273	00011
4	SC_R_4	−0.022678	0.003997	0.990592	−0.006679	0.005713	00100
5	SC_R_5	−0.001330	−0.001610	1.00189	0.005155	0.999831	00101
6	SC_R_6	0.091701	0.123207	0.9686	0.996738	−0.03991	00110
7	SC_R_7	0.063603	0.058364	1.01134	1.00321	0.972972	00111
8	SC_R_8	−0.086567	0.930572	−0.024035	0.003278	0.029575	01000
9	SC_R_9	−0.054049	0.948573	0.003266	−0.005873	1.03568	01001
10	SC_R_{10}	−0.056141	0.95309	−0.007465	0.99143	0.025770	01010
11	OC_R_1	−0.066707	0.955813	−0.033354	0.992657	1.02223	01011
12	OC_R_2	0.006070	1.00641	1.00248	0.003306	−0.00266	01100
13	OC_R_3	0.030508	1.05198	0.968742	0.040935	0.946518	01101
14	OC_R_4	−0.001143	0.999772	1.0003	1.00284	0.000054	01110
15	OC_R_5	−0.043673	0.945483	1.00606	0.978312	1.03961	01111
16	OC_R_6	0.834451	−0.038746	−0.116668	0.012697	−0.01778	10000
17	OC_R_7	1.0919	0.137313	−0.029459	−0.034003	0.981191	10001
18	OC_R_8	0.831157	−0.071293	−0.094656	1.01668	0.061815	10010
19	OC_R_9	1.10143	−0.000089	0.114098	1.01403	0.960284	10011
20	OC_R_{10}	0.683893	−0.284004	0.941717	−0.009732	0.131686	10100

Solution to Problem 7.11.3.3

We first perform circuit analysis of the fault-free amplifier given in order to establish its properties in the frequency domain. That was done by LTSpice circuit simulation. The results are shown in the following set of figures.

First, Fig. 7.11.29 depicts the frequency dependence of the power gain of the amplifier. It is obtained as: $A_P = V_{out\_all} \cdot i_{RD}/(v_{input1} \cdot i_{Iexcitation1})$ where i_{RD} is the current of the load resistor R_D. Enormous value of the gain is obtained which confirms the appropriateness of both the design method and the architecture chosen.

Similar conclusion may be drawn from Fig. 7.11.30, where the voltage gain is depicted.

Finally, Fig. 7.11.31 depicts the noise response of the amplifier.

For the bandwidth $B = (105 - 95)$ MHz, the total output rms noise voltage is 4.039 mV and the total rms noise voltage of the source resistor seen at the output

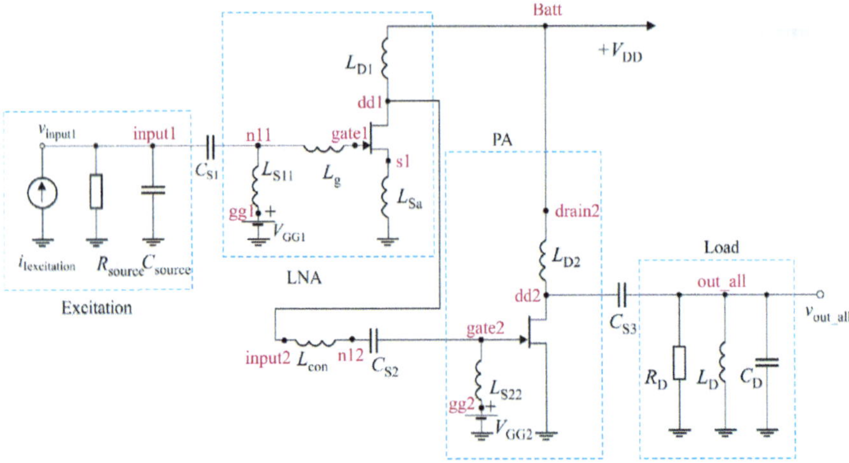

Fig. 7.11.27 Two-stage power RF amplifier tuned at 100 MHz

Fig. 7.11.28 Transfer characteristics: **a** of the GaAs MESFET and **b** of the GaN HEMT used

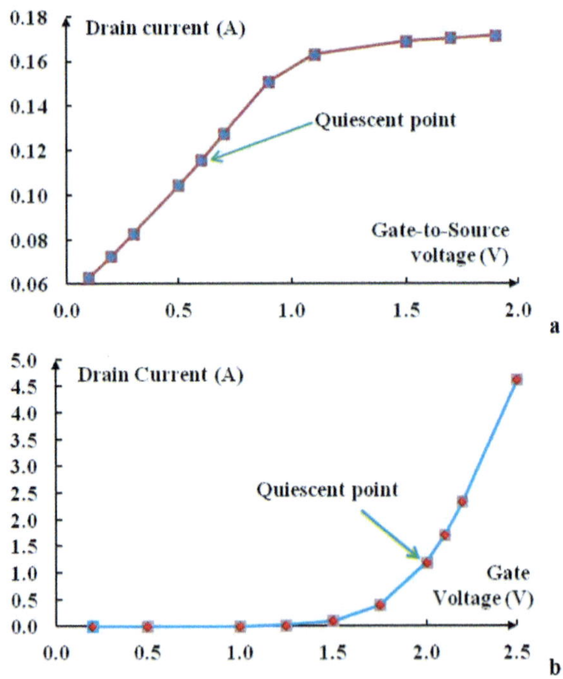

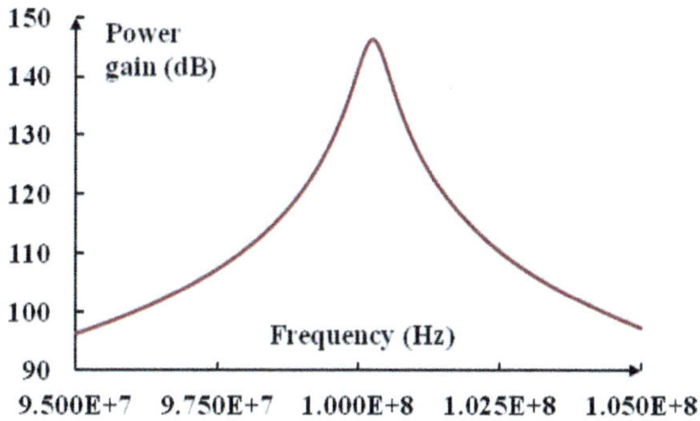

Fig. 7.11.29 Power gain obtained by AC analysis

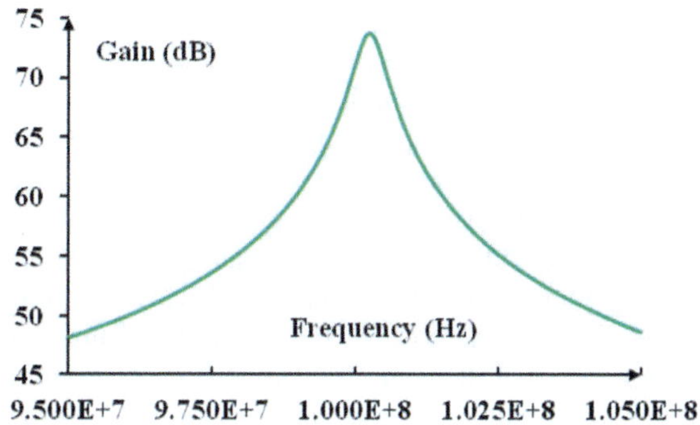

Fig. 7.11.30 Frequency dependence of the voltage gain

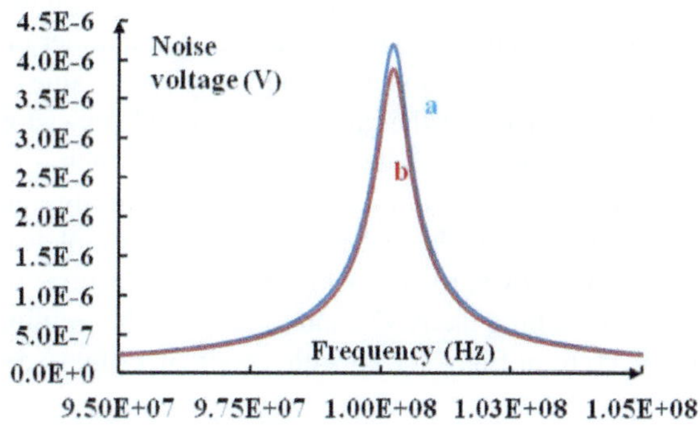

Fig. 7.11.31 Spectral amplitude of the **a** total output rms noise voltage and **b** rms noise voltage of the source resistor seen at the output

is 3.7507 mV (the spectral amplitude of a 50 Ω resistor is 0.92 nV/Hz$^{1/2}$ and its amplitude for the given frequency range—for a flat amplitude characteristic—is 2.9 μV). Based on this one may conclude that the contribution of the two-stage amplifier of Fig. 7.11.27 to the overall noise at the output is by serious margin less than the noise coming at the input. That, once more, confirms the design method and the schematic chosen.

Table 7.11.24 Ambiguity groups

Fault type	No.	V_{out_all}			I_{DD}	
		Amp (V)	Phase (°)	THD_v (%)	I_{DDQ} (A)	THD_i (%)
LD_CS	1	0	0	0	1.83199	0.715734
CD_CS		0	0	0	1.83199	0.715734
RD_CS		0	0	0	1.83199	0.715734
VGG2_OC	2	1.560e−03	−107.69	0.165106	0.116952	0.229492
LS22_OC		1.560e−03	−107.69	0.165106	0.116952	0.229492
CS2_OC	3	1.544e−10	70.35	10.967007	1.83237	0.937526
Lcon_OC		1.544e−10	70.35	10.967007	1.83237	0.937526

For diagnostic purposes time-domain analysis was performed. The circuit was excited by a 50 μA amplitude source at 100 MHz. After performing 73 simulations for the fault-free and the faulty circuits one gets a table containing circuit responses. Here for the purpose of diagnosis two test points were selected being most convenient for measurement while introducing minimum disturbance in the circuit's functionality. These are the output voltage v_{out_all} and the power supply current (I_{DD}).

The following quantities were selected to create a signature per fault: the amplitude of the output voltage (Amp), the phase of the output voltage (Phase), the THD of the output voltage THD_v, the DC component of the supply current I_{DDQ}, and the THD of the power supply current THD_i.

After selection of the ambiguity groups (as depicted in Table 7.11.24) the fault dictionary was created excerpts of which are given in Table 7.11.25. The numerical values in this table are shown with reduced number of figures for convenience. One is to pay attention to some exceptionally large values of I_{DDQ} signaling the fact that the fault is bringing the gate of the second transistor to a high-voltage value and forcing the transistor to behave as a closed switch.

Given the complete information on the circuit and the reduced fault dictionary from Table 7.11.25 the reader is expected to complete the fault dictionary by creating the missing rows, to select the ANN architecture, to install an open-source software for training the ANN, and to perform the training. To facilitate we suggest the architecture of Fig. 7.11.26 with a code of 7 binary digits.

Table 7.11.25 Part of the fault dictionary and codes assigned to the faults (here the number of significant figures is reduced for convenience)

Fault type	No.	V_{out_all}			I_{DD}		Binary code of the fault						
		Amp (V)	Phase (°)	THD_v (%)	I_{DDQ} (A)	THD_i (%)							
FF	1	3.702	103.2	0.382	1.794	0.974	0	1	0	0	0	0	0
$L_{Sa} \cdot 1.2$	2	3.054	88.13	0.404	1.822	1.041	0	1	1	0	1	0	1
$L_{Sa} \cdot 0.8$	3	4.635	127.3	0.444	1.828	1.793	0	0	0	1	0	1	1
$L_{D1} \cdot 1.2$	4	1.279	91.79	0.246	1.707	6.206	0	1	0	1	1	0	0
$L_{D1} \cdot 0.8$	5	2.304	−104.9	0.368	1.909	1.012	0	0	1	0	0	1	1
$V_{GG1} \cdot 1.2$	6	4.353	146.8	0.424	1.767	1.702	0	0	0	0	1	1	0
$V_{GG1} \cdot 0.8$	7	3.798	103.7	0.449	1.775	1.252	0	1	0	0	0	1	1
		...			...								
$L_{con} \cdot 0.8$	23	3.464	101.6	0.371	1.863	1.461	0	0	1	1	1	0	1
$C_{s3} \cdot 1.2$	24	3.330	−22.61	2.063	2.016	5.680	1	0	0	0	0	1	1
$C_{s3} \cdot 0.8$	25	2.655	168.1	0.154	1.694	2.521	0	1	0	1	1	0	1
$L_D \cdot 1.2$	26	0.4317	74.30	0.406	1.737	4.660	0	0	0	0	0	1	0
$L_D \cdot 0.8$	27	0.3484	−112.7	0.280	1.816	1.739	0	1	0	1	0	0	1
$C_D \cdot 1.2$	28	0.4032	74.18	0.384	1.727	22.49	0	0	1	1	1	0	0
$C_D \cdot 0.8$	29	0.5116	−113.9	0.233	1.842	0.591	0	0	1	1	1	1	0
$R_D \cdot 1.2$	30	11.01	111.4	0.759	1.727	4.035	0	1	0	0	1	0	1
$R_D \cdot 0.8$	31	2.770	133.7	0.318	1.810	1.870	0	0	1	1	0	1	0
L_{D1}_SC	32	5.5e−12	−71.68	11.31	1.832	0.935	0	1	1	0	0	1	0
L_{Sa}_SC	33	2.378	149.4	3.242	1.675	10.93	0	1	0	0	1	1	1
		...			...								
C_{S3}_CS	42	1.050	−36.79	47.85	6.00e3	9.816	0	0	0	0	1	0	1
L_D_CS	43	0	0	0	1.831	0.7157	0	0	1	0	1	0	1
T_1DG_SC	44	1.7e−03	−139.1	0.181	3.80e3	68.7e3	0	0	1	1	1	1	1
T_1DS_SC	45	1.0e−04	−131.0	0.202	6.00e3	6.12e3	0	0	0	0	0	0	1
		...			...								
C_{S2}_OC	58	1.5e−10	70.35	10.96	1.832	0.937	0	0	1	1	0	0	0
C_{S3}_OC	59	0	0	0	1.803	3.518	0	1	1	1	1	0	1
L_D_OC	60	7.8e−02	70.24	1.130	1.813	1.127	1	0	0	0	0	0	1
		...			...								
T_1G_OC	65	4.4e−8	−102.03	0.697	1.798	7.22e3	0	1	1	1	0	0	1
T_2D_OC	66	1.4e−14	179.30	60.68	0.115	0.843	0	1	0	1	0	0	0
T_2S_OC	67	7.4e−14	−164.26	7.811	0.116	0.625	1	0	0	0	1	0	0
T_2G_OC	68	1.5e−3	175.41	0.207	0.115	0.489	0	0	1	0	0	0	1

References

1. Andrejević M, Litovski V (2004) ANN application in electronic diagnosis—preliminary results. In: Proceedings of IEEE 24th international conference on microelectronics (MIEL2004), Niš, Serbia, pp 597–600
2. Litovski V (2023) LNAE, discrete and integrated large signal amplifiers. Springer Nature, Berlin
3. Litovski V (2024) LNAE, noise in electronic circuits and low noise amplifier design. Springer Nature, Berlin

7.12 A Short Review of Feed-Forward ANNs

Abstract One of the algorithms extensively used in diagnosis of electronic circuits and systems is the artificial neural networks (ANNs). These were proven on many occasions as a powerful tool to recognizing (separate) among the faulty responses of the system obtained by simulation the one which is the most similar to the measured. In this chapter, the very basics of analog feed-forward ANNs will be given with the goal to help the reader to start implementing the concept by her/him using already available open-source software solutions.

7.12.1 Introduction

We are all acquainted with the classical approximation theory based on algebraic polynomials. Examples of that are the Newton's interpolation formula (point matching) and the Chebyshev approximant. When non-monotonic phenomena have to be approximated including multiple-variable problems, it came out, better solutions are necessary. So the research went in two directions. The first one, which emerged in the early fifties of the last century (just since the digital computers became available to the academia), is related to use of exponential polynomials [1]. A specific approach to this concept was named the Grey Theory [2]. The second one, which became ubiquitous and omnipotent, was named artificial neural network. It uses exponential functions too but in a specific manner. The latter will be shortly described in the sequel since, as already shown in this book, it may be considered as a very convenient tool in diagnostic applications.

The fact that the human brain works in a completely different way than the conventional digital computer was crucial for the start of research on artificial neural networks, general known as *neural networks*. Struggle to understand how the brain works starts even from Ramon y Cajal (1911) who gave the idea of nerves as structural elements of the brain. If we compare the speed of nerves with the speed of operation of silicon chips in a computer, we will see that the nerves are much slower than chips: Events in a silicon chip have a duration of the order of nanoseconds (10^{-9} s) in contrast to events in the nervous system whose duration is on the order of

V. B. Litovski, *Lecture Notes in Analog Electronics*, Lecture Notes in Electrical Engineering 1265, https://doi.org/10.1007/978-981-97-8257-4_12

milliseconds (10^{-3} s). The explanation for the "slowness" of the human brain lies in the fact that the number of natural nerves (nerves cells) that exist in the execution of "operations" varies very often, as well as the number of realized couplings between them. Namely, it is estimated that in the human brain there are 10 billion nerves and 60,000 billion synapses, that is, connections among them [3]. In addition, the energy efficiency of the brain is 10^{-16} joules (J) per event per second, while the corresponding value, for the best computer, is about 10^{-6} J per operation per second [4]. Based on that, we conclude that the brain is an extremely energy efficient structure.

It can also be said that the human brain is an extremely complex, nonlinear, and parallel computer (information-processing system). It has the ability to organize nerves to perform certain calculations (e.g., character recognition, perception, and movement control) much faster than the fastest computer would do. If, for example, we are talking about the ability of human vision, functions take place in the brain that provide a representation of the environment around us, but also, what is much more important, supply us information that we need due to interaction with that environment [5]. More precisely, the brain routinely performs the recognition operation (e.g., recognition of a familiar person in an unknown event) in a certain situation, at a speed of 100–200 ms, while some tasks with insignificant complexity put in pain a conventional computer [5].

How, then, does the human brain manage to do that. At birth, the brain has a perfect structure and the ability to build its own rules according to our experience. Experience is built up over years, with the human brain developing dramatically in the first two years of life, but it continues to develop quite well. During that earliest development, as many as a million synapses are formed in the brain per second.

Synapses are elementary structural and functional units that establish a connection between nerves. The most common type of synapses is chemical synapses, which work as follows. Presynaptic processes release transmitters, substances that spread along the synaptic nodes between nerve cells, and they then act on postsynaptic processes. In this way, the synapse translates the presynaptic electrical signal into a chemical signal and then back into a postsynaptic electrical signal [6]. It can also be said, in electronic terminology, that it is a unilateral four-pole. The usual description of a neural network assumes that a synapse is a simple connection that, in a receptive nerve, can handle excitation or inhibition, but not both at the same time.

The nerve, in its development, can be compared to the plasticity of the brain, i.e., with its characteristic of being malleable. Plasticity enables such a development of the nervous system that it adapts to its environment [5]. In the "adult" brain, this shaping can be initiated in two ways: by creating new synapses between nerves or by modifying existing connections.

When we talk about the appearance of a typical nerve, that is, a nerve cell, we can say that it consists of a body, dendrites, and axons. According to the morphological origin, we distinguish two types of cell fibers: the axon, whose surface is flatter, branches very little at its end, and has a longer length, unlike dendrites (named for their resemblance to a tree trunk) which have an irregular surface and a large number of "branches." Namely, 90% of the connections that accept the signal from other nerves are made via dendrites.

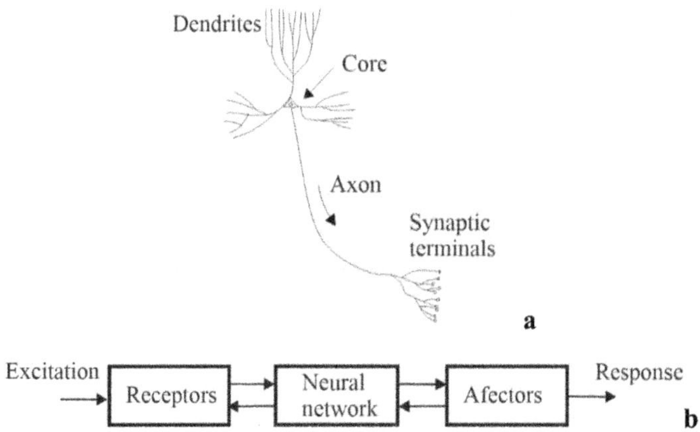

Fig. 7.12.1 **a** Pyramidal nerve cell, **b** block diagram of the nervous system of the brain

In the brain, nerves can be of different shapes and sizes. One example is pyramidal nerve cell that is shown in Fig. 7.12.1a. This type of nerve cell is the most abundant in the cortex of the cerebrum. Like other types of nerves, they receive signals from dendrites that extend through the spinal cord. The pyramidal cell can receive 10,000 or more synaptic signals and also it can forward them to several thousand target nerves.

The key features of a nerve in information processing include:

1. Dendrites: Nerves have branching structures called dendrites that receive signals from other nerves or sensory receptors. These signals are typically in the form of neurotransmitters released by neighboring nerves.
2. Cell Body (Soma): The cell body integrates incoming signals from the dendrites and, if the accumulated signals are strong enough, generates an action potential—a brief electrical impulse.
3. Axon: The action potential travels down the axon, a long, slender projection of the nerve. The axon is coated in a myelin sheath, which helps speed up the transmission of the electrical signal.
4. Axon Terminals: At the end of the axon, the signal is transmitted to other nerves or to muscle cells or glands at junctions called synapses. Neurotransmitters are released into the synapse, where they bind to receptors on the next nerve, transmitting the signal.
5. Action Potential: Nerves communicate through the generation and propagation of action potentials—rapid changes in the nerve's membrane potential. This electrochemical signal allows for the transmission of information over long distances in the nervous system.
6. Synaptic Transmission: The synapse is a crucial point where communication occurs between nerves. Neurotransmitters released into the synapse bind to receptors on the postsynaptic nerve, transmitting the signal from one nerve to the next.

Before we continue with the discussion about "artificial nerves," we will briefly look at the terminology that will be used. For natural nerves, as before, we will use "nerve" and for the artificial term "neuron" which comes from "νευρον." In this way, confusion will be avoided about the domain in question. We call the natural nerve system the "nervous system" and the artificial one "a neural network." The abbreviation ANN will be read as artificial neural network.

As ability to transmit and process signals is a basic feature for the functioning of nerves as information-processing individuals in the human brain, it is the same with neural networks that are composed of neurons. Generally speaking, an ANN is a machine designed to implement the way the brain performs certain tasks; the network is usually made of electronic components, or it is simulated by appropriate software on a computer. In order to achieve good performance, the ANN "employs" a large number of connections, which perform simple arithmetic operations between neurons i.e., between "working units." In the literature, the following definition of a neural network is offered, seen as an adaptable machine.

ANN is a parallel distributed processor that has an innate tendency toward storage experimental knowledge as well as its adaptation for use. There are two reasons why we compare the neural network with the brain:

- The network will gain knowledge and experience through the learning process.
- Interneuron connections, known as synaptic weights, are used to store knowledge.

The procedure that ensures the learning process is called the learning algorithm, and its task is to modify the synaptic weights within the ANN so that the desired end goal is achieved.

According to the previous presentation, it is obvious that ANN provides a large number of calculation operations thanks to the parallel organized structure, as well as the ability to generalize. These two characteristics allow us, by using neural networks, to solve very complex problems that are intractable in a different way. In practice, however, ANN cannot solve the problem by them. Namely, it is necessary to make a connection with the corresponding system proposed by the engineering design team. In particular, the problem we want to solve is usually broken down into simpler tasks like the ones that ANN solves in a way that matches its natural ability, e.g., interpolation, extrapolation, character recognition, associative memory, control, and the like. The use of neural networks has several useful advantages and capabilities:

1. Nonlinearity. Basically, the natural, and accordingly artificial, neuron is a nonlinear element. Therefore, an ANN consisting of connected neurons is inherently nonlinear. It is possible to say that the nonlinearity is specific in that it is distributed throughout the network. Nonlinearity is an extremely important feature, especially if the physical mechanism underlying the process which is responsible for generating the input signal by its nature, nonlinear.
2. Input–Output Mapping. A popular example for the process of creating neural networks, supervised learning, implies the modification of synaptic weights using a set of specific patterns for learning, that is, given examples. Each example contains a unique input signal and a desired one—the response, that is, the output

signal. Examples are brought to the input of the network by random selection from the set of input data, while the weights of connections (free parameters) are modified in such a way that the difference between the desired response and the actual value generated by the input signal is made as small as possible, according to selected statistical criteria. The creation of the network is repeated for several examples from the dataset, until the network reaches a constant state, that is, when there no longer exists significant changes in the value of the weights of connections. Examples that have already been used for learning can be used again with the fact that their order must be changed. We see that the network learns on examples so that it gradually constructs the input–output mapping based on the current problem that is being solved.

3. Adaptability. Neural networks have an innate ability to adjust their synapses weights in accordance with the environment. Practically, the neural network trained to act in a specific environment can be easily adapted to work (with minor changes) in the current environmental conditions. Moreover, when it works in a non-stationary environment (an environment that changes in time), neural the network can be designed so that it changes the values of the weights of connections in real time.

   The natural architecture of neural networks for pattern classification, signal processing, and control, in conjunction with the adaptability feature, creates an ideal tool for adaptive application for classification of samples, adaptive signal processing, and adaptive control.

4. Confirmed Response. In the context of pattern classification, a neural network can be designed in this way to provide information not only about which sample was selected, but also information about how much is that decision reliable. This information can later be used to reject unreliable, unclear, patterns that can occur, and thus improve the classification feature owned by the network.
5. Contextual Information. In the neural network, the acquired knowledge is presented the very structure and activation state of the network. During training, one acts on each neuron in the network by the global activity of the other neurons. Therefore, contextual information about the state of the network is in fact its inherent feature.
6. Error Tolerance. It is characteristic of a neural network, implemented as hardware, to be fault tolerant in the sense that its performance degrades slightly in unfavorable conditions. For example, if a neuron or one of its interconnections is damaged, during the exploitation of the network this is recognized as damage to the "memory" of the learned mapping. However, bearing in mind that the information is distributed over the entire network, in order for the overall response to be significantly impaired, it is necessary for the network to be significantly impaired as well. Essentially, the neural network exhibits a slight degradation of performance during operation rather than leading to complete failure.
7. Applicability in VLSI. The strongly parallel nature of neural networks makes them potentially fast for computing certain tasks. This same phenomenon makes neural networks ideally suited for application in the form of integrated circuits (VLSI technology).

8. Neurobiological Analogy. The analogy of neural networks with the human brain was the main motive for designing neural networks. Namely, the human brain is living proof that parallel, failure-resistant processing is not only physically possible, but also very fast and powerful. Neurobiologists look at artificial neural networks as a research tool with which to interpret neurobiological phenomena. For example, neural networks have been used to create insights into the development of premotor cortex in the oculomotor systems and the way they process signals [7].

Bearing in mind that the inspiration for designing artificial neural networks came from a neurobiological analogy, it seems appropriate to make a brief overview of the structural levels and organization of the human brain, which we will do in the next paragraph.

7.12.2 Structural Levels of Human Brain Organization

The human nervous system can be viewed through three levels, described in the block diagram in Fig. 7.12.1b [8]. The central part of the system is the brain, in Fig. 7.12.1b presented as a neural network, which continuously receives information, processes it, and makes appropriate decisions.

Two sets of arrows are also shown in Fig. 7.12.1b; those pointing from left to right indicate that the signal containing information is transmitted forward through the system, while arrows pointing from right to left indicate the presence of a feedback loop in the system. Receptors, in the same figure, convert the body or external stimuli into electrical impulses that transmit information to the nervous system (brain). Effectors, on the other hand, convert electrical impulses generated by the neural network into visible responses as outputs from the system.

7.12.3 Neuron Models

A neuron is a process-informational unit and as such, represents the basis for the work of the neural network. Figure 7.12.2 schematically represents the neuron model. We can distinguish three basic elements of the model, described in the following way:

1. *A set of synapses*, that is, connections, each of which is determined by its weight (strength). That means that the signal x_j, at the input of synapse j which is connected to neuron k, is multiplied by synaptic weight w_{kj}. It is important to pay attention to the way of marking the weights we work with. The first letter in the subscript refers to the neuron, while the second letter indicates the synapse to which refers the weighted relationships. Reversed notation may be found in some literature sources. The weight w_{kj} is positive if the corresponding synapse is excitatory, that is, it is negative if the synapse is inhibitory.

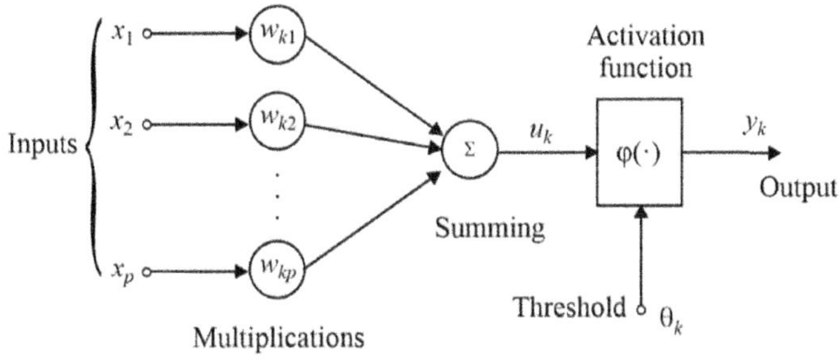

Fig. 7.12.2 Schematic representation of the neuron model

2. *The adder* is adding the input signals multiplied by the corresponding weights of the neuron synapses. As a result, linear combination of the input signals is obtained. The adder is often seen as if it performs scalar multiplication of the vector of excitations **x** and the vector of weights **w**.
3. *The activation function* limits the neuron's output amplitudes to some finite value. Usually, the normalized amplitude level of the output neuron is written in square brackets, where it refers to the segment [0, 1], or, alternatively, [−1, 1].

The neuron model shown in Fig. 7.12.2 also includes external influences, thresholds θ_k, that change the value of the input signal of the activation function. Mathematically, the neuron model can be described in the following ways:

$$u_k = \sum_{j=1}^{p} w_{kj} x_j \tag{7.12.1}$$

and

$$y_k = \varphi(u_k - \theta_k), \tag{7.12.2}$$

where $x_1, x_2, \ldots, x_p$ are the input signals; $w_{k1}, w_{k2}, \ldots, w_{kp}$ are the synaptic weights of the kth neuron; u_k is the output of the linear combination (the scalar product), θ_k is the threshold; $\varphi(\cdot)$ is the activation function; and y_k is the output signal of the neuron. The use of the threshold θ_k has the effect of applying an *affine transformation* to the output u_k of the linear combiner in the model of Fig. 7.12.2, which is expressed as

$$v_k = u_k - \theta_k. \tag{7.12.3}$$

Depending on whether the threshold θ_k is positive or negative, the relationship between the effective internal level of activity or the activation potential v_k of the neuron k and the linear combination of the inputs u_k is modified in the manner shown in Fig. 7.12.3. We notice that as a result of this affine mapping, the diagram $v_k - u_k$ does not pass through the origin.

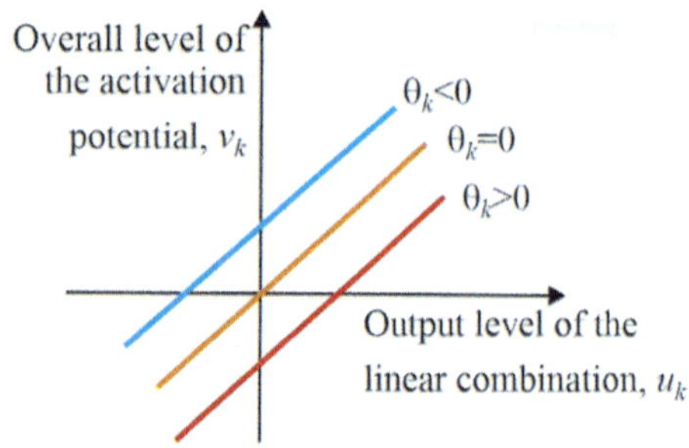

Fig. 7.12.3 Affine mapping due to the presence of a threshold

The threshold θ_k is the external parameter of the artificial neuron k. Its influence is introduced by relation (7.12.2).

The combination of (7.12.1) and (7.12.2) yields:

$$v_k = \sum_{j=0}^{p} w_{kj}x_j - \theta_k \tag{7.12.4}$$

and

$$y_k = \varphi(v_k). \tag{7.12.5}$$

In (7.12.4) we got a new synapse, whose input

$$x_0 = -1, \tag{7.12.6}$$

and its weight is

$$w_{k0} = \theta_k. \tag{7.12.7}$$

Accordingly, we can reformulate the model of the kth neuron as in Fig. 7.12.4. In this figure (Fig. 7.12.4a), the threshold action is represented in two ways: (1) by adding a new input signal that has constant value equal to -1, and (2) adding a new synaptic weight equal to the threshold θ_k. Alternatively, we can change the neuron model as in Fig. 7.12.4b, where the combination of fixed input $x_0 = +1$ and weight $w_{k0} = b_k$ is taken as biasing b_k. In any case, the models shown on Figs. 7.12.3 and 7.12.4 although they look different, they are mathematically the same.

The activation function, denoted with $(\cdot)$, will define the output of the neuron. There are three basic types of activation function:

1. *Binary step (Threshold) function.* For this type of function, which is graphically represented in Fig. 7.12.5a, we have

$$\varphi(v) = \begin{cases} 1 & v \geq 0 \\ 0 & v < 0 \end{cases}. \tag{7.12.8}$$

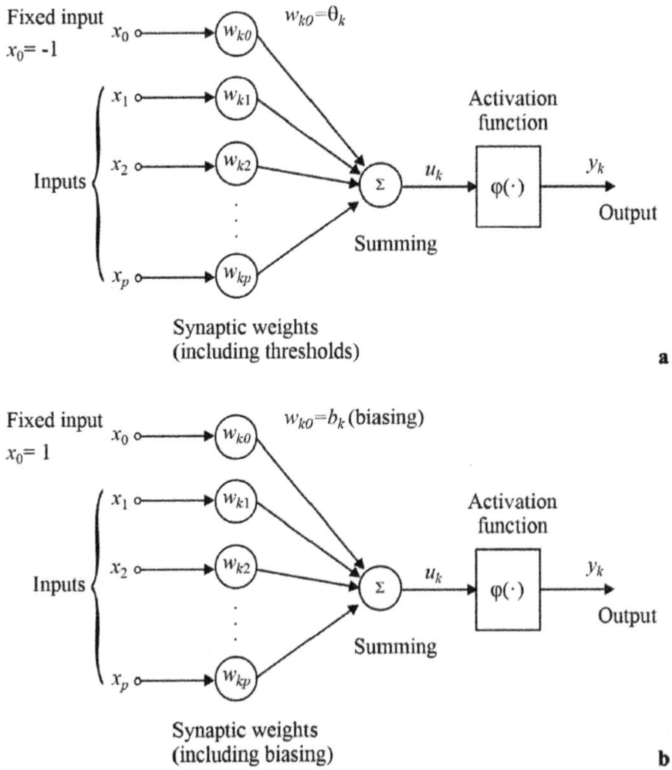

Fig. 7.12.4 Two nonlinear neuron models

Accordingly, the output of the kth neuron, which has the role of a threshold, is expressed as follows:

$$y_k = \begin{cases} 1 & v_k \geq 0 \\ 0 & v_k < 0 \end{cases}, \tag{7.12.9}$$

where v_k is the internal activation level of the neuron, that is

$$v_k = \sum_{j=1}^{p} w_{kj}x_j - \theta_k. \tag{7.12.10}$$

This type of neuron is presented in the literature as the McCulloch-Pitts model, in recognition of the dedicated pioneering work of McCulloch and Pitts, on the study of neural networks. In this model, the output of the neuron takes the value 1 (if the total internal activation potential of this neuron is non-negative) or equal to zero in the opposite. This statement expresses the all-or-none feature inherent to the McCulloch-Pitts model.

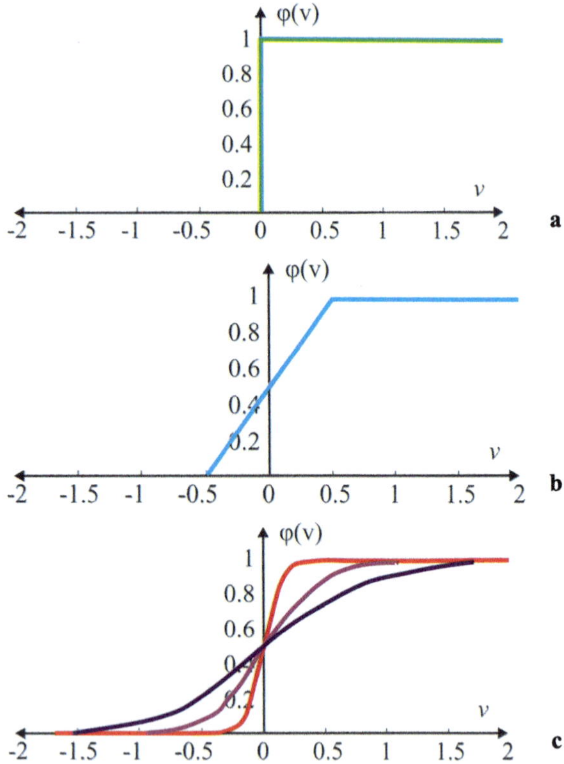

Fig. 7.12.5 Activation functions. **a** Threshold function, **b** piece-wise linear function, and **c** sigmoid function (three examples)

2. *Piece-wise linear function*. For this function, as depicted in Fig. 7.12.5b, we have

$$\varphi(v) = \begin{cases} 1, \; v \geq \frac{1}{2} \\ v, \; -\frac{1}{2} > v > -\frac{1}{2} \\ 0, \; v \leq -\frac{1}{2} \end{cases}, \tag{7.12.11}$$

where it is assumed that the amplification factor within the linear region is equal to unity. This form of activation functions can be considered as an approximation of a nonlinear amplifier. The following two examples can be characterized as special cases of piece-wise linear functions:

- As long as the linear operation is maintained, i.e., the activation function is not entering saturation, the linear combination of input signals is preserved.
- The piece-wise linear function reduces to a threshold function if the gain factor in the linear area becomes infinitely large.

3. *Sigmoid function*. The sigmoid function is by far the most widely used in ANN synthesis. It is characterized as a function that strictly increases. One example of a sigmoid function is the logistic function, defined by:

$$\varphi(v) = \frac{1}{1 + \exp(-a \cdot v)}, \tag{7.12.12}$$

where a is the slope of the sigmoid function. By changing the parameter a, a sigmoid function is created with a variable slope, as shown in Fig. 7.12.5c. In fact, the slope in the origin is $a/4$. In the limiting case, when the parameter a, i.e., the slope, is approaching infinity, the sigmoidal function would become a threshold function. While the threshold function only takes the values 1 or 0, the sigmoid function takes values in the continuous range from 0 to 1. Also, it should be noted that the sigmoid function is differentiable, while the threshold function is not (differentiability is an important property from the point of view of application of certain learning algorithms).

The activation function described in (7.12.8), (7.12.11), and (7.12.12) takes values from 0 to 1. Sometimes it is desirable, however, to have an activation function with values between −1 and +1, while it takes an antisymmetric form in relation to the origin. In this case, the threshold function from Eq. (7.12.8) will be reformulated as:

$$\varphi(v) = \begin{cases} 1 & v > 0 \\ 0 & v = 0 \\ -1 & v < 0 \end{cases}, \tag{7.12.13}$$

which is known as the *signum function*. For the sigmoid case we can use the hyperbolic tangent:

$$\varphi(v) = \tanh\left(\frac{v}{2}\right) = \frac{1 - \exp(-v)}{1 + \exp(-v)}. \tag{7.12.14}$$

The property of the activation function of the sigmoid type that it can also have negative values, as expressed in (7.12.14), can have certain analytical advantages.

Beyond the basic activation functions discussed above, there is a broad set of alternatives. One of them is the so-called radial basis function (RBF) which is one of the most popular neural network's activation function. These are strictly positive definite functions. One example of RBF would be:

$$\varphi(v) = e^{-(\varepsilon \cdot v)^2}, \tag{7.12.15}$$

where ε is an adjustable constant defining the width, around θ_k, of the "bell" graphically representing the function.

Recently, as large set of variations of the Rectified Linear Unit function (ReLU) is in use. "Rectified" means that the value of the function given by (7.12.3), for negative argument u_k, is equal to zero.

7.12.4 Network Architecture

In general, we can identify four different classes of neural network architectures.

7.12.4.1 Single-Layer Feed-Forward ANN

A layered neural network is a network where neurons are organized to form layers. The simplest form of a layered network is when we have an input layer of excitation nodes that act on neurons (nodes) in the output layer but not in the opposite direction. In other words, this network is strictly feed-forward (non-recurrent) type. This is shown in Fig. 7.12.6 for the case when we have four nodes in both the input and the output layer. At the same time, the excitation layer does not perform any calculations; it just distributes the input signals to neurons located in the output layer. This type of network is called single-layer feed-forward network, where the adjective "single-layer" refers to the output layer of the computing nodes (neurons). It is important to note that there are connections between every input node and all output neurons. It is said for such a network that it is fully connected.

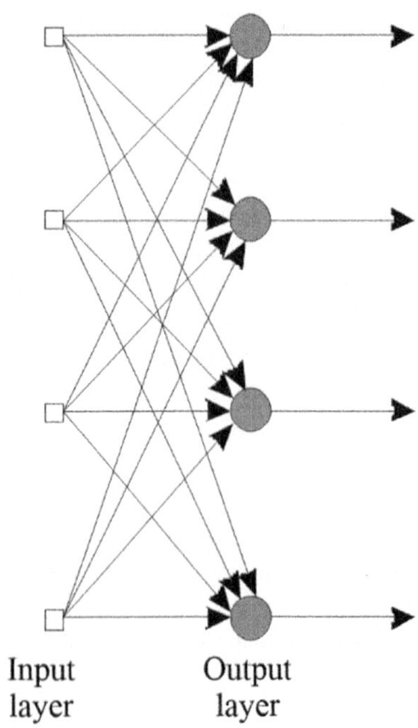

Fig. 7.12.6 Single-layer feed-forward fully-connected neural network

7.12.4.2 Multilayer Feed-Forward ANN

The second class of feed-forward ANNs can be divided into those with single and those with several hidden layers, in which the computing nodes are called hidden neurons or hidden units. The role of the hidden neurons is to act between the external inputs and outputs of the network. By adding one or more hidden layers, the network is expected to become more capable of performing more complex calculations.

The excitation vortices in the input layer of the neural network form elements of the input vector which constitute input signals that excite the neurons in the second layer (the first hidden layer). The output signals in the second layer are used as inputs for the third layer and so on for the rest of the network. Accordingly, the neurons in each layer of the network have input signals only coming from the previous layers. The set of output signals of neurons in the output (end) layer of the network establishes the general response of the network to the activation pattern applied to the excitation vortices in the input (first) layer. Figure 7.12.7 illustrates the layout of a multilayer feed-forward neural network for the case when we have one hidden layer. The network shown in Fig. 7.12.7 can be characterized by the abbreviation: "10-4-2 network," which means that there are ten excitation vortices, four hidden neurons, and two output neurons. As another example, for a feed-forward neural network with p excitation vortices, h_1 neurons in the first hidden layer, h_2 neurons in the second hidden layer, and q neurons in the output layer, we say that it is a p–h_1–h_2–q network.

For the neural network shown in Fig. 7.12.7, we say that it is fully connected in the sense that every cell in every layer of the network is connected to every cell in the next adjacent layer. If, however, some of the synaptic connections are missing in the network, we can say that the network is partially connected.

7.12.4.3 Recurrent Networks

The feature of a recurrent ANN is feedback that can be realized in one or more loops. For example, a recurrent network can contain one layer of neurons, where the output of each neuron leads to the inputs of all others, as illustrated in Fig. 7.12.8. The structure described on this picture has no self-feedback loops, where self-feedback refers to a neuron which brings its own output signal to its own input. The recurrent network of Fig. 7.12.8 has no hidden neurons, too.

Figure 7.12.9 shows a class of recurrent networks that, now, contain hidden neurons. The feedback connections, shown in Fig. 7.12.9, originate from both hidden and output neurons. Feedback, regardless of whether it is of architecture, like in Figs. 7.12.8 or 7.12.9, has a profound impact on network's learning capability and on its behavior as well. Further, the feedback includes use of special branches that are composed of elements of a unit delay (marked with z^{-1}) what as a result has a nonlinear dynamic behavior based on the nonlinear nature of neurons. Dynamic nonlinearity plays an important role in the memory function of a neural network.

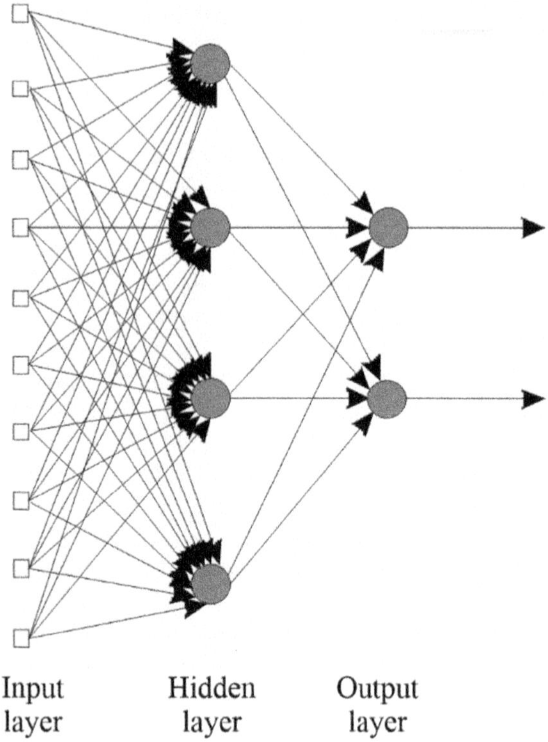

Fig. 7.12.7 Multilayer fully-connected feed-forward neural network (one hidden layer)

7.12.5 Presentation of Knowledge

According to the literature [9], the following definition is used:

> Knowledge refers to memorized information or models used by man or machine to interpret, predict, and respond appropriately to the external world.

The basic characteristics of knowledge representation are twofold: first, what information is currently being observed; and second, how that information is physically remembered for subsequent use. In the real world of application of intelligent devices, it can be said that obtaining a good solution depends on well-presented knowledge [10]. It is the same with the ANN, which represents a special class of intelligent machines.

The main task of the neural network is to learn the model of the environment in which it is embedded, then to maintain the model consistent enough with the environment, and in this way to achieve the set of goals that are important for the application. Environmental knowledge (external knowledge) contains two types of information:

1. The state of the environment, represented by facts about what is now known, and what was known. This form of knowledge is called prior information.

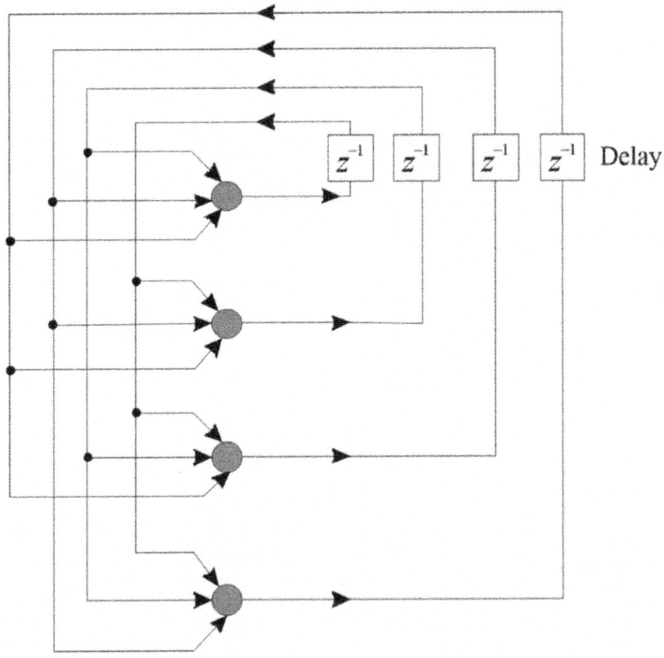

Fig. 7.12.8 Recurrent network containing one layer of neurons

2. Observations (measurements) related to the environment, obtained from sensors designed to examine the environment in which the neural network is assumed to operate. Typically, these measurements are inherently noisy, so observations are subject to sensor errors. In any case, the measurements obtained in this way provide a wealth of information about the examples used to train the neural network.

Each example contains one input–output pair: the input signal and the corresponding desired ANN response. Therefore, a set of examples represents the knowledge about the environment that is of interest to the neural network. Let us discuss, for example, the problem of recognizing handwritten numbers, where the input contains one image with black or white dots, and each image represents one of ten digits, but so that they are clearly visible against the background. In this example, the desired response is determined by the identity of a particular digit whose image is presented to the network as an input signal. The set of examples used to train the network contains a wide range of handwritten numbers, which, in fact, represent the real situation, given the indefinite number of authors of the numbers. With such a given set of examples, designing a neural network can be done as follows:

First, an appropriate architecture for the neural network is chosen, with one input layer containing input nodes whose number is equal to the number of points in the input image, and one output layer containing 10 neurons (one for each digit). A

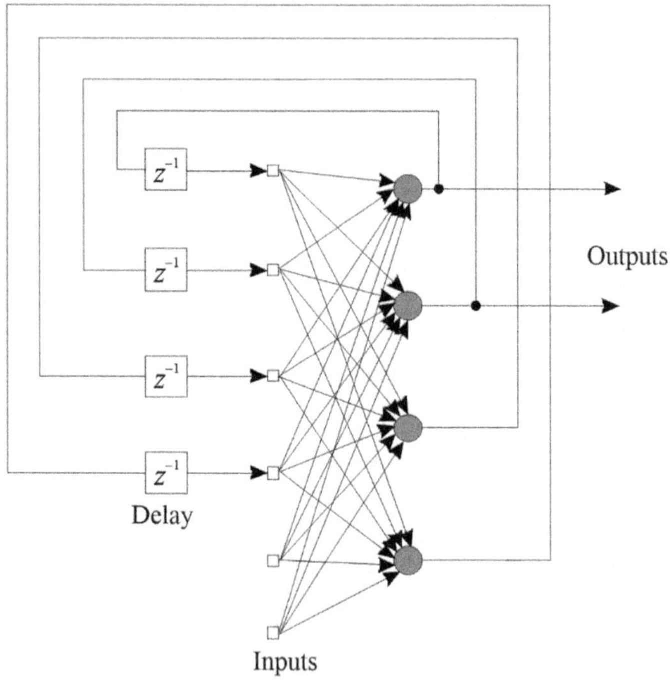

Fig. 7.12.9 Recurrent network that also contains hidden neurons

subset of the examples is used to train the network *using the appropriate algorithm*. This phase of network design is called learning.

Second, the recognition properties of the network thus learned are tested with data that have not been used before. Namely, the input image is presented to the network, but this time the identity of the digit found in that image is not detected. The characteristic of the network is, now, in evaluating the comparison of the digit recognized by the network with the real identity of the requested digit. The second stage of network operation is called generalization, a term borrowed from psychology.

This is where the fundamental difference between ANN design and classic information processing for sample classification lies. Thus, when we talk about classical data processing, we first formulate a mathematical model of the environment, and then we proceed to check the model with real data and only then do we design based on the model. In contrast to this approach, the design of ANN is directly based on real, actual, data. Therefore, neural networks not only provide an implicit model of the environment in which they are embedded, but also provide the function of processing information that is of interest to the operation of the neural network.

In a neural network of a given architecture, knowledge about the environment is represented by defined values of free parameters (weights and thresholds) of the network. The form of presentation of this knowledge determines the very design of ANN and, further, contains the key to its characteristics.

The topic of knowledge representation within a nervous system is, however, very complex. It becomes even more complex when we have multiple sources of information that activate the network, all the more so if these sources interact with each other. Modern research results in artificial intelligence are in fact a breakthrough in knowledge representation.

7.12.6 A Very Short Introduction to the Process of Learning

Given the ANN's expected responses, $\mathbf{y}_n$, as functions of the input signals of the nth "lesson" (somewhere called epoch) $\mathbf{x}_n$, and the actual response of the ANN $\hat{\mathbf{y}}_n(\mathbf{x}_n, \mathbf{w}, \boldsymbol{\theta})$ to the same excitation, one may define the discrepancy, here named nth error, as the Euclidean distance:

$$d_n = \|\mathbf{y}_n - \hat{\mathbf{y}}_n\|. \tag{7.12.16}$$

The overall error accumulated after N lessons will be

$$D = \frac{1}{N}\sqrt{\sum_{n=1}^{N}(d_n^2)}. \tag{7.12.17}$$

This quantity is a function of the network's parameters $\mathbf{w}$ and $\boldsymbol{\theta}$ which are unknown in advance. It is the extraction of their values what we call learning or training the ANN. The goal is to get:

$$D = min. \tag{7.12.18}$$

This means that the learning process is in fact an iterative minimization procedure.

There are many procedures which may lead to the solution. Some are deterministic based on the derivatives of D with respect to the ANN's parameters, and others are stochastic searching the parameter space for the global minimum.

In the first group we may mention the Newton–Raphson method, the steepest descent (used throughout this book based on [11]) method, and the most popular one, the back-propagation algorithm.

In the second group we find the simulated annealing, the genetic algorithms, and similar.

The reader, if interested, is advised to consult the literature, especially [12, 13].

References

1. Angeloni L, Costarelli D (2024) Approximation by exponential-type polynomials. J Math Anal Appl 532(1):127927
2. Julong D (1989) Introduction to grey system theory. J Gray Syst 1:1–24
3. Shepherd GM, Koch C (1990) Introduction to synaptic circuits. In: Shepherd GM (ed) The synaptic organization of the brain. Oxford University Press, New York, pp 3–31
4. Wilamowski B, Binged J, Kaynak O (2000) VLSI implementation of neural networks. Int J Neural Syst 10(3):191–197
5. Churchland PS, Sejnowski TJ (1992) The computational brain. MIT Press, Cambridge
6. Freeman WJ (1975) Mass action in the nervous system. Academic Press, New York
7. Robinson DA (1992) Signal processing by neural networks in the control of eye movements. In: Proceedings of the computational neuroscience symposium, Indiana University/Purdue University, Indianapolis, pp 73–78
8. Arbib MA (1987) Brain, machines, and mathematics, 2nd edn. Springer, New York
9. Fischler MA, Firschein O (1987) Intelligence: the eye, the brain, and the computer. Addison Wesley, Reading
10. Woods WA (1986) Important issues in knowledge representation. Proc IEEE 74(10):1322–1334
11. Zografski Z (1991) A novel machine learning algorithm and its use in modeling and simulation of dynamical systems. In: Proceedings of 5th annual European computer conference, IEEE COMPEURO'91, Bologna, Italy, pp 860–864
12. Haykin S (2008) Neural networks and learning machines, 3rd edn. Pearson
13. Jain AK, Mao J, Mohiuddin KM (1996) Artificial neural networks: a tutorial. Computer 29(3):31–44. https://doi.org/10.1109/2.485891

7.13 Further Reading

Abstract When writing a book the author is following his own thoughts, knowledge, preferences, and ideas. That imposes frames to the parts of the book being it a text or literature. Here we offer additional literature which, we think, will allow the reader to communicate with ideas which are extensions the ones given above. Some of them are proposing fully alternative approaches of which the reader should be aware.

7.13.1 Literature

1. Liao W, Tian F, Liu J (2009) Efficient fault diagnosis method in nonlinear circuits based on neural network. Comput Eng Appl 45(19):228–231
2. Yunyan H, Minfang P, Chenglai T, Hu T (2012) Analog circuit fault diagnosis using multi-wavelet transform and SVM. In: Third international conference on digital manufacturing and automation (ICDMA), pp 214–217
3. Kumar A, Singh AP (2013) Neural network based fault diagnosis in analog electronic circuit using polynomial curve fitting. Int J Comput Appl 61(16):28–34. ISSN 0975-8887
4. Gao X (2014) A novel method to optimum test-nodes selection in analog-circuit fault diagnosis. In: Fifth IEEE international conference on intelligent control and information processing (ICICIP), Dalian, pp 424–429
5. Kavithamani A (2014) Certain new approaches to fault diagnosis of analog electronic circuits using frequency response methods. Anna University, Faculty of Electrical and Electronics Engineering. URL: https://hdl.handle.net/10603/24732
6. Guoming S, Qi L, Gang L, Shuyan J, Houjun W (2015) Analog circuit fault diagnosis using wavelet feature optimization approach. In: 12th IEEE international conference on electronic measurement & instruments (ICEMI), pp 119–124
7. Yu W, He Y (2015) Analog circuit fault diagnosis via sensitivity computation. J Electron Test 31(1):119–122

V. B. Litovski, *Lecture Notes in Analog Electronics*, Lecture Notes in Electrical Engineering 1265, https://doi.org/10.1007/978-981-97-8257-4_13

8. Luo H, Lu W, Wang Y, Wang L, Zhao X (2016) A novel approach for analog fault diagnosis based on stochastic signal analysis and improved GHMM. Measurement 81:26–35
9. Mosin S (2016) An approach to design-for-testability automation of analogue integrated circuits using OBIST strategy. In: 5th Mediterranean conference on embedded computing, MECO 2016, Bar, Montenegro, pp 211–214
10. Tang X, Xu A (2016) Practical analog circuit diagnosis based on fault features with minimum ambiguities. J Electron Test 32(1):83–95
11. Yu W, He Y, Gao K (2016) A compute method of analog circuit fault diagnosis via particle swarm optimization. J Autom Control Eng 4(6):439–442
12. Tadeusiewicz M, Hałgas S (2017) Diagnosis of a soft short and local variations of parameters occurring simultaneously in analog CMOS circuits. Microelectron Reliab 72:90–97
13. Tang X, Xu A, Li R, Zhu M, Dai J (2018) Simulation-based diagnostic model for automatic testability analysis of analog circuits. IEEE Trans Comput-Aided Des Integr Circuits Syst 37(7):1483
14. Srimani S, Ghosh K, Rahaman H (2020) Wavelet transform based fault diagnosis in analog circuits with SVM classifier. In: IEEE international test conference, Bangalore, India, pp 1–10
15. Chatterjee N, Ortega J, Meric I, Xiao P, Tsameret I (2021) Machine learning on transistor aging data: test time reduction and modeling for novel devices. In: IEEE international reliability physics symposium (IRPS), Monterey, CA, USA, pp 1–9
16. Cloete JB, Stander T, Wilke DN (2022) Parametric circuit fault diagnosis through oscillation-based testing in analogue circuits: statistical and deep learning approaches. IEEE Access 10:15671–15680
17. Li Y, Chen H, Lu N, Jiang B, Zio E (2022) Data-driven optimal test selection design for fault detection and isolation based on CCVKL method and PSO. IEEE Trans Instrum Meas 71:1–10
18. Cloete JB, Stander T, Wilke DN (2022) Parametric circuit fault diagnosis through oscillation-based testing in analogue circuits: statistical and deep learning approaches. IEEE Access 10:15671–15680

Index

V. B. Litovski, *Lecture Notes in Analog Electronics*, Lecture Notes in Electrical Engineering 1265, https://doi.org/10.1007/978-981-97-8257-4

The manufacturer's authorised representative in the EU is Springer Nature Customer Service Centre GmbH, Europaplatz 3, 69115 Heidelberg, Germany. If you have any concerns regarding our products, please contact ProductSafety@springernature.com

Printed and bound by CPI Group (UK) Ltd, Croydon, CR0 4YY

30/04/2025

01856994-0001